Progress in Inflammation Research

Series Editor

Prof. Michael J. Parnham PhD
Senior Scientific Advisor
PLIVA Research Institute Ltd.
Prilaz baruna Filipovića 29
HR-10000 Zagreb
Croatia

Advisory Board

Forthcoming titles:

Antirheumatic Therapy: Actions and Outcomes,
R.O. Day, D.E. Furst, P.L. van Riel, B. Bresnihan (Editors), 2005
NPY Family of Peptides in Immune Disorders, Inflammation, Angiogenesis and Cancer,
G.Z. Feuerstein, Z. Zukowska (Editors), 2005
Turning up the Heat on Pain: Vanilloid Receptors in Pain and Inflammation,
A.B Malmberg, K.R. Bley (Editors), 2005
Regulatory T-Cells in Inflammation, L. Taams, A.N. Akbar, M.H.M. Wauben (Editors), 2005
Sodium Channels, Pain, and Analgesia, K. Coward, M. Baker (Editors), 2005
Complement and Kidney Disease, P.F. Zipfel (Editor), 2005

(Already published titles see last page.)

Antibiotics as Anti-Inflammatory and Immunomodulatory Agents

Bruce K. Rubin
Jun Tamaoki

Editors

Birkhäuser Verlag
Basel · Boston · Berlin

Editors

Bruce K. Rubin
Department of Pediatrics
School of Medicine
Wake Forest University
Medical Center Boulevard
Winston-Salem, NC 27157-1081
USA

Jun Tamaoki
First Department of Medicine
Tokyo Women's Medical University
8-1 Kawada-Cho, Shinjuku
Tokyo 162-8666
Japan

A CIP catalogue record for this book is available from the Library of Congress, Washington D.C., USA

Bibliographic information published by Die Deutsche Bibliothek
Die Deutsche Bibliothek lists this publication in the Deutsche Nationalbibliografie;
detailed bibliographic data is available in the internet at http://dnb.ddb.de

ISBN 3-7643-5925-0 Birkhäuser Verlag, Basel – Boston – Berlin

Part of Springer Science+Business Media
Printed on acid-free paper produced from chlorine-free pulp. TCF ∞
Cover design: Markus Etterich, Basel
Cover illustration: Inhibitory effect of clarithromycin on LPS-induced MAC5AC gene expression and I-kappa-B-alpha phosphorylation in human airway epithelial cells. With the friendly permission of Jun Tamaoki.
Printed in Germany
ISBN 3-7643-5925-0

9 8 7 6 5 4 3 2 1

www.birkhauser.ch

Contents

List of contributors

Ronald Anderson, MRC Unit for Inflammation and Immunity, Department of Immunology, University of Pretoria, Pretoria, and Tshwane Academic Division of the National Health Laboratory Service, South Africa;
e-mail: randerso@medic.up.ac.za

Arata Azuma, Fourth Department of Internal Medicine, Nippon Medical School, 1-1-5 Sendagi, Bunkyo-Ku, Tokyo 113-8602, Japan; e-mail: a-azuma@nms.ac.jp

Andrew Bush, Department of Paediatric Respiratory Medicine, Royal Brompton and Harefield NHS Trust, Sydney Street, London SW3 6NP, UK;
e-mail: abush@rbh.nthames.nhs.com

Axel Dalhoff, Bayer AG, Aprather Weg, 42096 Wuppertal, Germany;
e-mail: axel.dalhoff@bayerhealthcare.com

Larry H. Danziger, Department of Pharmacy Practice, University of Illinois at Chicago, USA; e-mail: danziger@uic.edu

Charles Feldman, Department of Medicine, University of Witwatersrand, Medical School, 7 York Road, Parktown, 2193, Johannesburg, South Africa;
e-mail: feldmanc@medicine.wits.ac.za

Mark H. Gotfried, Department of Medicine, University of Arizona, Phoenix, Arizona; and Department of Pharmacy Practice, University of Illinois at Chicago, Chicago, USA

Qutayba Hamid, McGill University, Canada; e-mail: qutayba.al_heialy@staff.mcgill.ca

Markus O. Henke, Department of Pulmonary Medicine, Universität Marburg, Baldingerstrasse 1, 35043 Marburg, Germany;
e-mail: markus.henke@staff.uni-marburg.de

Adam Jaffé, Portex Respiratory Medicine Group, Great Ormond Street Hospital for Children NHS Trust & Institute of Child Health, Great Ormond Street, London WC1N 3JH, UK; e-mail: a.jaffe@ich.ucl.ac.uk

Rose Jung, Department of Clinical Pharmacy, University of Colorado Health Science Center, Denver, USA

Jun-ichi Kadota, Division of Pathogenesis and Disease Control, Department of Infectious Diseases, Oita University Faculty of Medicine, 1-1 Hasama, Oita 879-5593, Japan; e-mail: kadota@med.oita-u.ac.jp

Kei Kasahara, Department of Medicine II, Nara Medical University Hospital, Nara Medical University, 840 Shijyocho, Kashihara, Nara 634-8521, Japan

Eiji Kita, Department of Bacteriology, Nara Medical University Hospital, Nara Medical University, 840 Shijyocho, Kashihara, Nara 634-8521, Japan;
e-mail: eijikita@nmu-gw.naramed-u.ac.jp

Shoji Kudoh, Fourth Department of Internal Medicine, 1-1-5 Sendagi, Bunkyo-Ku, Tokyo 113-8602, Japan; e-mail: kuntonjp@nms.ac.jp

Marie-Thérèse Labro, INSERM U479, CHU X. Bichat, 16 rue Henri Huchard, 75018 Paris, France; e-mail: labro@bichat.inserm.fr

Yuichi Majima, Department of Otorhinolaryngology, Mie University School of Medicine, 2-174 Edobashi, Tsu, Mie 514-8507, Japan;
e-mail: majima@clin.medic.mie-u.ac.jp

Keiichi Mikasa, Center for Infectious Diseases, Nara Medical University Hospital, Nara Medical University, 840 Shijyocho, Kashihara, Nara 634-8521, Japan

Michael J. Parnham, PLIVA Research Institute Ltd, Prilaz baruna Filipovića 29, 10000 Zagreb, Croatia; e-mail: michael.parnham@pliva.hr

Bruce K. Rubin, Department of Pediatrics, School of Medicine, Wake Forest University, Medical Center Boulevard, Winston-Salem, NC 27157-1081, USA;
e-mail: brubin@wfubmc.edu

Theodore J. Standiford, Pulmonary and Critical Care Medicine, University of Michigan Medical School, Ann Arbor, MI 48109-0360, USA

Kazuhiko Takeuchi, Department of Otorhinolaryngology, Mie University School of Medicine, 2-174 Edobashi, Tsu, Mie 514-8507, Japan;
e-mail: kazuhiko@clin.medic.mie-u.ac.jp

Kiyoshi Takeyama, First Department of Medicine, Tokyo Women's Medical University School of Medicine, 8-1 Kawada-cho, Shinjuku-ku, Tokyo 162-8666, Japan;
e-mail: kiyot@kj8.so-net.ne.jp

Hajime Takizawa, Department of Respiratory Medicine, University of Tokyo, Graduate School of Medicine, 7-3-1 Hongo, Bunkyo-ku, Tokyo 113-8655, Japan;
e-mail: takizawa-phy@h.u-tokyo.ac.jp

Jun Tamaoki, First Department of Medicine, Tokyo Women's Medical University, 8-1 Kawada-Cho, Shinjuku, Tokyo 162-8666, Japan;
e-mail: jtamaoki@chi.twmu.ac.jp

Kazuhiro Tateda, Department of Microbiology and Infectious Disease, Toho University School of Medicine, 5-21-16 Ohmorinishi, Ohtaku, Tokyo 143-8540, Japan;
e-mail: kazu@med.toho-u.ac.jp

Keizo Yamaguchi, Department of Microbiology and Infectious Disease, Toho University School of Medicine, 5-21-16 Ohmorinishi, Ohtaku, Tokyo 143-8540, Japan

Preface

The antibiotic era began in earnest during World War II with the "miracle of penicillin". Following the introduction of penicillin, the quest was on to discover similar antimicrobial agents. In the late 1940s, erythromycin A was isolated from a soil sample found in the Philippine island of Iloilo, and in 1952 erythromycin was introduced by Eli Lilly Company under the name of Ilosone, as an alternative to penicillin for emerging penicillin-resistance bacteria. It was recognized early on that the gastrointestinal side effects of erythromycin A could be modified by altering the chemical structure of the agent, and in the early 1990s clarithromycin and azithromycin were developed to be more acid-stable and with fewer side effects. Not long after this, it was shown that the macrolide antibiotics had immunomodulatory effects separate from antimicrobial properties.

The "steroid sparing" properties of the 14-member macrolides troleandomycin and oleandomycin, were first described in patients with severe, steroid-dependent asthma. Erythromycin was also found to reduce the need for corticosteroids in patients with asthma and, as described by Rose Jung, Mark H. Gotfried and Larry H. Danziger, in these trials some severe, steroid-dependent asthmatics were able to discontinue systemic corticosteroids with the use of macrolide antibiotics. Although it was speculated that the mechanism of macrolide action for severe asthma was by interfering with corticosteroids metabolism, in the clinical trials the reduction in steroid side effects, dosage, and in some cases discontinuation of steroids suggested a different effect on the underlying disease.

This was exploited in the 1980s in Japan for the treatment of the nearly uniformly fatal airway disease diffuse panbronchiolitis (DPB), as described by Arata Azuma and Shoji Kudoh. Since that time, many investigators in Japan – and now around the world – have studied these immunomodulatory properties not only of macrolide antibiotics but also of other classes of antimicrobials. Studies in the last 5 years have confirmed these effects, not only for the treatment of DPB but for also cystic fibrosis (CF) as discussed by Adam Jaffé and Andrew Bush. With the widespread adoption of macrolide therapy for the treatment of CF there has been an explosion of interest and publications in the field. A literature search conducted in

June 2004 from the PubMed database shows that there have been nearly 300 references to the immunomodulatory or anti-inflammatory properties of antibiotics since 1976.

This book is divided into two sections; the first, on basic research, evaluates the effects of macrolide antibiotics on bacteria other than by ribosomally-mediated bacteriostasis. Specifically the macrolide antibiotics have been shown to influence the expression of virulence factors in gram-negative organisms and decrease the ability of these bacteria to form biofilms as detailed in the chapters by Kazuhiro Tateda, Theodore J Standiford, and Keizo Yamaguchi. A series of six chapters then follow detailing the various anti-inflammatory and immunomodulary effects of these antibiotics. Immunomodulation in this sense refers to the ability to downregulate deleterious hyperimmunity leading to airway damage as opposed to anti-inflammatory properties, which refers to the suppression of all inflammatory responses whether beneficial or not. Thus immunomodulation should not impair the normal host defense but will prevent an acute inflammatory response from becoming chronic and destructive inflammation. Michael Parnham gives a superb overview of the role of inflammation and its resolution with antibiotics. This is then followed by chapters that document the effect of macrolide antibiotics on cell membrane protection and epithelial stabilization (Charles Feldman and Ronald Anderson), neutrophil activation and chemotaxis (Jun-ichi Kadota), reduction of proinflammatory cytokine expression and release (HajimeTakizawa), the oxidative burst (Marie-Thérèse Labro), and immune activation (Jun-ichi Kadota).

Related to these immunomodulatory effects are the effects on mucus secretion. It is well established that mucus secretion is beneficial to the airway preventing bacterial infection, airway desiccation, and aiding particle clearance; however mucus hypersecretion can lead to airflow obstruction and entrap microorganisms as seen in patients with chronic airway inflammation. Many chronic inflammatory airway diseases such as COPD, asthma, sinusitis, DPB, bronchiectasis and CF are associated with hyperinflammation and airway obstruction with secretions. Kiyoshi Takeyama discusses the role of macrolides in mucus production and secretion and Jun Tamaoki reviews the related data on the regulation of ion channels and how this relates to macrolide antibiotics and mucus secretion.

The second part of the book discusses the clinical results using antibiotics as mucoregulatory agents in a variety of diseases. Shoji Kudoh, who was the first to describe the role of macrolides in the treatment of DPB, and Arata Azuma provide a superbly updated overview of DPB including the current Japanese recommendations for the use of macrolides in treating this disease. These guidelines have proven useful for establishing appropriate therapy for Adam Jaffé and Andrew Bush, who discuss not only their landmark studies of azithromycin for the treatment of CF but also the results of recent large-scale studies that have led to wide acceptance of this therapy. This is followed by a chapter by Kazuhiko Takeuchi, Yuichi Majima, and Qutayba Hamid that reviews the use of macrolides in the therapy chronic upper airway diseases including sinusitis and nasal polyposis. Rose Jung, Mark H. Gotfried, and Larry H. Danziger then summarize the use of macrolides and the treatment of chronic asthma; in particular for persons with neutrophil-predominant, steroid dependent asthma. The role of immunomodulatory antibiotics in the treatment of lung injury is reviewed by Arata Azuma.

Eiji Kita, Keiichi Mikasa and Kei Kasahara give a superb review of the data suggesting a possible role of immunomodulatory antibiotics that can decrease proinflammatory cytokines for the therapy of nonpulmonary disorders including arthritis, inflammatory bowel disease, and cancer. The final chapter by Markus O. Henke, Axel Dalhoff, and Bruce K. Rubin reviews the immunomodulatory properties of antibiotics other than macrolides with the special emphasis on the quinolones, where data now support the ability of these agents to affect the immune systems.

This is an exciting and a rapidly changing field and we are delighted to have the opportunity to summarize the state of the art as of 2004. Thus it is timely that this book be published summarizing these data and it is appropriate that half of the authors are from Japan. We personally believe it is likely that we will see a more widespread use of these antibiotics for their immunomodulatory properties as well as the development of derivatives of these medications that have no antibacterial properties but that do have more potent and directed immunomodulatory activity. This may permit more precise therapy for preventing biofilm diseases or chronic inflammation while reducing the risk of developing antimicrobial resistance to the macrolide class of antibiotics. The editors would like to thank Michael Parnham, the PIR series editor, for suggesting this book and for agreeing to write the overview chapter. We would also like to thank our editors at Birkhäuser Publishing including Karin Neidhart and Hans Detlef Klüber for their outstanding support. Finally the Editors of this monograph would like to thankfully acknowledge the many students and postdoctoral investigators who have worked with us over the years and enriched both our research laboratories and our lives.

Winston-Salem/Tokyo, July 2004

Bruce K. Rubin
Jun Tamaoki

I. Basic research

Indirect antimicrobial effects

Effects of antibiotics on *Pseudomonas aeruginosa* virulence factors and quorum-sensing system

Kazuhiro Tateda[1], Theodore J. Standiford[2] and Keizo Yamaguchi[1]

[1]Department of Microbiology and Infectious Disease, Toho University School of Medicine, 5-21-16 Ohmorinishi, Ohtaku, Tokyo 143-8540, Japan
[2]Pulmonary and Critical Care Medicine, University of Michigan Medical School, Ann Arbor, Michigan, USA

Introduction

Pseudomonas aeruginosa is an opportunistic pathogen that causes a wide range of acute and chronic infections, including sepsis, wound and pulmonary infections [1]. In particular, this organism is a major cause of pulmonary damage and mortality in patients with cystic fibrosis (CF), diffuse panbronchiolitis (DPB) and other forms of bronchiectasis [2, 3].

P. aeruginosa is known to produce a variety of virulence factors, such as pigment and exotoxins. The synthesis and expression of these factors is regulated by a cell-to-cell signaling mechanism referred to as quorum sensing [4, 5]. Two major quorum-sensing components in *P. aeruginosa*, *Las* and *Rhl*, enables bacteria to coordinately turn on and off genes in a density-dependent manner by the production of small diffusible molecules called autoinducers [6, 7]. The expression of these autoinducer-regulated virulence factors directly contributes to the course and outcome of individuals infected with *P. aeruginosa*.

A breakthrough in chemotherapy for patients with chronic *P. aeruginosa* pulmonary infection was realized when a patient with DPB was treated with erythromycin for a prolonged period. This resulted in a dramatic improvement in clinical symptoms, respiratory function and radiographic findings [8]. This astute observation, made by Dr. Shoji Kudoh, lead to a subsequent open trial study which established the clinical effectiveness of long-term erythromycin therapy in DPB patients [9]. Clinical experience in DPB has lead to the use of long-term macrolide therapy in patients with chronic sinusitis, bronchiectasis and CF. While there is mounting evidence of clinical efficacy, the mechanisms of action are still unknown. Currently, investigators are working on two major research directions; 1) macrolide effects on host inflammatory and immune systems, and 2) specific effects of macrolides on the bacteria themselves, including the expression of bacterial virulence factors.

In this chapter, we will review the effects of sub-MIC of macrolides on *P. aeruginosa*, particularly activity of these antibiotics on the bacterial quorum-sensing sys-

Antibiotics as Anti-Inflammatory and Immunomodulatory Agents, edited by Bruce K. Rubin and Jun Tamaoki

tem; a system that may be crucial in the pathogenesis of chronic *P. aeruginosa* infection. Immunomodulatory properties on host responses and clinical efficacy of macrolides will be more comprehensively addressed in other chapters.

An overview of macrolide antibiotics

The macrolide class of antimicrobials is characterized by a multi-membered lactone ring with one or more amino sugars attached. Macrolides are grouped according to the number of atoms comprising the lactone ring, such as 12-, 14-, 15- and 16-membered rings. The 14-membered ring group includes erythromycin, clarithromycin, roxithromycin and oleandomycin, whereas the 16-membered group contains josamycin, kitasamycin and rokitamycin. The only 15-membered ring is azithromycin, which is characterized by a higher degree of intracellular accumulation within leukocytes and more potent antibacterial activity against gram-negative organisms [10].

Macrolides inhibit bacterial protein synthesis by binding to the 50S ribosomal subunit causing an inhibition of translocation of peptidyl-tRNA and the initial steps of 50S subunit assembly. The spectrum of activity of macrolides includes aerobic gram-positive bacteria, especially *Staphylococcus* spp., and *Streptococcus* spp. A few gram-negative bacteria (e.g., *Campylobacter* spp., *Helicobacter* spp., and *Legionella* spp.), and other atypical pathogens including *Mycoplasma* spp. and *Chlamydia* spp., are also susceptible to this class of antibiotics. In contrast, *P. aeruginosa*, as well as other enteric microorganisms, are intrinsically resistant owing to the exclusion of the macrolide from the cytoplasm by the outer membrane architecture.

Generally, the mode of therapeutic efficacy of antibiotics is attributed to the inhibition of bacterial growth *in vivo* when antibiotic concentrations (usually in serum) exceed the minimum inhibitory concentration (MIC), measured on a short exposure time (generally 24 h) to planktonic forms of the bacteria. However, concentrations below the MIC can still attenuate growth and the expression of a variety of bacterial virulence factors, compromising the ability of the pathogen to cause disease. This activity of antibiotics is referred to as sub-MIC effects. The MIC of macrolides for most *P. aeruginosa* strains is in the range of 128–512 μg/ml (our laboratory data). Peak serum concentrations of erythromycin after a 250 mg oral dose are, however, only 1.0–1.5 μg/ml and the mean sputum concentration after an intravenous dose of 1 g every 12 h was 2–3 μg/ml [11, 12]. Thus, judged by conventional criteria, *P. aeruginosa* is fully resistant to macrolide antibiotics. However, there is increasing evidence of a role of sub-MIC macrolides in suppressing virulence factors of this organism.

A characteristic of macrolides that augment their efficacy is that they can concentrate within leukocytes and can enhance the function of aspects of the cellular immune system [13, 14]. For example, intracellular macrolides may be transported to the site of an infection, where they are partially released [15]. These data may explain, in part, how relatively higher concentrations of macrolides can occur at the site of infection, as compared to lower levels observed in serum. Furthermore, macrolide accumulation has been demonstrated to occur not only in host cells, but also within bacteria, especially after a prolonged incubation period [16], which may account for sub-MIC effects on pathogens and perhaps clinical efficacy. These data suggests that macrolide antibiotics have the potential for antibacterial activity, not only through direct bactericidal and bacteriostatic effects, but also through suppression of virulence factors.

Macrolide effects on bacteria

The cellular and molecular mechanisms accounting for the dramatic effect of macrolides in DPB patients has been the subject of intensive research. To summarize a large body of work, the clinical efficacy of macrolides in DPB and CF patients is likely attributable to modulation of host inflammatory and immunological pathways and modulation of bacterial virulence factors, such as suppression of exoproducts (e.g., toxins, pigments, alginate) and bacterial cell components (e.g., flagella, pili, lipopolysaccharide [LPS]). In the discussion to follow, we focus on macrolides effects on bacteria, especially sub-MIC macrolide effects on virulence factors of *P. aeruginosa* and its "quorum-sensing" regulatory system.

Sub-MIC effect of macrolides on bacteria and its virulence factors

Suppression of bacterial exoproducts

P. aeruginosa produces a variety of extracellular products, such as pigment, toxins and exopolysaccharide, which contribute to the pathogenesis through cell/tissue destruction, inflammation and other local and systemic effects [17]. Molinari and associates demonstrated that erythromycin, clarithromycin, and azithromycin differed in their ability to inhibit various *P. aeruginosa* virulence factors. Specifically, azithromycin reduced the synthesis of elastase, protease, lecithinase, and DNase to a greater degree than the other macrolides tested, and was the only agent to suppress pyocyanin production [18, 19]. Sato et al. have reported that erythromycin suppresses the production of pyocyanin dose-dependently *in vitro* [20]. Kita and collaborators have reported that erythromycin over a concentration range of 0.1–10 μg/ml suppressed production of elastase, protease and leucocidin in *P. aeruginosa*; although growth of bacteria was not affected significantly during 24 h culture [21]. Sakata and associates have reported that elastase production was inhibited completely by erythromycin in 27 (79.4%) of 34 strains at concentrations between 0.125 and 64 μg/ml [22]. Likewise, Hirakata and colleagues reported that erythromycin suppressed the *in vitro* production of exotoxin A, total protease, elastase, and phospholipase C by *P. aeruginosa* D4 in a dose-dependent manner [23]. A similar investigation confirmed the greater sub-MIC inhibitory activity of azithromycin, as compared to erythromycin, roxithromycin, and rokitamycin against *P. aeruginosa* exoenzymes and exotoxin A [24].

Strains of *P. aeruginosa* involved in chronic lung infection in DPB and CF develop a mucoid phenotype which is attributable to hyperproduction of alginate. These strains transform into a biofilm coating airway surfaces [25]. Within biofilms, bacteria are protected from antibiotics and the host immune system. Sub-MIC of macrolides have been shown to inhibit both the production of alginate and the formation and stability of biofilms [26–28].

Kobayashi has reported that 14- and 15-membered macrolides specifically inhibited the enzyme guanosine diphosphomannose dehydrogenase (GMD), which is involved in the biosynthesis of alginate, but that the 16-membered macrolide midecamycin was ineffective [29]. It is also notable that macrolides can inhibit α-dornase (recombinant human DNase I) with azithromycin displaying greater activity than erythromycin [30].

Several explanations have been proposed for the sub-MIC effects of macrolides on the expression of *P. aeruginosa* exoproducts. This effect may be due to direct inhibition of translation at the ribosomal level, although it is difficult to imagine how the inhibition of enzymes to as low as 30% of normal function would not substantially impact bacterial growth. It has also been suggested that short peptide chains are preferentially more susceptible to macrolides and this would allow for differential inhibition of enzymes [31]. Regardless of mechanisms involved, it does appear that certain macrolides, but not all family members, are active in suppressing virulence factors of *P. aeruginosa*, and this effect is closely linked with those macrolides that demonstrate clinical efficacy, including erythromycin, clarithromycin, roxithromycin and azithromycin.

Bacterial cell surface components and adherence to host cells

The bacterial cell surface components of LPS and outer membrane proteins of *P. aeruginosa* were disrupted when bacteria were grown at sub-MIC of erythromycin or clarithromycin, but not kitasamycin, josamycin, rokitamycin or oleandomycin [32] (Figure 2).

Erythromycin treatment induced reduction of LPS amounts, as determined by the amount of 2-keto-3-deoxyoctulosonic acid, which is a conserved portion of the LPS molecule. Additionally, a reduction of amount of a 38 kDa protein and a concomitant increase of a 41 kDa protein, which are considered to be *Pseudomonas* outer membrane proteins, were demonstrated. Sub-MIC of erythromycin and clarithromycin also rendered *P. aeruginosa* more susceptible to serum bactericidal activity [33]. These alterations of cell surface structures, such as LPS and outer mema 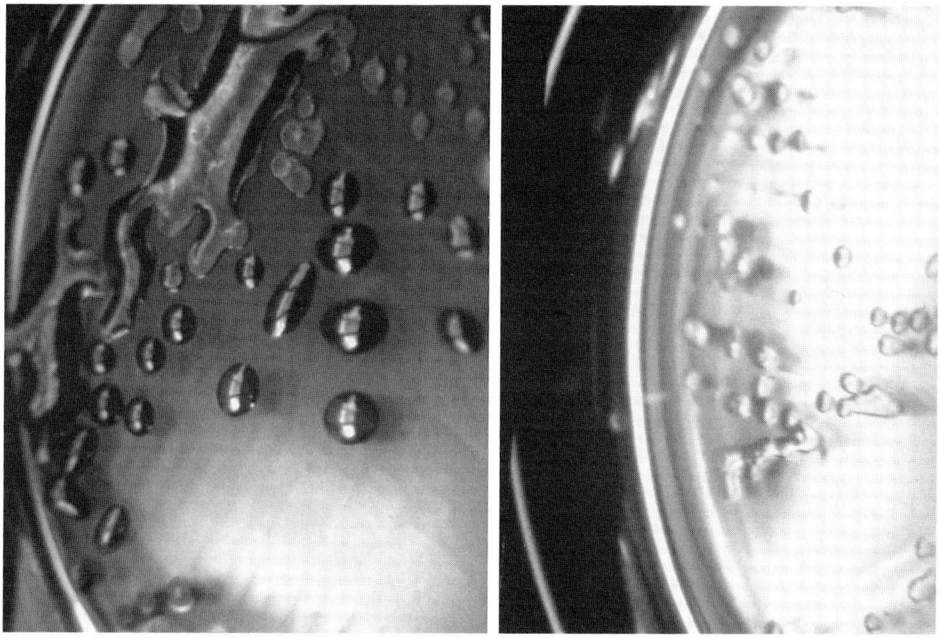 b

Figure 1
Colony of mucoid-type P. aeruginosa *grown in agar with (b) or without (a) sub-MIC of erythromycin (10 µg/ml). Smooth colony has changed to rough in the presence of erythromycin, that suggests suppression of exopolysaccharide alginate.*

brane proteins, may facilitate the access of complement to the outer surface, thus increasing bacterial susceptibility.

Tissue invasion requires the attachment of the microorganism to the host cell. Depending on the host site, the microbe will encounter mucosal or epithelial cells to which it must adhere or be eliminated. Gram-negative bacteria attach primarily by means of proteinaceous appendages known as fimbriae and pili, which extend through the mucus layer to bind to the appropriate host receptor. A number of antibiotics have been shown to impair bacterial adherence [34]. Yamasaki and collaborators have provided compelling evidence that exposure of *P. aeruginosa* to erythromycin at 1/4 MIC for only 4 h significantly reduced the number of pili and hence adherence [35]. Another important cell surface structure is flagella, which facilitates bacterial motility and adherence, and enables bacteria to establish a colony in a more hospitable environment. Molinari and associates have reported that erythromycin, clarithromycin and azithromycin inhibited *P. aeruginosa* motility at sub-MIC [18, 19]. Moreover, Kawamura-Sato and collaborators have report-

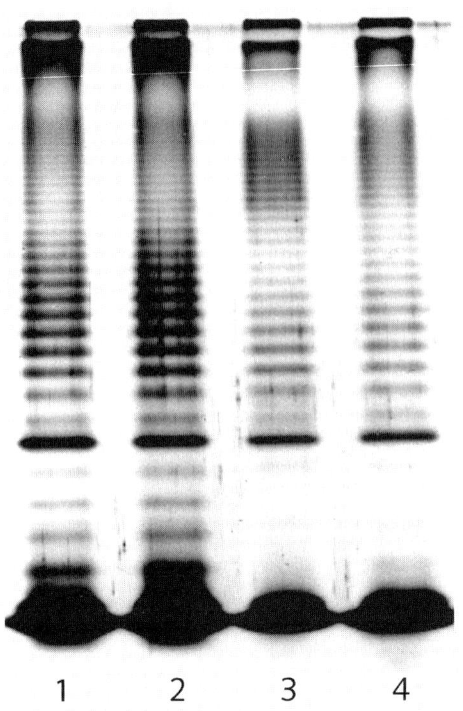

Figure 2
Changes of LPS of P. aeruginosa *grown in agar with sub-MICs of macrolide antibiotics. Lane 1: no antibiotic. Lane 2: josamycin 16 µg/ml. Lane 3: erythromycin 16 µg/ml. Lane 4: azithromycin 4 µg/ml. Change of LPS pattern, especially reduction of lower molecular weight LPS bands, was observed in bacteria grown in the presence of sub-MICs of erythromycin, azithromycin, but not josamycin [32].*

ed that azithromycin can inhibit flagellin expression more effectively than either erythromycin or clarithromycin at concentrations as low as 1/8 MIC [36]. This activity may disrupt biofilm formation in *P. aeruginosa* through inhibition of flagellin expression even at concentrations below the MIC.

Direct killing effects of macrolides with longer incubation

The macrolides do not exhibit intrinsic activity against *P. aeruginosa* based on conventional antimicrobial testing procedures, although appreciable additive and synergistic activities have been observed when macrolides were paired with other antibiotics [37–39]. However, we have reported reduction of viability of *P. aerugi-*

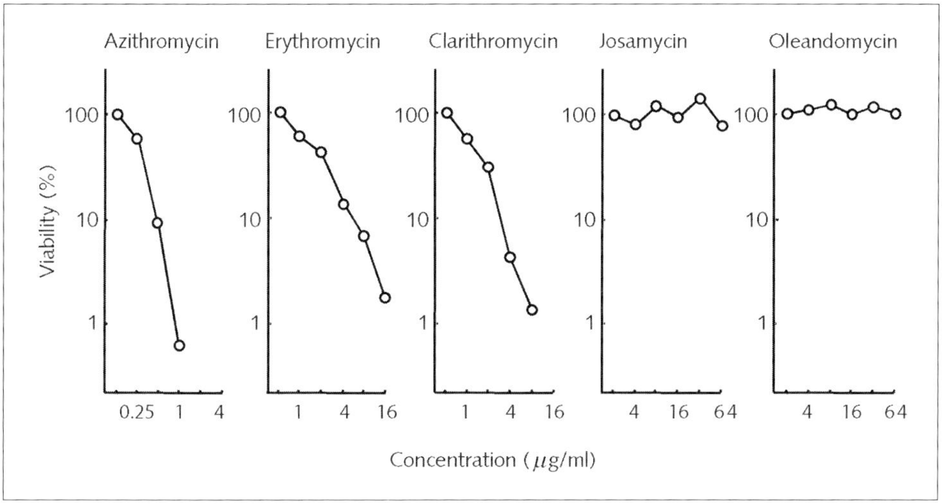

Figure 3
Bactericidal activity of macrolides against P. aeruginosa *after longer incubation*
P. aeruginosa *was incubated on agar with various concentrations of macrolides for 48 hours, and then bacterial viability was compared to that of control bacteria [16].*

nosa when the bacteria were incubated with macrolides for a prolonged time [16]. Exposure to azithromycin for 48 h or more significantly decreased viability of *P. aeruginosa* PAO1 in a concentration-dependent manner, whereas no effect on viability was observed with 24 h or less of incubation. As shown in Figure 3, this time-dependent bactericidal activity was observed with erythromycin, clarithromycin, and azithromycin, but not with josamycin, oleandomycin or other classes of antibiotics (ceftazidime, tobramycin, minocycline, ofloxacin). This reduction in organism viability correlated with a decline in bacterial protein synthesis, which was associated with time-dependent intracellular accumulation of the antibiotic (Fig. 4). Moreover, it is likely that the macrolides may sensitize bacteria to stresses, as these antibiotics induced alterations in a major stress protein, Gro-EL, in both resting and inducible states [40]. These data suggest that conventional antimicrobial susceptibility testing, which is done against planktonic organisms, may not reflect antimicrobial effects of macrolides on *P. aeruginosa* at the site of infection, which may account for discrepancies between clinical efficacy and MIC values.

Figure 5 shows a schematic representing potential effect of macrolides on *P. aeruginosa*. In the respiratory tract or alveolar spaces of patients with persistent *P. aeruginosa* infections, bacteria live on the surface of respiratory cells, where they exist within secreted mucus and host-cell debris in the form of microcolonies or biofilm [41, 42]. As the bacterium multiply, they express virulence factors that may

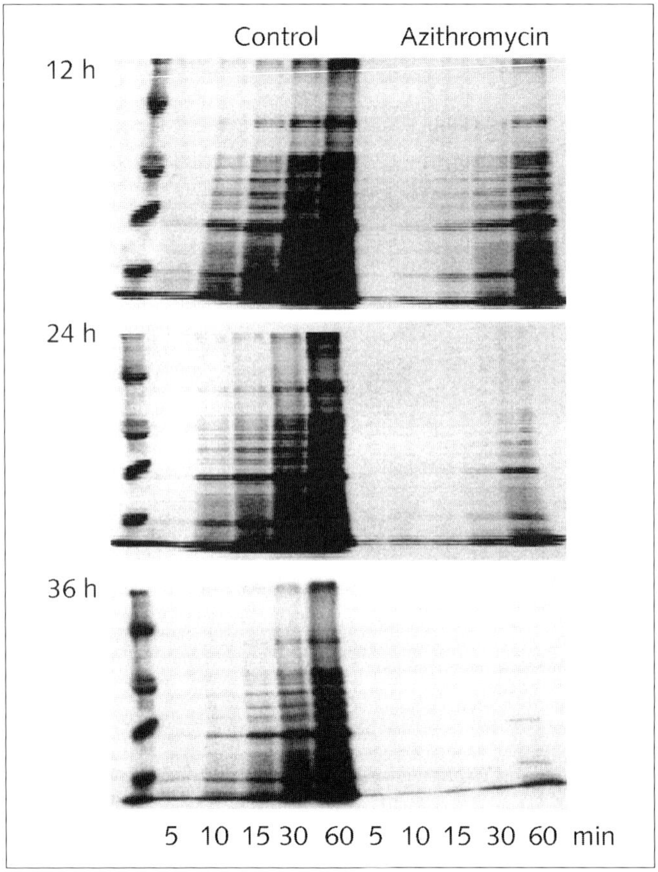

Figure 4
Effects of sub-MIC of azithromycin on protein synthesis of P. aeruginosa
Bacteria was grown on agar with or without azithromycin (4 μg/ml) for 12, 24 or 36 h, and then protein synthesis was examined in a pulse-chase method using ^{35}S-methionine. Significant suppression of protein synthesis was observed in the presence of azithromycin in a time-dependent manner [16].

injure host cells and induce local host responses, including the production of inflammatory mediators, increases in vascular permeability, and leukocyte accumulation. Bacterial populations directly adhering to epithelial cells may be exposed to high macrolide concentrations due to the generation of antibiotic concentration gradients. Under these conditions, sub-MICs of the drug may suppress the virulence of *P. aeruginosa*. Moreover, in patients undergoing macrolide therapy for prolonged periods, bacteria continuously exposed to the antibiotic may be sensitized to the

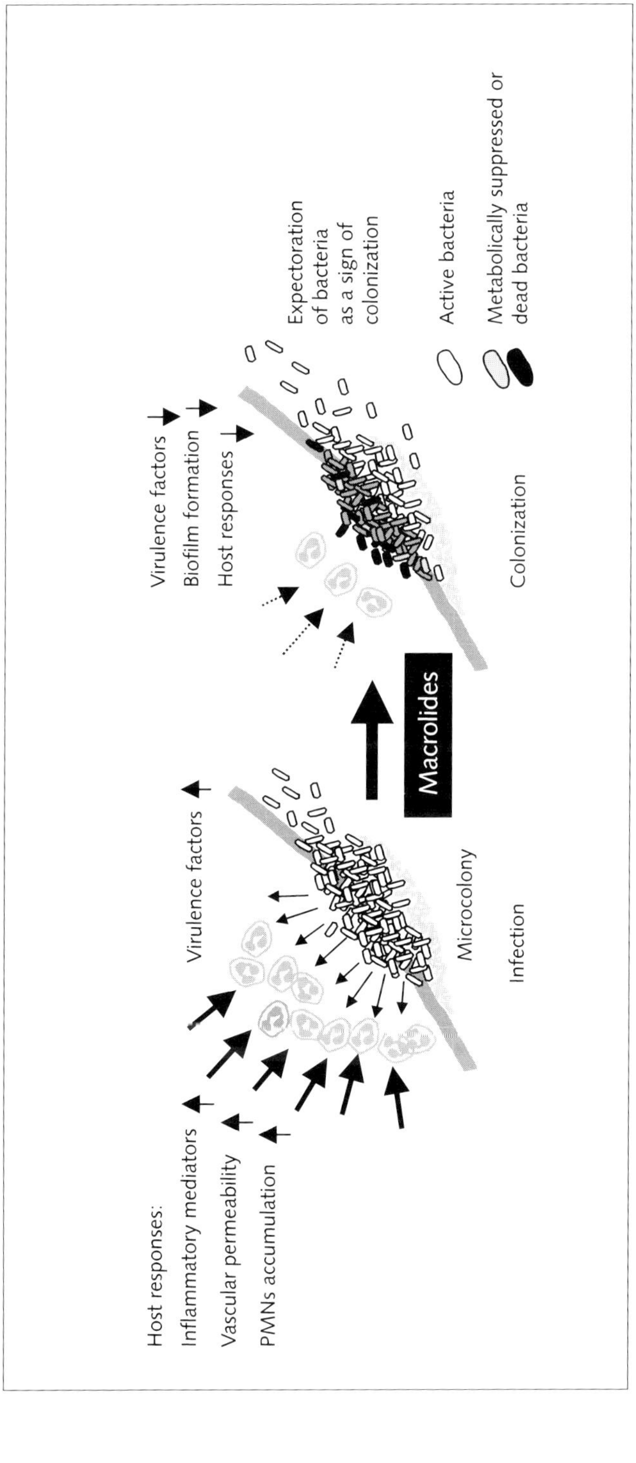

Figure 5: Possible mechanisms of macrolide effects on bacteria

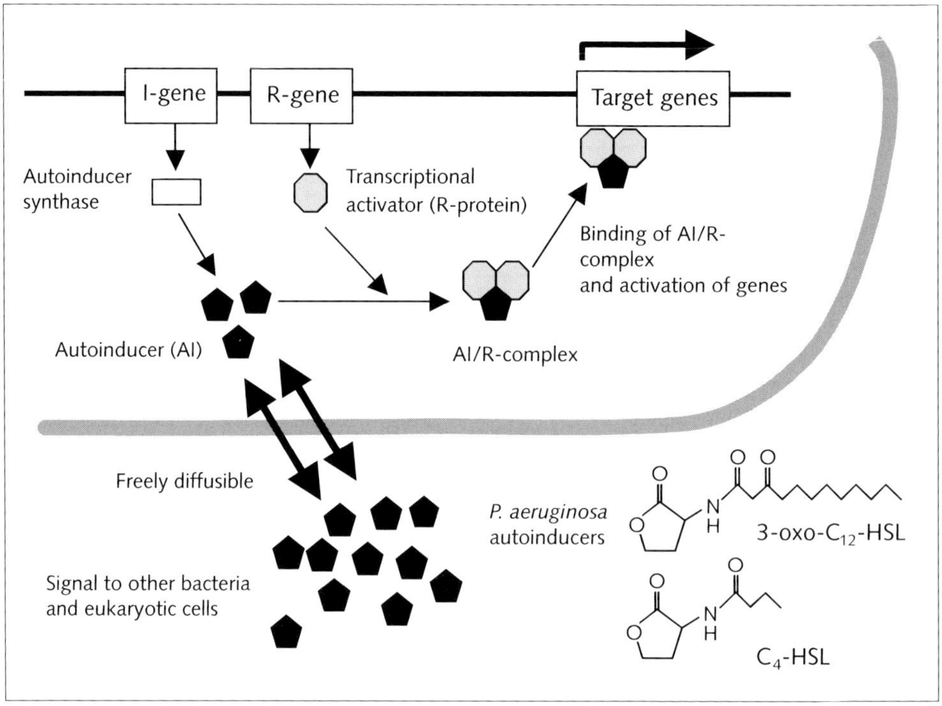

Figure 6
HSL-mediated quorum-sensing systems in bacteria

serum bactericidal effect. Bacteria closely associated with host cells may gradually lose their viability as a consequence of the direct anti-pseudomonal bactericidal activities of these medications. In addition, macrolides may disrupt biofilm attachment to host epithelium. Thus, we speculate that long-term macrolide therapy may shift the host-pathogen interaction from infection to a relatively benign colonization state and possibly even to eradication in some patients. This hypothesis is consistent with the common clinical observation that long-term macrolide therapy leads to improvements in clinical symptoms and laboratory data before any observable bacteriological response.

Quorum-sensing systems as new therapeutic targets

Role of quorum-sensing systems in chronic pulmonary P. aeruginosa infection

P. aeruginosa possesses at least two separate but interrelated quorum-sensing systems, *las* and *rhl* [43, 44]. As the bacterial population increases, the autoinducer signal molecules, 3-oxo-C_{12}-homoserine lactone (HSL) and C_4-HSL, accumulate in the environment. When the concentration of autoinducer reached to a threshold in bacteria, these molecules bind to and activate their cognate transcriptional regulators (Fig. 6). Both systems have been found to regulate multiple virulence factors, such as extracellular toxins (e.g., elastase, alkaline protease, exotoxin A), rhamnolipid and pyocyanin. To investigate the effects of quorum-sensing systems during infections, strains of *P. aeruginosa* that contain deletions in one or more of the quorum-sensing genes were tested in various infection models, including a burn injury mouse model, a murine model of acute pneumonia and a rat model of chronic lung infection [45–48]. A general observation obtained from these models indicates that strains containing a mutation in quorum-sensing genes were less virulent as compared with wild-type *P. aeruginosa*. Another interesting aspect in quorum-sensing research is the contribution and association of this system in biofilm formation. Accumulating data demonstrated that quorum-sensing systems are essential for differentiation and maturation within biofilm in *P. aeruginosa* infection [49–53].

Quorum-sensing is functionally active during *P. aeruginosa* infections in humans. Sputum samples obtained from CF patients chronically infected with *P. aeruginosa* contain mRNA transcripts for the quorum-sensing genes [54]. Sputum from *P. aeruginosa*-infected CF patients also contains the autoinducer molecules 3-oxo-C_{12}-HSL and C_4-HSL [49]. These autoinducer molecules were directly extracted and measured in CF sputum [55]. These samples contained approximately 20 nM 3-oxo-C_{12}-HSL and 5 nM C_4-HSL. In contrast, when bacteria were grown in a biofilm, considerably higher concentrations (300–600 μM) of 3-oxo-C_{12}-HSL were measured [56]. Although it is difficult to define exact concentrations of autoinducer molecules at the site of infection, particularly in biofilm, these results demonstrate that quorum-sensing systems may be active during *P. aeruginosa* infection and potentially regulate the expression of various genes *in vivo*.

Accumulating evidence suggests that the quorum sensing signal molecule 3-oxo-C_{12}-HSL is also a potent stimulator of multiple eukaryotic cells and thus may modulate the host inflammatory response during *P. aeruginosa* infection. *In vitro* experiments have shown that 3-oxo-C_{12}-HSL stimulates the production of the inflammatory chemokine IL-8 from human lung bronchial epithelial cells [57, 58]. In addition, Smith et al. have reported that 3-oxo-C_{12}-HSL could stimulate a complex response *in vivo* by inducing several inflammatory cytokines and chemokines [47]. More recently, we have reported that 3-oxo-C_{12}-HSL from a concentration of 12 μM induces apoptosis in certain types of cells, such as macrophages and neutrophils, but not in epithelial cells [59] (Fig. 7). Taken together, these data suggest that the quorum-sensing molecules have a critical role in the pathogenesis of *P. aeruginosa* infection, not only in the induction of bacterial virulence factors but also in the modulation of host responses. The role of bacterial quorum-sensing systems and their regulation in infection have been reviewed elsewhere [60–63].

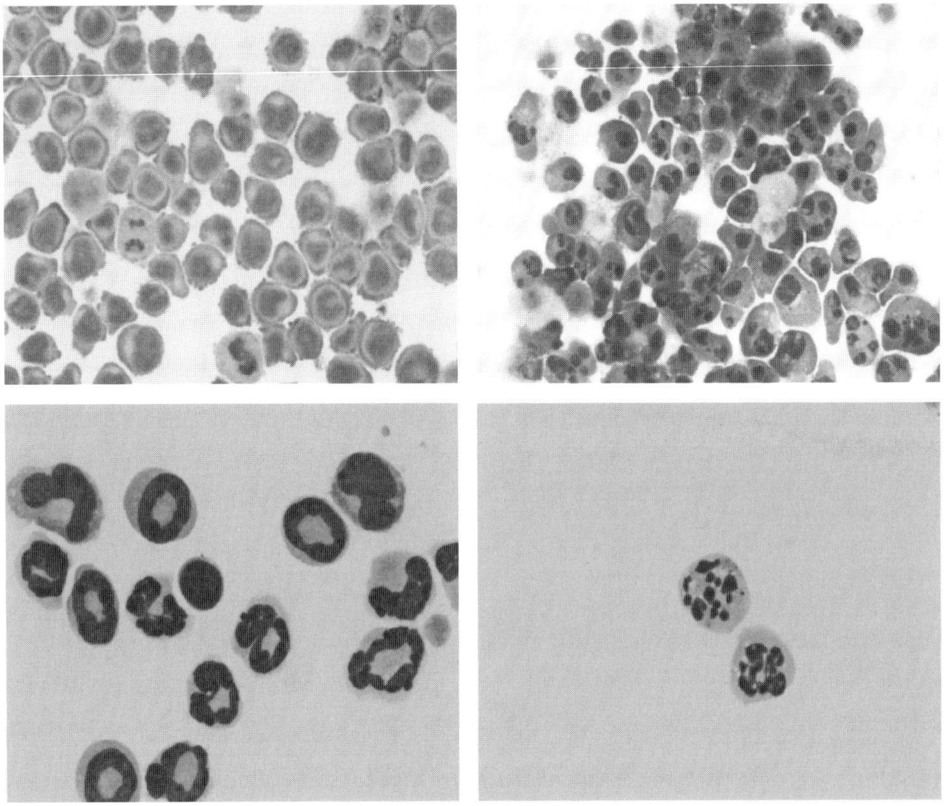

Figure 7
Induction of apoptosis by Pseudomonas *3-oxo-C_{12}-HSL in macrophage and neutrophil*
Macrophage cell line U-937 and mouse neutrophil were incubated with or without 3-oxo-C_{12}-HSL, and then morphology of cells was examined at 4 h after incubation.
a: U-937 cell, control. b: U-937 cell, 3-oxo-C_{12}-HSL. c: neutrophil, control. d: neutrophil, 3-oxo-C_{12}-HSL [59].

Potential of macrolides as quorum-sensing inhibitors

The discovery that gram-negative bacteria employ HSL autoinducer molecules to globally regulate the production of virulence determinants has identified a novel target for therapeutic intervention. The ability to interfere with bacterial virulence by jamming signal generation or signal transduction is intellectually seductive and pharmaceutically appealing, and may also be of considerable clinical importance. Strategies to inhibit quorum-sensing systems include chemical antagonists and specific antibody to inhibit the autoinducers, HSL-destroying enzyme lactonase, and

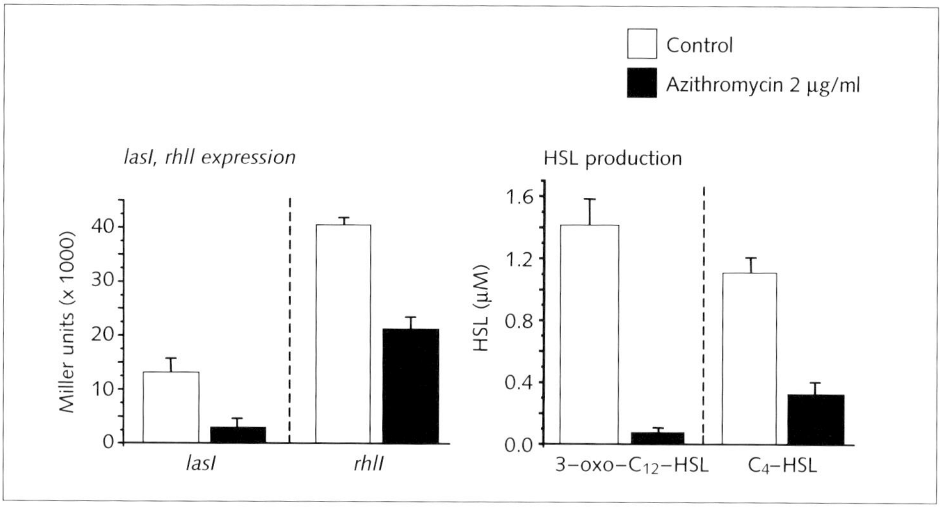

Figure 8
Effects of azithromycin on quorum-sensing systems of P. aeruginosa
P. aeruginosa *was incubated with or without azithromycin 2 μg/ml for 10 hours, and then autoinducer synthase expression (*lasI, rhlI*) and HSL production were examined [67].*

suppression of quorum-sensing by interfering with associated genes and gene products. Several investigators have reported the feasibility of HSL-analogues [64, 65] and synthetic derivatives of natural furanone as means to inhibit bacterial quorum-sensing systems [66].

Clinical and experimental data described above provided a hint that certain macrolides and their analogues may function as *Pseudomonas* quorum-sensing inhibitors. As shown in Figure 8, 2 μg/ml of azithromycin significantly suppressed transcription of *lasI* by 80% and *rhlI* by 50% in *P. aeruginosa* PAO1 [67]. Additionally, the production of 3-oxo-C_{12}-HSL and C_4-HSL was inhibited to approximately 6% and 28% of the control, respectively. In contrast, azithromycin treatment did not alter the expression of the *xcpR* gene, which codes for a structural protein belonging to the type II secretion pathway. These data suggested that azithromycin suppressed quorum-sensing systems in *P. aeruginosa*, and azithromycin's effects on these bacteria are somewhat selective in nature. Importantly, we have observed suppression of *lasI* gene expression by erythromycin, clarithromycin and roxithromycin, but not by oleandomycin and josamycin. These results suggested that the clinically effective macrolides are also the macrolides that are active in suppressing quorum-sensing system, and are consistent with the notion that macrolides might reduce the production of *Pseudomonas* virulence factors by inhibiting the synthesis of the autoinducer molecules.

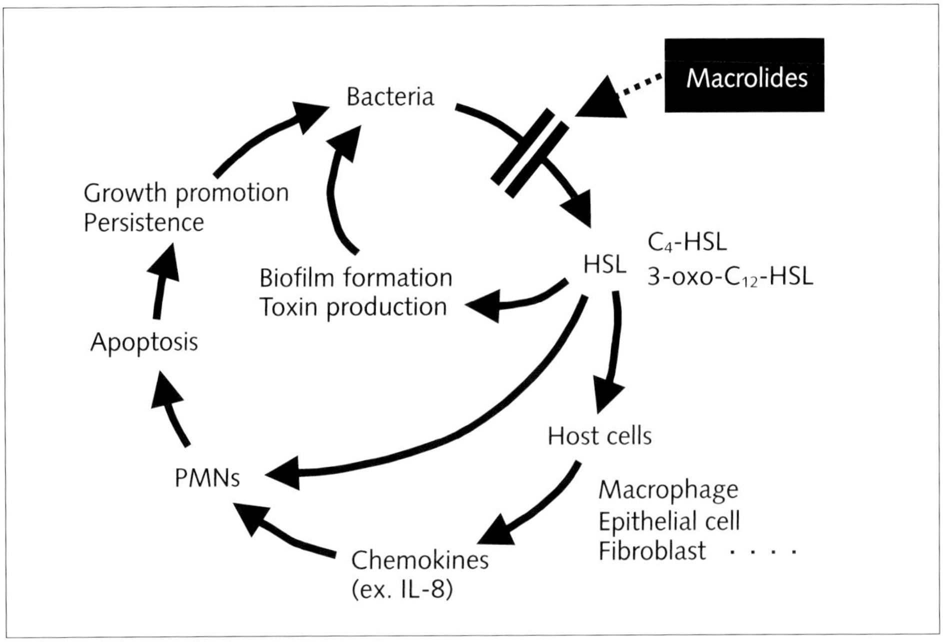

Figure 9
Inhibition of HSL production by macrolides and its impact on pathogenesis of chronic P. aeruginosa *pulmonary infection [59].*

Figure 9 demonstrates several potential mechanisms by which macrolide antibiotics may suppress quorum-sensing systems and highlight their contribution to clinical efficacy in chronic *P. aeruginosa* pulmonary infections. Activation of the quorum-sensing cascade promotes biofilm formation at the site of infection, which make conditions more favorable for bacterial persistence in the lung. Bacterial autoinducers, especially 3-oxo-C_{12}-HSL, stimulates several types of cells, such as epithelial cells, fibroblasts, and macrophages, to induce production of neutrophil chemotactic factors (IL-8 in humans and MIP-2 in mice). Migrated neutrophils are triggered to produce several toxic substances for killing of bacteria, but these molecules, in conjunction with bacterial virulence factors, promote tissue destruction that is a hallmark of the lungs of CF patients. In sites where bacteria are actively producing autoinducers and autoinducer-regulated virulence factors, host cells come in contact with these bacterial factors. In these sites, neutrophils begin to undergo apoptosis, and this process may be accelerated by the presence of bacterial factors, such as 3-oxo-C_{12}-HSL. Apoptotic neutrophils, in addition to secreted mucus and other cell debris, may serve as nutrients for the growth of bacteria and a niche for their survival. Macrolide antibiotics strongly suppress *Pseudomonas* quorum-sensing systems, particularly autoinducer production, which may contribute to suppression of virulence factor expression and biofilm formation. Additionally, macrolides may alter pathogen-driven host responses, such as IL-8 production and apoptosis in neutrophil. Taken together, this evidence supports a potential role of certain macrolides as *Pseudomonas* quorum-sensing inhibitors, which may explain at least in part clinical efficacy of this class of antibiotics in chronic *P. aeruginosa* pulmonary infections. Further research regarding the mechanisms of action and putative target molecules of bacterial quorum-sensing systems, is warranted.

Conclusions

Clinical and basic science data summarized in this review suggests the potential of macrolides as a prototypic inhibitor of bacterial quorum-sensing systems. Given that clinical efficacy of macrolides is associated with suppression of bacterial virulence, including quorum-sensing activity, further investigation aimed at characterizing molecular mechanisms involved may prove fruitful in identifying novel strategies of antimicrobial chemotherapy against antibiotic resistant organisms and biofilm disease.

Acknowledgement
We thank Y. Ishii, S. Kimura and E. Tuzuki (Toho University) for their helpful assistance and discussion. We also express our appreciation to H. Hashimoto, S. Miyairi, M. Horikawa, N. Gotoh, M. Ishiguro (Quorum-sensing group member) and J.C. Pechere, C. Van Delden (University of Geneva) for their helpful suggestion and critical discussion.

References

1 Richards MJ, Edwards JR, Culver DH, Gaynes RP (1999) Nosocomial infections in medical intensive care units in the United States. National Nosocomial Infections Surveillance System. *Crit Care Med* 27(5): 887–92
2 Hoiby N (1994) Diffuse panbronchiolitis and cystic fibrosis: East meets West. *Thorax* 49(6): 531–2
3 Wilson R, Dowling RB (1998) Lung infections. 3. *Pseudomonas aeruginosa* and other related species. *Thorax* 53(3): 213–19
4 Passador L, Cook JM, Gambello MJ, Rust L, Iglewski BH (1993) Expression of *Pseudomonas aeruginosa* virulence genes requires cell-to-cell communication. *Science* 260(5111): 1127–30

5 Kaplan HB, Greenberg EP (1985) Diffusion of autoinducer is involved in regulation of the *Vibrio fischeri* luminescence system. *J Bacteriol* 163(3): 1210–14
6 Pearson JP, Gray KM, Passador L, Tucker KD, Eberhard A, Iglewski BH, Greenberg EP (1994) Structure of the autoinducer required for expression of *Pseudomonas aeruginosa* virulence genes. *Proc Natl Acad Sci USA* 91(1): 197–201
7 Pearson JP, Passador L, Iglewski BH, Greenberg EP (1995) A second N-acylhomoserine lactone signal produced by *Pseudomonas aeruginosa. Proc Natl Acad Sci USA* 92(5): 1490–4
8 Kudoh S, Kimura H (1984) Clinical effect of low-dose long-term administration of macrolides on diffuse panbronchiolits. *Jpn J Thorac Dis* 22: 254
9 Kudoh S, Uetake T, Hagiwara K, Hirayama M, Hus LH, Kimura H, Sugiyama Y (1987) Clinical effects of low-dose long-term erythromycin chemotherapy on diffuse panbronchiolitis. *Jpn J Thorac Dis* 25(6): 632–42
10 Peters DH, Friedel HA, McTavish D (1992) Azithromycin: A review of its antimicrobial activity, pharmacokinetic properties and clinical efficacy. *Drugs* 44(5): 750–99
11 Wilson JT, van Boxtel CJ (1978) Pharmacokinetics of erythromycin in man. *Antibiot Chemother* 25: 181–203
12 Kirst HA, Sides GD (1989) New directions for macrolide antibiotics: pharmacokinetics and clinical efficacy. *Antimicrob Agents Chemother* 33(9): 1419–22
13 Butts JD (1994) Intracellular concentrations of antibacterial agents and related clinical implications. *Clin Pharmacokinet* 27(1): 63–84
14 Tulkens PM (1991) Intracellular distribution and activity of antibiotics. *Eur J Clin Microbiol Infect Dis* 10(2): 100–106
15 Gladue RP, Bright GM, Isaacson RE, Newborg MF (1989) *In vitro* and *in vivo* uptake of azithromycin (CP-62,993) by phagocytic cells: possible mechanism of delivery and release at sites of infection. *Antimicrob Agents Chemother* 33(3): 277–82
16 Tateda K, Ishii Y, Matsumoto T, Furuya N, Nagashima M, Matsunaga T, Ohno A, Miyazaki S, Yamaguchi K (1996) Direct evidence for antipseudomonal activity of macrolides: exposure-dependent bactericidal activity and inhibition of protein synthesis by erythromycin, clarithromycin, and azithromycin. *Antimicrob Agents Chemother* 40(10): 2271–5
17 Pollack M (1984) The virulence of *Pseudomonas aeruginosa. Rev Infect Dis* 6 (Suppl 3): S617–626
18 Molinari G, Paglia P, Schito GC (1992) Inhibition of motility of *Pseudomonas aeruginosa* and *Proteus mirabilis* by subinhibitory concentrations of azithromycin. *Eur J Clin Microbiol Infect Dis* 11(5): 469–71
19 Molinari G, Guzman CA, Pesce A, Schito GC (1993) Inhibition of *Pseudomonas aeruginosa* virulence factors by subinhibitory concentrations of azithromycin and other macrolide antibiotics. *J Antimicrob Chemother* 31(5): 681–8
20 Sato K, Suga M, Nishimura J, Kushima Y, Muranaka H, Ando M (1997) Pyocyanine synthesis by *Pseudomonas aeruginosa* in chronic airway infection and the effect of erythromycin on its biological activity. *Jpn J Antibiot* 50 (Suppl): 89–91

21 Kita E, Sawaki M, Oku D, Hamuro A, Mikasa K, Konishi M, Emoto M, Takeuchi S, Narita N, Kashiba S (1991) Suppression of virulence factors of *Pseudomonas aeruginosa* by erythromycin. *J Antimicrob Chemother* 27(3): 273–84
22 Sakata K, Yajima H, Tanaka K, Sakamoto Y, Yamamoto K, Yoshida A, Dohi Y (1993) Erythromycin inhibits the production of elastase by *Pseudomonas aeruginosa* without affecting its proliferation *in vitro*. *Am Rev Respir Dis* 148(4 Pt 1): 1061–5
23 Hirakata Y, Kaku M, Mizukane R, Ishida K, Furuya N, Matsumoto T, Tateda K, Yamaguchi K (1992) Potential effects of erythromycin on host defense systems and virulence of *Pseudomonas aeruginosa*. *Antimicrob Agents Chemother* 36(9): 1922–7
24 Mizukane R, Hirakata Y, Kaku M, Ishii Y, Furuya N, Ishida K, Koga H, Kohno S, Yamaguchi K (1994) Comparative *in vitro* exoenzyme-suppressing activities of azithromycin and other macrolide antibiotics against *Pseudomonas aeruginosa*. *Antimicrob Agents Chemother* 38(3): 528–33
25 Govan JR, Deretic V (1996) Microbial pathogenesis in cystic fibrosis: mucoid *Pseudomonas aeruginosa* and *Burkholderia cepacia*. *Microbiol Rev* 60 (3): 539–74
26 Yasuda H, Ajiki Y, Koga T, Kawada H, Yokota T (1993) Interaction between biofilms formed by *Pseudomonas aeruginosa* and clarithromycin. *Antimicrob Agents Chemother* 37(9): 1749–55
27 Ichimiya T, Yamasaki T, Nasu M (1994) *In-vitro* effects of antimicrobial agents on *Pseudomonas aeruginosa* biofilm formation. *J Antimicrob Chemother* 34(3): 331–41
28 Ichimiya T, Takeoka K, Hiramatsu K, Hirai K, Yamasaki T, Nasu M (1996) The influence of azithromycin on the biofilm formation of *Pseudomonas aeruginosa in vitro*. Chemotherapy 42(3): 186–91
29 Kobayashi H (1995) Biofilm disease: its clinical manifestation and therapeutic possibilities of macrolides. *Am J Med* 99(6A): 26S–30S
30 Ripoll L, Reinert P, Pepin LF, Lagrange PH (1996) Interaction of macrolides with alpha dornase during DNA hydrolysis. *J Antimicrob Chemother*37(5): 987–91
31 Menninger JR, Coleman RA, Tsai LN (1994) Erythromycin, lincosamides, peptidyl-tRNA dissociation, and ribosome editing. *Mol Gen Genet* 243(2): 225–33
32 Tateda K, Ishii Y, Hirakata Y, Matsumoto T, Ohno A, Yamaguchi K (1994) Profiles of outer membrane proteins and lipopolysaccharide of *Pseudomonas aeruginosa* grown in the presence of sub-MICs of macrolide antibiotics and their relation to enhanced serum sensitivity. *J Antimicrob Chemother* 34(6): 931–42
33 Tateda K, Hirakata Y, Furuya N, Ohno A, Yamaguchi K (1993) Effects of sub-MICs of erythromycin and other macrolide antibiotics on serum sensitivity of *Pseudomonas aeruginosa*. *Antimicrob Agents Chemother* 37(4): 675–80
34 Shibl AM (1985) Effect of antibiotics on adherence of microorganisms to epithelial cell surfaces. *Rev Infect Dis* 7(1): 51–65
35 Yamasaki T, Ichimiya T, Hirai K, Hiramatsu K, Nasu M (1997) Effect of antimicrobial agents on the piliation of *Pseudomonas aeruginosa* and adherence to mouse tracheal epithelium. *J Chemother* 9(1): 32–7
36 Kawamura-Sato K, Iinuma Y, Hasegawa T, Horii T, Yamashino T, Ohta M (2000) Effect

of subinhibitory concentrations of macrolides on expression of flagellin in *Pseudomonas aeruginosa* and *Proteus mirabilis*. *Antimicrob Agents Chemother* 44(10): 2869–72

37 Saiman L, Chen Y, Gabriel PS, Knirsch C (2002) Synergistic activities of macrolide antibiotics against *Pseudomonas aeruginosa*, *Burkholderia cepacia*, *Stenotrophomonas maltophilia*, and *Alcaligenes xylosoxidans* isolated from patients with cystic fibrosis. *Antimicrob Agents Chemother* 46(4): 1105–7

38 Bui KQ, Banevicius MA, Nightingale CA, Quintiliani R, Nicolau DP (2000) *In vitro* and *in vivo* influence of adjunct clarithromycin on the treatment of mucoid *Pseudomonas aeruginosa*. *J Antimicrob Chemother*45(1): 57–62

39 Yanagihara K, Tomono K, Sawai T, Kuroki M, Kaneko Y, Ohno H, Higashiyama Y, Miyazaki Y, Hirakata Y, Maesaki S et al (2000) Combination therapy for chronic *Pseudomonas aeruginosa* respiratory infection associated with biofilm formation. *J Antimicrob Chemother* 46(1): 69–72

40 Tateda K, Ishii Y, Matsumoto T, Kobayashi T, Miyazaki S, Yamaguchi K (2000) Potential of macrolide antibiotics to inhibit protein synthesis of *Pseudomonas aeruginosa*: suppression of virulence factors and stress response. *J Infect Chemother* 6(1): 1–7

41 Lam J, Chan R, Lam K, Costerton JW (1980) Production of mucoid microcolonies by *Pseudomonas aeruginosa* within infected lungs in cystic fibrosis. *Infect Immun* 28(2): 546–56

42 Gilligan PH (1991) Microbiology of airway disease in patients with cystic fibrosis. *Clin Microbiol Rev* 4(1): 35–51

43 Gambello MJ, Iglewski BH (1991) Cloning and characterization of the *Pseudomonas aeruginosa* lasR gene, a transcriptional activator of elastase expression. *J Bacteriol* 173(9): 3000–9

44 Ochsner UA, Koch AK, Fiechter A, Reiser J (1994) Isolation and characterization of a regulatory gene affecting rhamnolipid biosurfactant synthesis in *Pseudomonas aeruginosa*. *J Bacteriol* 176(7): 2044–54

45 Rumbaugh KP, Griswold JA, Iglewski BH, Hamood AN (1999) Contribution of quorum sensing to the virulence of *Pseudomonas aeruginosa* in burn wound infections. *Infect Immun* 67(11): 5854–62

46 Pearson JP, Feldman M, Iglewski BH, Prince A (2000) *Pseudomonas aeruginosa* cell-to-cell signaling is required for virulence in a model of acute pulmonary infection. *Infect Immun* 68(7): 4331–4

47 Smith RS, Harris SG, Phipps R, Iglewski BH (2002) The *Pseudomonas aeruginosa* quorum-sensing molecule N-(3-oxododecanoyl)homoserine lactone contributes to virulence and induces inflammation *in vivo*. *J Bacteriol* 184(4): 1132–9

48 Wu H, Song Z, Givskov M, Doring G, Worlitzsch D, Mathee K, Rygaard J, Hoiby N (2001) *Pseudomonas aeruginosa* mutations in lasI and rhlI quorum sensing systems result in milder chronic lung infection. *Microbiology* 147(Pt 5): 1105–13

49 Singh PK, Schaefer AL, Parsek MR, Moninger TO, Welsh MJ, Greenberg EP (2000) Quorum-sensing signals indicate that cystic fibrosis lungs are infected with bacterial biofilms. *Nature* 407(6805): 762–4

50 Parsek MR, Greenberg EP (1999) Quorum sensing signals in development of *Pseudomonas aeruginosa* biofilms. Methods Enzymol 310: 43–55
51 Parsek MR, Greenberg EP (2000) Acyl-homoserine lactone quorum sensing in gram-negative bacteria: a signaling mechanism involved in associations with higher organisms. *Proc Natl Acad Sci USA* 97(16): 8789–93
52 De Kievit TR, Iglewski BH (1999) Quorum sensing, gene expression, and Pseudomonas biofilms. *Methods Enzymol* 310: 117–28
53 Sauer K, Camper AK, Ehrlich GD, Costerton JW, Davies DG (2002) *Pseudomonas aeruginosa* displays multiple phenotypes during development as a biofilm. *J Bacteriol* 184(4): 1140–54
54 Storey DG, Ujack EE, Rabin HR, Mitchell I (1998) *Pseudomonas aeruginosa* lasR transcription correlates with the transcription of lasA, lasB, and toxA in chronic lung infections associated with cystic fibrosis. *Infect Immun* 66(6): 2521–8
55 Erickson DL, Endersby R, Kirkham A, Stuber K, Vollman DD, Rabin HR, Mitchell I, Storey DG (2002) *Pseudomonas aeruginosa* quorum-sensing systems may control virulence factor expression in the lungs of patients with cystic fibrosis. *Infect Immun* 70(4): 1783–90
56 Charlton TS, de Nys R, Netting A, Kumar N, Hentzer M, Givskov M, Kjelleberg S (2000) A novel and sensitive method for the quantification of N-3-oxoacyl homoserine lactones using gas chromatography-mass spectrometry: application to a model bacterial biofilm. *Environ Microbiol* 2(5): 530–41
57 DiMango E, Zar HJ, Bryan R, Prince A (1995) Diverse *Pseudomonas aeruginosa* gene products stimulate respiratory epithelial cells to produce interleukin-8. *J Clin Invest* 96(5): 2204–10
58 Smith RS, Fedyk ER, Springer TA, Mukaida N, Iglewski BH, Phipps RP (2001) IL-8 production in human lung fibroblasts and epithelial cells activated by the *Pseudomonas* autoinducer N-3-oxododecanoyl homoserine lactone is transcriptionally regulated by NF-kappa B and activator protein-2. *J Immunol* 167(1): 366–74
59 Tateda K, Ishii Y, Horikawa M, Matsumoto T, Miyairi S, Pechere JC, Standiford TJ, Ishiguro M, Yamaguchi K (2003) The *Pseudomonas aeruginosa* autoinducer N-3-oxododecanoyl homoserine lactone accelerates apoptosis in macrophages and neutrophils. *Infect Immun* 71(10): 5785–93
60 de Kievit TR, Iglewski BH (2000) Bacterial quorum sensing in pathogenic relationships. *Infect Immun* 68(9): 4839–49
61 Miller MB, Bassler BL (2001) Quorum sensing in bacteria. *Annu Rev Microbiol* 55: 165–99
62 Whitehead NA, Barnard AM, Slater H, Simpson NJ, Salmond GP (2001) Quorum-sensing in Gram-negative bacteria. *FEMS Microbiol Rev* 25(4): 365–404
63 Schauder S, Bassler BL (2001) The languages of bacteria. Genes Dev 15(12): 1468–80
64 Reverchon S, Chantegrel B, Deshayes C, Doutheau A, Cotte-Pattat N (2002) New synthetic analogues of N-acyl homoserine lactones as agonists or antagonists of transcrip-

tional regulators involved in bacterial quorum sensing. *Bioorg Med Chem Lett* 12(8): 1153–7

65 Smith KM, Bu Y, Suga H (2003) Induction and inhibition of *Pseudomonas aeruginosa* quorum sensing by synthetic autoinducer analogs. *Chem Biol* 10(1): 81–9

66 Hentzer M, Wu H, Andersen JB, Riedel K, Rasmussen TB, Bagge N, Kumar N, Schembri MA, Song Z, Kristoffersen P et al (2003) Attenuation of *Pseudomonas aeruginosa* virulence by quorum sensing inhibitors. *Embo J* 22(15): 3803–15

67 Tateda K, Comte R, Pechere JC, Kohler T, Yamaguchi K, Van Delden C (2001) Azithromycin inhibits quorum sensing in *Pseudomonas aeruginosa. Antimicrob Agents Chemother* 45(6): 1930–3

Anti-inflammatory effects

Antibiotics, inflammation and its resolution: An overview

Michael J. Parnham

PLIVA Research Institute Ltd, Prilaz baruna Filipovića 29, HR-10 000 Zagreb, Croatia

Introduction

Inflammation is a dynamic process that involves chronological changes. Initially, the acute response with plasma exudation and vasodilation facilitates the infiltration of blood-borne leukocytes and the release of chemotactic agents, such as complement factor C5a, at the site of tissue injury or infection. The neutrophilic granulocytes are the first cells to respond to the tissue alarm signals. Neutrophils are vital for host defence, particularly against bacteria and compromise of this defence is hazardous. Their release of proteinases and other inflammatory mediators, together with increased production of oxygen species contributes to the killing of bacteria, but also damages the surrounding tissue [1, 2]. Consequently, resolution of the acute inflammatory response is crucial to avoid excessive injury to structural tissue. Recent investigations indicate that locally released lipids such as prostaglandin D_2 derivatives play an important role in this process of resolution of inflammation [3]. They contribute towards the induction of programmed cell death (apoptosis) of neutrophils, thereby curtailing the continued release of inflammatory agents [2, 4]. The apoptotic neutrophils are phagocytosed by macrophages, which further stimulate the healing process by clearing tissue debris, releasing growth factors and stimulating formation of replacement connective tissue [4]. Failure to kill microorganisms or sustained immune responses to local (auto) antigens leads to prolonged inflammatory responses, macrophages and lymphocytes releasing cytokines and other inflammatory products that contribute towards severe tissue damage.

Thus, while stimulation of the acute inflammatory response – including the activity of neutrophils – can be beneficial in facilitating removal of bacteria, subsequent stimulation of leukocyte apoptosis and of inflammatory mediator release can be crucial in preventing undesirable tissue damage, either in infectious diseases or in non-infectious chronic inflammatory conditions. The outcome of pharmacological modulation of inflammation is therefore dependent on the timing of treatment as well as the ultimate indication. Early stimulation of the acute inflammatory response to inflammation may be beneficial in infections, but facilitated resolution

Antibiotics as Anti-Inflammatory and Immunomodulatory Agents, edited by Bruce K. Rubin and Jun Tamaoki

of the response is needed to limit tissue damage. On the other hand, inhibition of unresolved inflammation, either by antibiotics or specifically anti-inflammatory agents, is needed to relieve patients with chronic inflammatory disorders. This chapter will review some of the recent evidence for the modulation of this dynamic inflammatory process by antibacterial drugs. The reader is referred to later chapters for detailed discussion of the effects of these drugs on leukocyte chemotaxis, oxidative burst and cytokine release, as well as effects on immune responses.

Modulation of proinflammatory processes

Many antibacterial drugs have been shown to exert effects on leukocytes, particularly neutrophils, and some of these agents have been found to affect experimental inflammation in animals. The most promising drugs have been administered to patients with inflammatory disorders, a topic discussed in a later chapter. The antibacterial agents that have been most investigated in this respect are the macrolides, quinolones and tetracyclines.

Accumulation of antibiotics in inflammatory cells

Macrolide antibiotics accumulate in inflammatory cells at concentrations up to several hundred-fold higher than those in extracellular fluid [5, 6] enabling phagocytic cells to deliver concentrated active drug to sites of infection. The mechanism of intracellular accumulation is not clear, but exhibits characteristics of an active (protein-mediated) process [5]. Concentration occurs in the cytoplasm and azurophilic granules of neutrophils, thus favouring antibiotic delivery to bacteria phagocytosed by leukocytes. Cytokines stimulate *in vitro* accumulation of macrolides into macrophages, suggesting that at the site of inflammation (infection), cells may accumulate even more macrolide antibiotics than under physiological conditions [7].

Efflux or release of macrolides from leukocytes varies among macrolides, being very fast with erythromycin and clarithromycin, but very slow with azithromycin [8, 9], so that the latter agent is retained much longer in the cells. This offers the possibility of both prolonged activity against invading bacteria and extended modulation of leukocyte function, beyond that which might be observed in short-term cell cultures *in vitro*.

Other antibacterials can also accumulate to some degree in cells, but nowhere near the extent of that of the macrolides. For instance, uptake *via* the nucleoside transport system may explain the approximate 20-fold cellular accumulation of clindamycin into alveolar macrophages [10]. Apart from erythromycin, the only other antibiotics that showed some selective accumulation (2–5-fold) were the lipid-soluble chloramphenicol, rifampin, tetracycline, and lincomycin. Neutrophil uptake of the quinolone, ciprofloxacin, is also approximately 5-fold that of the extracellular fluid [10a].

Effects on plasma exudation and infiltration of leukocytes

Leukocyte adhesion is an initial hallmark of the inflammatory process. The recruitment of these cells to a site of inflammation occurs through a sequence of events involving the specific arrest of leukocytes on the vascular endothelium and their transmigration across the endothelial cell barrier. Four phases are involved in this adhesion process – margination, capture, rolling and adhesion – mediated by cell adhesion molecules of the selectin and integrin families, their expression being stimulated by locally released inflammatory cytokines [11]. The directed migration of the leukocytes into the tissue is further stimulated by locally generated chemotactic factors, such as chemokines and complement anaphylatoxins, accompanied by plasma exudation and swelling that is facilitated by the release of vasodilatory factors, such as prostaglandin E_2. Several classes of antibacterials have been shown to modulate various aspects of this initial acute inflammatory response, effects on chemotaxis being discussed in a later chapter.

The macrolide, erythromycin, has been reported to be capable of downregulating expression of integrins CD11b/CD18 and of Mac-1 on leukocytes after short-term incubation [12, 13]. Erythromycin treatment for 2 weeks of rats with experimental otitis media led to a downregulation of L-selectin and Mac-1 expression on peripheral blood neutrophils and inhibited macrophage and neutrophil accumulation in middle ear effusions [14, 15]. The macrolide roxithromycin was ineffective on whole blood cells *in vitro* [12], but was found to reduce Mac-1 expression on neutrophils after treatment of patients with chronic lower respiratory tract disease, including diffuse panbronchiolitis [16], suggesting that prolonged contact is required to cause inhibition. Roxithromycin also inhibited neutrophil adhesion to bronchial epithelial cells *in vitro* [17]. Similarly, in human bronchial epithelial and synovial (fibroblast-like) cells, clarithromycin markedly inhibited expression of several adhesion molecules, such as intercellular adhesion molecule-1 (ICAM-1), lymphocyte function-associated antigen-3 (LFA-3) and vascular cell adhesion molecule-1 (VCAM-1) [18]. Clearly, inhibition of adhesion molecule expression makes a notable contribution to the anti-inflammatory effects of macrolides [19].

Erythromycin, but not clarithromycin, also ameliorates neutrophil-induced endothelial cell damage, at least partially by stimulating endothelial NO synthetase (eNOS)-mediated NO production by a protein kinase A-dependent mechanism and/or by enhancing NOS expression [20, 21]. This NO generation could either enhance vasodilation or modify the function of migrating leukocytes.

Used for the treatment of leprosy on the basis of its weak activity against *M. leprae*, clofazimine has been shown to be of benefit in a number of other skin diseases including cutaneous discoid, pyoderma gangrenosum and pustular psoriasis [22]. A possible mechanism was proposed to be inhibition of the expression of ICAM-1 and HLA-DR molecules as seen in dermal biopsies from patients with *erythema dyschromicum perstans* lesions [23]. To what extent this contributes to other clinical effects of the antibiotic is unclear.

Following adherence to the vascular endothelium, leukocytes move between the endothelial cell junctions and enter the tissue along the concentration gradient of chemotactic mediators. Inhibitory effects of macrolides on leukocyte chemotaxis were documented some time ago *in vitro* [24] as well as *in vivo* [25]. All quinolones modestly but significantly impair rat macrophage chemotaxis, in a concentration-dependent manner [26], while clofazimine has also been shown to inhibit neutrophil motility *ex vivo* [27]. Effects of antibiotics on chemotaxis will be discussed in detail in a later chapter.

The ability of macrolides to inhibit plasma exudation and cell infiltration *in vivo* is illustrated by the fact that several of these antibiotics were found to be effective in carrageenan-induced paw oedema, the standard animal model used for evaluating anti-inflammatory drugs [28]. Rats pretreated with erythromycin or roxithromycin were also protected from airway inflammatory reactions, including vascular leakage, caused by injection of *E. coli* endotoxin lipopolysaccharide [29]. Importantly, no protection was observed in neutropenic rats, indicating that the main target for the anti-inflammatory activity of the macrolides was the neutrophil. This conclusion is supported by the results of another study showing that clarithromycin and erythromycin inhibit endotoxin lipopolysaccharide-induced recruitment of neutrophils into guinea pig trachea [30]. A similar anti-inflammatory action, targeting the neutrophil, was seen in the rat model of immune complex-induced lung injury. Erythromycin and josamycin both inhibited neutrophil accumulation and reduced the concentration of NO in exhaled air [31].

Enhancement of initial cellular defence reactions

The stimulation of leukocyte, especially neutrophil activity is a crucial aspect of defence against infection. Lysozyme released from neutrophilic granules, together with other degradative enzymes, is directly injurious to bacteria. Following opsonisation by complement or immunoglobulin, phagocytosis of opsonised bacteria leads to the stimulation of the oxidative burst that generates reactive oxygen species capable of breaking down bacterial membranes and proteins. Chemokines, such as interleukin-8 (IL-8), further stimulate the cells, also generating cytokines that activate other inflammatory processes.

Macrolides directly stimulate exocytosis (degranulation) by human neutrophils *in vitro* [26]. With the exception of roxithromycin, these agents also stimulate macrophage chemotaxis, phagocytosis and cytocidal activity against *Candida albi-*

cans [32]. In this way, macrolides facilitate their own direct antibacterial activity by stimulating host defense reactions against bacteria and other microorganisms. This stimulatory activity of macrolides can also be seen after repeated administration to otherwise healthy animals. In healthy mice, a 28-day (but not a 7-day) treatment with erythromycin or roxithromycin (10 mg/kg) resulted in increased production of proinflammatory cytokines by isolated macrophages and IL-2 by isolated splenocytes [33, 34]. It should be noted that stimulatory effects of macrolides on host defence reactions in healthy animals differ markedly from inhibitory effects in experimental inflammatory models, as discussed below. Although macrolide antibiotics generally inhibit neutrophil responses *in vitro* [6, 26], in the *healthy* guinea pig, roxithromycin given for 14 days enhanced the oxidative burst of neutrophils in these animals [35]. It has been suggested that macrolides may stimulate non-activated leukocytes, but their reactivity may be reversed following priming by cytokines [6]. In support of this proposal, macrolides have recently been shown to stimulate cyclic AMP in lipopolysaccharide (LPS)-primed peripheral blood human leukocytes, but not in unstimulated leukocytes [6a].

Some cephalosporins, β-lactams and quinolones have also been reported to enhance neutrophil bacterial killing and/or phagocytosis and the phagocyte oxidative burst *in vitro* [26]. These effects are discussed in more detail in a later chapter. Quinolones, however, at clinically achievable concentrations, generally do not affect granulocyte functions [35a]. Most quinolone antibacterials, particularly ciprofloxacin, have been shown to superinduce proinflammatory cytokine gene transcription (IL-2 and interferon-γ) production by mitogen-activated human T lymphocytes *in vitro*, apparently by activation of the nuclear factor AP-1 [35a]. This has lead to their study as immunomodulators, as discussed elsewhere in this volume.

Recently, the effect of the quinolone, moxifloxacin, on THP-1 monocytic cells, stimulated *in vitro* with zymogen A or *S. aureus*, has been shown to be biphasic [36]. Within the first hour, moxifloxacin increased the release of NO and hydrogen peroxide, but after 4 h lipid peroxidation, lysosomal enzyme release and the release of proinflammatory cytokines was inhibited. Such a biphasic action could potentially enhance initial antibacterial activity, while subsequently facilitating resolution of inflammation and tissue healing. This biphasic activity has also been proposed for the macrolide, azithromycin, on the basis of *in vivo* data obtained by administering the antibacterial (500 mg/day) to healthy human subjects for three consecutive days [37]. An initial neutrophil degranulating effect of azithromycin, 2.5–24 h after the last dose, was reflected in rapid decreases in azurophilic granule enzyme activities in cells and corresponding increases in serum. The oxidative response to a particulate stimulus (opsonised zymosan) *ex vivo* was also acutely enhanced. These actions were associated with high plasma and neutrophil drug concentrations. A continuous fall in chemokine and interleukin-6 serum concentrations, within the non-pathological range, accompanied a delayed downregulation of the oxidative burst and an increase in apoptosis of neutrophils up to 28 days after the last azithromycin dose.

Consequently, azithromycin – and perhaps some other antibiotic agents, such as quinolones – may complement their direct antibacterial actions by enhancing cellular defence mechanisms and then facilitate resolution of undesirably prolonged inflammation.

Inhibition of inflammatory responses

Considerable evidence has accumulated for the inhibitory effects of antibiotics, particularly of macrolides, tetracyclines and quinolones, on the generation of inflammatory mediators, including reactive oxygen species and cytokines, as well as for the their inhibitory effects on immune responses. These anti-inflammatory effects are discussed in detail in later chapters. In-keeping with these inhibitory actions, anti-inflammatory effects of several antibiotics in experimental animal models have been reported.

Macrolides

In general, macrolides inhibit synthesis of reactive oxygen species and/or secretion of proinflammatory cytokines *in vitro* while exerting variable effects on the release of anti-inflammatory cytokines. Other inflammatory mediators are also inhibited. Thus, in contrast to the stimulatory effects of erythromycin on eNOS, discussed above in relation to interactions between neutrophils and endothelial cells, inducible NO synthetase (iNOS) expression by stimulated alveolar macrophages is reduced by treatment with erythromycin, clarithromycin and josamycin *in vitro* [38, 39]. In rat acute carrageenan-induced pleurisy, NO production, TNF-α levels, and prostaglandin E_2 were significantly reduced by pretreatment with roxithromycin, clarithromycin, and erythromycin [40]. The same three macrolides (but not the 16-membered josamycin) given for 4 weeks to mice have also been shown to reduce plasma total NO, IL-1β, IL-6 and TNF-α levels and lung iNOS mRNA after i.p. injection of LPS [41]. Using a similar *in vivo* i.p. LPS test system in rats, a series of modified macrolides, devoid of antibacterial activity, have been described that inhibit neutrophilia, as well as cytokine release *in vitro* [42]. This suggests that non-antibacterial compounds can be developed with anti-inflammatory activity, based on the macrolide structure.

Erythromycin administration also caused anti-inflammatory effects in zymosan-induced peritonitis in rats, the anti-inflammatory effect being most obvious after 28 days of (pre) treatment [43, 44]. This suggests that the anti-inflammatory effect of macrolides is a rather slow process that needs a prolonged period to become fully developed. Similar findings were obtained with roxithromycin on mouse endotoxin

Table 1 - Effects of macrolides on experimental inflammatory models in vivo. *From [6]*

Model	Species	Macrolide	Change
Healthy	guinea pig	roxithromycin	↑ ciliary activity ↑ neutrophil oxidative burst
	mice	roxithromycin erythromycin	↑ macrophage interleukin-1 ↑ splenocyte interleukin-2
Carrageenan pleurisy	rat	roxithromycin erythromycin clarithromycin	↓ NO, prostaglandin E_2, TNF-α
Zymosan peritonitis	rat	erythromycin	↓ inflammation
Lipopolysaccharide inflammation	mouse	roxithromycin	↓ interleukin-1, TNF-α
Adjuvant arthritis	rat	erythromycin azithromycin	↓ lysosomal enzymes
Lipopolysaccharide airway inflammation	rat	erythromycin roxithromycin	↓ vascular leakage
	guinea pig	erythromycin clarithromycin	↓ neutrophil accumulation
Immune complex lung inflammation	rat	erythromycin josamycin	↓ neutrophil accumulation ↓ NO in exhalate
Otitis media	rat	erythromycin	↓ leukotriene B_4, leukotriene C_4, ↓ prostaglandin E_2, neutrophil adhesion

lipopolysaccharide-induced inflammation [45] In this case, 7 weeks treatment was required in order to obtain pronounced suppression of cytokine secretion (interleukin-1β and TNF-α). The importance of neutrophils for the anti-inflammatory actions of macrolides was further supported by studies on the rat model of otitis media. The anti-inflammatory effects of macrolides in experimental models *in vivo* are summarized in Table 1.

It should be pointed out that macrolides are also inhibitors of mucus secretion *in vitro* and *in vivo*, an action that contributes to their beneficial effects on upper airway inflammation [46, 47]. This activity will be discussed in detail in a later chapter. Here it is worth noting that clarithromycin, at least, has been shown recently to inhibit the gene expression of the major mucin protein, muc5ac, in a *Pseudomonas aeruginosa* lung inflammation model of diffuse panbronchiolitis in mice [48]. Both clarithromycin and erythromycin (but not the 16-membered macrolide josamycin, nor ampicillin) inhibited mucus production and neutrophil infiltration induced in rats by intranasal ovalbumen and lipopolysaccharide [49]. The mucus and neutrophil-inhibiting activities of macrolides probably account for the long-standing therapeutic usefulness of these antibacterial agents, particularly erythromycin, in the treatment of human diffuse panbronchiolitis and in the renewed interest in their use for the treatment of asthma [46, 47, 50].

Macrolides may inhibit other chronic types of inflammation as well. Erythromycin and azithromycin were shown to be anti-inflammatory in reducing circulating lysosomal enzyme activities in adjuvant-induced arthritis in rats *in vivo* [51] and roxithromycin has been reported to exert an antiangiogenic effect through inhibition of TNF-α-mediated vascular endothelial growth factor (VEGF) induction [52].

The transcription factors NF-κB and AP-1 mediate a wide variety of cellular inflammatory responses and are under intense investigation as potential targets for anti-inflammatory drugs [53–55]. Both NF-κB and AP-1 seem to be important intracellular mediators of the anti-inflammatory actions of macrolides [6]. In an elegant series of experiments on LPS-primed human peripheral blood leukocytes and THP-1 cells *in vitro*, Abeyama et al. [6a] have recently shown that erythromycin, clarithromycin and roxithromycin inhibit the reactive oxygen intermediate-induced activation of NF-κB in a cyclic AMP-dependent manner. The LPS-induced *transcription*, but not the rapid TNF-α-induced *translocation* of the transcription factor, NF-κB was inhibited. In addition, these macrolides stimulated the generation of anti-inflammatory IL-10 in a cyclic AMP- and CREB-dependent manner.

Cyclines

Cyclines interfere with bacterial protein synthesis, and have been widely reported to inhibit various phagocyte functions, including cytokine release, at therapeutic concentrations [26]. Initially investigated in periodontal disease [56], these drugs were found to exhibit anti-inflammatory and bone resorption-inhibiting effects independent of their antibacterial activity. This led to the investigation of minocycline in the treatment of rheumatoid arthritis by several groups. Analysis of the results of these studies has confirmed the modest, but significant improvement in disease activity with minocycline with no absolute increased risk of side effects [57].

The anti-inflammatory properties of tetracyclines have been reviewed and their spectrum of anti-inflammatory activity proposed to make them attractive candidates for use in the prevention of acute lung injury [58]. As these authors point out, the most prominent action of tetracyclines is the downregulation of the expression of the metalloproteinases MMP-2 and MMP-3, an action that protects α1-proteinase inhibitor from inactivation. In this way, activation of neutrophil elastase is prevented. This protease-inhibiting activity can account for many of the beneficial actions of tetracyclines observed clinically. Removal of the dimethylamino group at position C4 of the tetracyclines abolishes antibacterial activity, but in at least one derivative,

Table 2 - Anti-inflammatory actions of tetracyclines [58, 60–63, 96]

Inhibition of metalloproteinases (MMP-2, MMP-8, MMP-9)
Prevention of inactivation of α1-proteinase inhibitor
Inhibition of expression of inducible NO synthetase (iNOS)
Reactive oxygen scavenging
Inhibition of TNF-α release (inhibition of TNF-α converting enzyme – TACE)
Inhibition of induction of IL-1β converting enzyme
Inhibition of expression of cyclooxygenase-2 (COX-2)
Inhibition of apoptosis
Inhibition of T lymphocyte proliferation
Inhibition of murine B lymphocyte immunoglobulin secretion and class switching
Reduction of mortality in murine endotoxin-induced shock
Inhibition of occlusion-induced rat cerebral ischemic damage
Decrease in incidence of adjuvant arthritis

CMT-3, inhibitory activity against metalloproteinases is retained [58]. CMT-3 has also been shown to improve the biomechanical properties of femoral bones in rats with adjuvant arthritis, but without any effect on joint inflammation [59]. Inhibition of T lymphocyte proliferation and of immunoglobulin synthesis may also contribute to inhibitory effects of other tetracyclines on systemic arthritic responses [60, 61]. A variety of anti-inflammatory actions of classical tetracyclines have been reported and these are summarised in Table 2. Inhibition of reactive oxygen species, prevention of inducible NO synthetase (iNOS) expression and inhibition of apoptosis may all contribute towards inhibitory effects of minocycline on experimental neuroinflammatory disorders [62, 63]. So far, the potential of tetracyclines and minocycline in particular, in treatment of chronic inflammation has not been realised, possibly because of concerns about side effects and the possible increase in bacterial resistance.

Quinolones

Inhibitory effects of 4-quinolones on cytokine production by human monocytes *in vitro* have been described in several studies and they have been shown to inhibit phagocytosis, adhesion and the oxidative burst of macrophages *in vitro* [26]. In addition to their stimulatory actions on proinflammatory cytokines *in vitro*, discussed earlier, they also selectively modify T lymphocyte functions [64]. Quinolones were recently proposed as anti-inflammatory agents, based on their somewhat different modulation of cytokine responses *in vivo*. In mice injected with a lethal dose of LPS, trovafloxacin, ciprofloxacin and tosufloxacin significantly protected mice against death and diminished serum levels of TNF-α and IL-6 [65]. Inhibition of collagen type II arthritis in the rat has also been reported with fluoroquinolones. However, this inhibitory activity *in vivo* may be indirect, as the inhibitory effect of ciprofloxacin on collagen type II arthritis was reversed in adrenalectomised rats [65a].

Fosfomycin

Fosfomycin (1-cis-1,2-epoxypropylphosphoric acid) is a broad-spectrum bactericidal antibiotic unrelated to any other known antibacterial agent. Like macrolides, it appears to inhibit cytokine production in association with inhibition of NF-κB activation [66, 67]. At least in T lymphocytes, the activity of fosfomycin as an inhibitor of cytokine release is less than that of the macrolide, clarithromycin [68]. In mice injected with LPS, fosfomycin significantly lowered peak serum levels of TNF-α and IL-1β and in the rat carrageenan air-pouch model, fosfomycin also reduced local PGE2 and TNF-α concentrations, as well as mRNA for cyclooxygenase-2 [26]. Fosfomycin has immunomodulatory activity on B and T lymphocyte functions and inhibits histamine release from basophils [26]. The immunomodulatory activity of fosfomycin (and of its enantiomer, which lacks antimicrobial activity) has been demonstrated in various animal models and it has also been shown to improve symptoms in patients with severe bronchial asthma [26, 69].

Other antibiotics

Fusidic acid, mainly used as an antistaphyloccocal agent, decreases granulocyte functions *in vitro*, without markedly altering monocyte functions. It also protects mice from LPS- and staphylococcal enterotoxin B-induced lethality, and suppresses TNF-α and IFN-γ release *in vivo* [26, 69]. Clindamycin also exerts an inhibitory effect in LPS-induced septic shock, through inhibition of proinflammatory cytokine release *in vitro* and *in vivo* [70, 71]. In a model of concanavalin A (Con A)-induced liver damage, prophylactic administration of fusidic acid protected mice from Con A-induced hepatitis and this was accompanied by markedly reduced plasma levels of proinflammatory cytokines [26]. In experimental autoimmune neuritis in rats (a model of Guillain-Barré syndrome), fusidic acid also alleviated symptoms and decreased proinflammatory cytokine release [72]. Similar inhibitory effects of fusidic acid have been observed in rats with experimental allergic encephalomyelitis (EAE), a model associated with mononuclear cell infiltration of the central nervous system (CNS) [73].

Dapsone (4,4' diaminophenyl sulfone), initially developed as an antitubercular drug, is used to treat leprosy. It was later tested in malaria and some inflammatory diseases. Dapsone inhibits neutrophil functions such as chemotaxis and oxidant production [26]. In addition, it irreversibly inhibits myeloperoxidase (MPO), by converting the enzyme into its inactive (ferryl) form and inhibits the production of prostaglandin E_2 by neutrophils [26]. Clofazimine, like dapsone, is also an inhibitor of MPO, but appears to act differently from dapsone in that it scavenges chlorinating oxidants [74].

Sulfonamides inhibit phagocyte functions [75], and sulfasalazine, found to be effective in rheumatoid arthritis in the late 1940s, is still widely used for the treatment of early rheumatoid arthritis and for inflammatory bowel disease [76, 77]. While a number of possible mechanisms of action have been proposed for this anti-inflammatory agent, it has recently been suggested to act *via* inhibition of NF-κB [78].

Pro-apoptotic effects and resolution of inflammation

In addition to the modulatory effects of antibiotics on the processes leading to the development of the inflammatory response, there is also evidence for their ability to facilitate the resolution of inflammation through stimulation of apoptosis. Indeed, therapeutic induction of apoptosis (programmed cell death) as a means to resolve chronic inflammation is gaining increasing interest [79, 80] and macrolides could be blueprints for this approach.

Neutrophils are normally extremely short-lived cells, with a circulating life-time of only 6–7 h [81, 82]. This means that normal individuals make (and destroy) about 50 billion neutrophils per day, and many more in inflammatory states [83]. In other words, at least 50 g of neutrophils are destroyed by apoptosis each day! Importantly, phagocytosis of bacteria also induces apoptosis in neutrophils and this is accompanied by specific gene-mediated attenuation of many functional aspects of these cells [2]. In contrast to necrotic neutrophils, apoptotic neutrophils are ingested by macrophages [4]. Thus, granulocyte-induced tissue injury and chronic inflammation may result not only from excessive leukocyte recruitment but also inhibition of normal apoptosis-based clearance mechanisms.

Several reports have described the pro-apoptotic effects of erythromycin. It was reported to accelerate apoptosis of neutrophils through a mechanism that is at least partially cAMP-dependent [84]. This action of erythromycin has been confirmed in isolated human neutrophils and in guinea-pig eosinophils stimulated with IL-5 and extended also to roxithromycin [85, 86]. Azithromycin was also shown to stimulate apoptosis of neutrophils, without releasing proinflammatory IL-8 or inducing the oxidative burst, but in the presence of *S. pneumoniae* this effect was abolished [87], probably due to the fact that phagocytosis of the bacteria had already induced apoptosis [2]. Both erythromycin and azithromycin showed pro-apoptotic potential in a whole blood model, as determined by flow cytometry [88]. Tilmicosin, which reduces pulmonary inflammation in calves, was also shown to significantly stimulate apoptosis of peripheral neutrophils when isolated cells were incubated with this macrolide for 2 h [89]. The tilmicosin-induced apoptosis, in contrast to that described above for azithromycin, occurred regardless of the presence or absence of

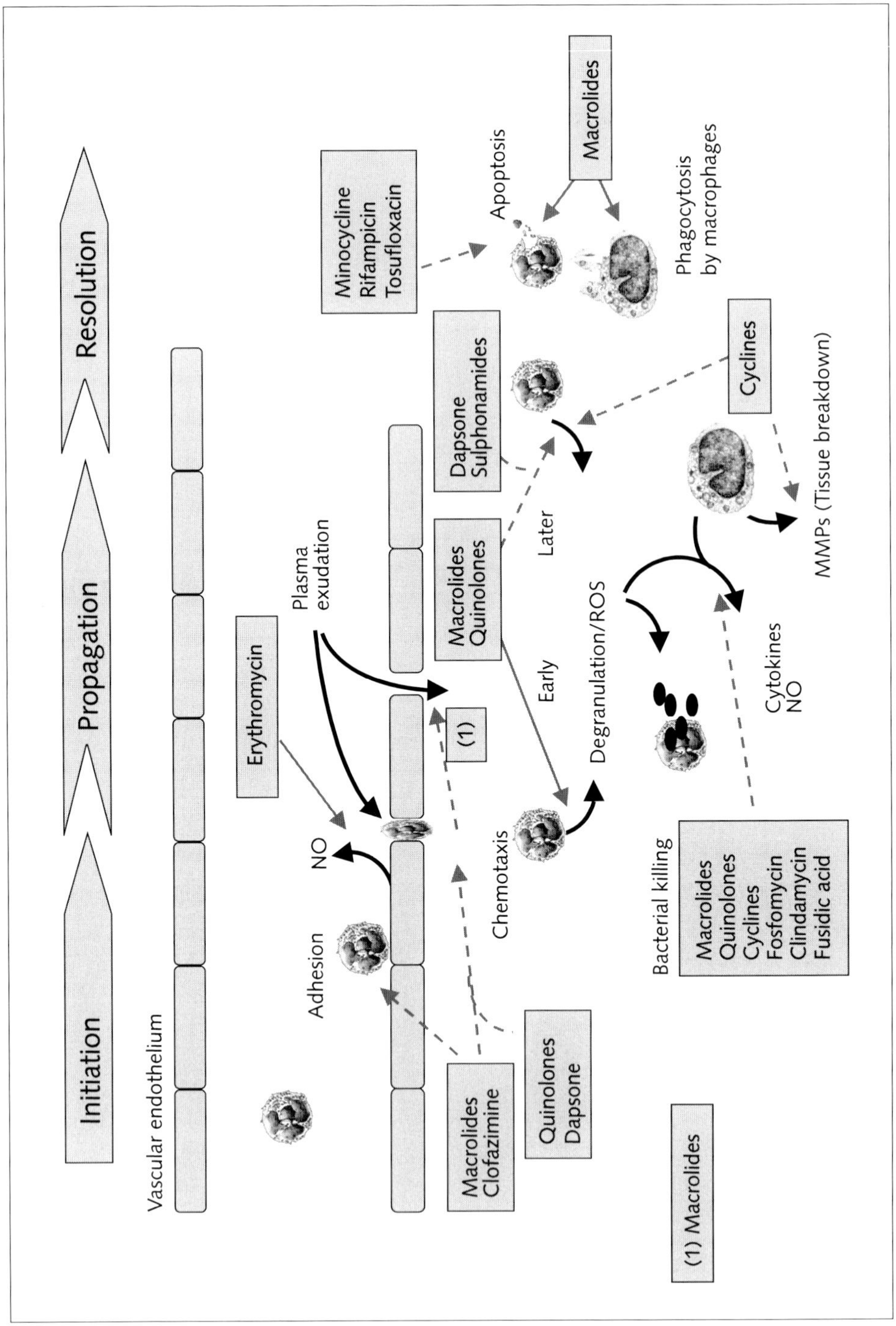
Initiation
Propagation
Resolution
Vascular endothelium
Adhesion
NO
Erythromycin
Plasma exudation
(1)
Chemotaxis
Macrolides Clofazimine
Quinolones Dapsone
Macrolides Quinolones
Early
Later
Degranulation/ROS
Dapsone Sulphonamides
Minocycline Rifampicin Tosufloxacin
Apoptosis
Macrolides
Phagocytosis by macrophages
Cyclines
MMPs (Tissue breakdown)
Cytokines NO
Bacterial killing
Macrolides Quinolones Cyclines Fosfomycin Clindamycin Fusidic acid
(1) Macrolides

bacteria (*Pasteurella haemolytica*). It would appear that either bacteria differentially modulate this process or that differences in pro-apoptotic effects exist between macrolides. In this respect, a 17-membered tylosine derivative was reported to cause apoptosis in several different cell lines [90] but the 16-membered macrolide, josamycin, had no effect on human neutrophil apoptosis [86]. Inhibition of the NF-kB pathway generally stimulates apoptosis in granulocytes *in vitro* [79]. However, a recent *in vivo* study has shown that activation of NF-κB in leukocytes recruited during the *onset* of inflammation leads to proinflammatory gene expression, while during *resolution* of inflammation, activation of NF-κB is associated with anti-inflammatory gene expression and apoptosis [91]. This is reminiscent of the stimulatory effects of macrolides in healthy animals and inhibitory effects in inflamed animals, described above in an earlier section. In the human volunteer study, in which azithromycin was administered for 3 days and caused initial stimulation of neutrophil degranulation, azithromycin was detectable for up to 28 days in circulating neutrophils that showed an increasing rate of apoptosis (and therefore neutrophil death) over the 28 days after stopping the treatment [37]. The findings described here thus suggest that the actions of macrolides on apoptosis may be time-dependent and associated with altered reactivity to activation of NF-κB.

A recent study has also opened a further facet to the ability of macrolides to facilitate the resolution of neutrophilic inflammation. Treatment of human alveolar macrophages *in vitro* with the 14- and 15-membered macrolides, erythromycin, clarithromycin or azithromycin (but not 16-membered macrolides, clindamycin or beta-lactam antibiotics) stimulated, in a phosphatidylserine receptor-dependent manner the phagocytosis of apoptotic neutrophils by the macrophages [92]. A sim-

Figure 1
Summary of main actions of antibiotics on different phases of the inflammatory response. Inhibitory actions are indicated by dashed and stimulatory actions by unbroken lines.
Initiation of inflammation *includes vasodilation and adhesion, chemotaxis and transendothelial migration of leukocytes, associated with plasma exudation. Several antibiotics inhibit aspects of this phase, while erythromycin, at least, has beneficial effects on endothelial cells. During* inflammation propagation, *leukocytes are activated, inflammatory mediators and degradative enzymes released and any responsible bacteria destroyed. Plasma exudation is further promoted and tissue in the immediate vicinity may also be destroyed. Many antibiotics exhibit inhibitory actions on different components of this phase. At least the macrolide azithromycin and the quinolone minocycline exert differential effects on leukocytes during the initiation and propagation phases.*
Resolution of inflammation *is associated with the release of anti-inflammatory cytokines and leukocyte apoptosis. This is facilitated by macrolides but inhibited by some other antibiotics. Effects of macrolide antibiotics on inflammation are predominantly restricted to 14- and 15-membered macrolides. MMP, metalloproteinase.*

ilar action of tilmicosin has been reported in promoting phagocytosis of neutrophils by macrophages [89]. Thus, in keeping with other effects of macrolides on inflammation, 14- and 15-membered, but not 16-membered macrolides, are able to clear neutrophils from inflammatory sites both by direct stimulation of apoptosis and their phagocytic removal by macrophages.

Effects of other antibiotics on apoptosis have been observed as well. However, in these cases, the apoptotic process was inhibited or delayed. Inhibition of apoptosis has been proposed to contribute towards inhibitory effects of minocycline on experimental neuroinflammatory disorders [63] and rifampicin has been reported to inhibit antiCD95-mediated apoptosis of Jurkat T cells and peripheral blood lymphocytes, at least partly *via* glucocorticoid receptor activation and the NF-κB signalling pathway [93, 94]. Recently, tosufloxacin, but not other quinolone antibiotic was found to delay neutrophil apoptosis *in vitro*, an action that was attenuated by a p38 mitogen-activated protein kinase (MAPK) inhibitor [95]. Such apoptosis-inhibiting actions, however, are likely to be of more relevance to immuno-enhancing effects of these drugs or possibly to inhibitory effects on autoimmune disorders. Whether inhibition of apoptosis may interfere with bacterial-killing effects of the antibiotics remains unclear.

Conclusions

The inflammatory response is modulated by a variety of different antibiotics (Fig. 1). The classes of antibiotics that have clearly the most effects on host defence mechanisms are macrolides, cyclones and quinolones, though others may affect the adaptive immune system, as discussed in a later chapter. Macrolides (and possibly clofazimine) have inhibitory effects on adhesion and transepithelial migration of leukocytes, but early migrating leukocytes appear to be stimulated by this class of antibacterials, as well as by the quinolone moxifloxacin. These stimulatory effects may facilitate bacterial killing in association with their therapeutic indication as antibiotics. Subsequently, macrolides and some quinolones inhibit leukocyte and other inflammatory responses, leading to dampening of the inflammatory process *in vivo*. In fact, a relatively broad number of antibiotics exhibit inhibitory effects on proinflammatory cytokine release. Inhibitory actions on inflammation could offer the possibility for additional therapeutic effects of antibiotics, such as the long-standing use of macrolides for the treatment of diffuse panbronchiolitis. Cyclines also inhibit metalloproteinase release, among other actions, and inhibit connective tissue breakdown. In fact, minocycline has shown beneficial effects in periodontal and rheumatic diseases.

Apart from direct inhibitory effects on inflammatory responses, macrolides, in particular, stimulate leukocyte apoptosis and may therefore assist the resolution of the inflammatory response. On the one hand, this can be beneficial in reducing "collateral damage" to surrounding tissues during bacterial infections, but also contributes to the growing understanding that macrolides have potential as anti-inflammatory agents. Their inhibitory actions on mucus secretion suggest that respiratory conditions are the most promising. Cyclines and quinolones also may offer new structural approaches to the development of anti-inflammatory agents, but their clinical application in this area has not been extensively investigated. Sulphasalazine, of course, represents the best example of an antibiotic that became an anti-inflammatory drug. Perhaps others may follow.

Dedicated to the memory of Professor Derek A. Willoughby.

References

1 Witko-Sarsat V, Rieu P, Descamps-Latscha B, Lesavre P, Halbwachs-Mecarelli L (2000) Neutrophils: molecules, functions and pathophysiological aspects. *Lab Invest* 80: 617–53

2 Kobayashi SD, Voyich JM, DeLeo FR (2003) Regulation of the neutrophil-mediated inflammatory response to infection. *Microbes Infection* 5: 1337–44

3 Gilroy DW, Colville-Nash PR, McMaster S, Sawatzky DA, Willoughby DA, Lawrence T (2003) Inducible cyclooxygenase-derived 15-deoxy(Delta)12-14 PGJ2 brings about acute inflammatory resolution in rat pleurisy by inducing neutrophil and macrophage apoptosis. *FASEB J* 17: 2269–71

4 Savill J, Haslett C (1999) Granulocytes. In: JD Winkler (ed): *Apoptosis and inflammation*. Birkhäuser Verlag, Basel, 53–84

5 Labro M-T (2002) Cellular accumulation of macrolide antibiotics. Intracellular bioactivity. In: W Schönfeld, HA Kirst (eds): *Macrolide antibiotics*. Birkhäuser Verlag, Basel, 37–52

6 Čulić O, Eraković V, Parnham MJ (2001) Anti-inflammatory effects of macrolide antibiotics. *Eur J Pharmacol* 429: 209–29

6a Abeyama K, Kawahara K, Iino S, Hamada, T Arimura S, Matsushita K, Nakajima T, Maruyama I (2003) Antibiotic cyclic AMP signalling by "primed" leukocytes confers anti-inflammatory cytoprotection. *J Leuk Biol* 74: 908–15

7 Bermudez LE, Inderlied C, Young LS (1991) Stimulation with cytokines enhances penetration of azithromycin into human macrophages. *Antimicrob Agents Chemother* 35: 2625–9

8 Gladue RP, Bright GM, Isaacson E, Newborg MF (1989) *In vitro* and *in vivo* uptake of azithromycin (CP-62,993) by phagocytic cells: Possible mechanism of delivery and release at sites of infection. *Antimicrob Agents Chemother* 33: 277–82

9 Fieta A, Merlini C, Grassi GC (1997) Requirements for intracellular accumulation and release of clarithromycin and azithromycin by human phagocytes. *J Chemother* 9: 23–31

10 Hand WL, Corwin RW, Steinberg TH, Grossman GD (1984) Uptake of antibiotics by human alveolar macrophages. *Am Rev Respir Dis* 129: 933–7
10a Easmon CS, Crane JP (1985) Uptake of ciprofloxacin by human neutrophils. *J Antimicrob Chemother* 16: 67–73
11 Muller WA (2003) Leukocyte-endothelial-cell interactions in leukocyte transmigration and the inflammatory response. *Trends Immunol* 24: 327–34
12 Okubo J (1997) Macrolides reduce the expression of surface Mac-1 molecule on neutrophil. *Kurume Med J* 44: 115–23
13 Lin HC, Wang CH, Liu CY, Yu CT, Kuo HP (2000) Erythromycin inhibits beta2-integrins (CD11b/CD18) expression, interleukin-8 release and intracellular oxidative metabolism in neutrophils. *Respir Med* 94: 654–60
14 Enomoto F, Ichikawa G, Nagaoka I, Yamashita T (1996) Evaluation of the effect of erythromycin on otitis media with effusion in experimental rat models. *Nippon Jibiinkoka Gakkai Kaiho* 99: 1126–35
15 Enomoto F, Ichikawa G, Nagaoka I, Yamashita T (1998) Effect of erythromycin on otitis media with effusion in experimental rat model. *Acta Otolaryngol* (Suppl) 539: 57–60
16 Kusano S, Kadota J, Kohno S, Iida K, Kawakami K, Morikawa T, Hara K (1995) Effect of roxithromycin on peripheral neutrophil adhesion molecules in patients with chronic lower respiratory tract disease. *Respiration* 62: 217–22
17 Kawasaki S, Takizawa H, Ohtoshi T, Takeuchi N, Kohyama T, Nakama K, Kasama T, Kobayashi K, Nakahara K, Morita Y et al (1998) Roxithromycin inhibits cytokine production by and neutrophil attachment to human bronchial epithelial cells *in vitro*. *Antimicrob Agents Chemother* 42: 1499–502
18 Matsuoka N, Eguchi K, Kawakami A, Tsuboi M, Kawabe Y, Aoyagi T, Nagataki S (1996) Inhibitory effect of clarithromycin on costimulatory molecule expression and cytokine production by synovial fibroblast-like cells. *Clin Exp Immunol* 104: 501–8
19 Suzuki H, Ikeda K (2002) Mode of action of long-term low-dose macrolide therapy for chronic sinusitis in the light of neutrophil recruitment. *Curr Drug Targets Inflamm Allergy* 1: 117–26
20 Mitsuyama T, Hidaka K, Furuno T, Hara N (1997) Neutrophil-induced endothelial cell damage: inhibition by a 14-membered ring macrolide through the action of nitric oxide. *Int Arch Allergy Immunol* 114: 111–15
21 Mitsuyama T, Hidaka K, Furuno T, Hara N (1998) Release of nitric oxide and expression of constitutive nitric oxide synthase of human endothelial cells: enhancement by a 14-membered ring macrolide. *Mol Cell Biochem* 181: 157–61
22 Lee SJ, Wegner SA, McGarigle CJ, Bierer BE, Antin JH (1997) Treatment of chronic graft-*versus*-host disease with clofazimine. *Blood* 89: 2298–302
23 Baranda L, Torres-Alvarez B, Cortes-Franco R, Moncada B, Portales-Perez DP, Gonzalez-Amaro R (1997) Involvement of cell adhesion and activation molecules in the pathogenesis of erythema dyschromicum perstans (ashy dermatitis). The effect of clofazimine therapy. *Arch Dermatol* 133: 325–29

24 Esterly NB, Furey NL, Flanagan LE (1978) The effect of antimicrobial agents on leukocyte chemotaxis. *J Invest Dermatol* 70: 51–5

25 Nelson S, Summer WR, Terry PB, Warr GA, Jakab GJ (1987) Erythromycin-induced suppression of pulmonary antibacterial defenses: a potential mechanism of superinfection in the lung. *Am Rev Respir Dis* 136: 1207–12

26 Labro M-T (2000) Interference of antibacterial agents with phagocytic functions: Immunomodulation or "immuno-fairy tales"? *Clin Microbiol Rev* 13: 615–50

27 Van Rensburg CEJ, Gatner EMS, Inkamp FMJH, Anderson R (1982) Effects of clozamine alone or combined with dapsone on neutrophil and lymphocyte functions in normal individuals and patients with lepromatous leprosy. *Antimicrob Agents Chemother* 21: 693–8

28 Scaglione F, Rossoni G (1998) Comparative anti-inflammatory effects of roxithromycin azithromycin and clarithromycin. *J Antimicrob Chemother* 41 Suppl B: 47–50

29 Tamaoki J, Sakai N, Tagaya E, Konno K (1994) Macrolide antibiotics protect against endotoxin-induced vascular leakage and neutrophil accumulation in rat trachea. *Antimicrob Agents Chemother* 38: 1641–3

30 Tamaoki J, Takeyama K, Yamawaki I, Kondo M, Konno K (1997) Lipopolysaccharide-induced goblet cell hypersecretion in the guinea pig trachea: inhibition by macrolides. *Am J Physiol* 272: L15–L19

31 Tamaoki J, Kondo M, Kohri K, Aoshiba K, Tagaya E, Nagai A (1999) Macrolide antibiotics protect against immune complex-induced lung injury in rats: role of nitric oxide from alveolar macrophages. *J Immunol* 163: 2909–15

32 Xu G, Fujita J, Negayama K, Yuube K, Hojo S, Yamaji Y, Kawanishi K, Takahara J (1996) Effects of macrolide antibiotics on macrophage functions. *Microbiol Immunol* 40: 473–9

33 Kita E, Sawaki M, Nishikawa F, Mikasa K, Yagyu Y, Takeuchi S, Yasui K, Narita N, Kashiba S (1990) Enhanced interleukin production after long-term administration of erythromycin stearate. *Pharmacology* 41: 177–83

34 Kita E, Sawaki M, Mikasa K, Hamada K, Takeuchi S, Maeda K, Narita N (1993) Alterations of host response by a long-term treatment of roxithromycin. *J Antimicrob Chemother* 32: 285–94

35 Sugiura Y, Ohashi Y, Nakai Y (1997) Roxithromycin stimulates the mucociliary activity of the Eustachian tube and modulates neutrophil activity in the healthy guinea pig. *Acta Otolaryngol* (Stockh) (Suppl) 531: 34–8

35a Riesbeck K (2002) Immunomodulating activity of quinolones: review. *J Chemother* 14: 3–12

36 Hall IH, Schwab UE, Ward ES, Ives TJ (2003) Effects of moxifloxacin in zymogen A or S. aureus stimulated human THP-1 monocytes on the inflammatory process and the spread of infection. Life Sci 73: 2675–85

37 Čulić O, Eraković V, Čepelak I, Barišić K, Brajša K, Ferenčić Ž, Galović R, Glojnarić I, Manojlović Z, Munić V et al (2002) Azithromycin modulates neutrophil function and

circulating inflammatory mediators in healthy human subjects. *Eur J Pharmacol* 450: 277–89
38 Tamaoki J, Gunwa H, Kondo M, Isono K, Nishimura K, Nagai A (1998) Effects of macrolide antibiotics on iNOS gene expression and NO production by alveolar macrophages. *Jpn J Antibiot* 51 (Suppl. 1): 12–14
39 Kohri K, Tamaoki J, Kondo M, Aoshiba K, Tagaya E, Nagai A (2000) Macrolide antibiotics inhibit nitric oxide generation by rat pulmonary alveolar macrophages. *Eur Respir J* 15: 62–7
40 Ianaro A, Ialenti A, Maffia P, Sautebin L, Rombola L, Carnuccio R, D'Acquisto F, Di Rosa M (2000) Anti-inflammatory activity of macrolide antibiotics. *J Pharmacol Exp Ther* 292: 156–63
41 Terao H, Asano K, Kanai K, Kyo Y, Watanabe S, Hisamitsu T, Suzaki H (2003) Suppressive activity of macrolide antibiotics on nitric oxide production by lipopolysaccharide stimulation in mice. *Mediators Inflamm* 12: 195–202
42 Pellacini F, Botta D, Romagnano S, Moriggi E, Pradella L (2002) Macrolides with anti-inflammatory activity. US Patent No. 6,455,576
43 Mikasa K, Kita E, Sawaki M, Kunimatsu M, Hamada K, Konishi M, Kashiba S, Narita N (1992) The anti-inflammatory effects of erythromycin in zymosan-induced peritonitis of mice. *J Antimicrob Chemother* 30: 339–48
44 Agen C, Danesi R, Blandizzi C, Costa M, Stacchini B, Favini P, Del Tacca M (1993) Macrolide antibiotics as anti-inflammatory agents: roxithromycin in an unexpected role. *Agents Actions* 38: 85–90
45 Suzaki H, Asano K, Ohki S, Kanai K, Mizutani T, Hisamitsu T (1999) Suppressive activity of a macrolide antibiotic, roxithromycin, on proinflammatory cytokine production *in vitro* and *in vivo*. *Mediators Inflamm* 8: 199–204
46 Majima Y (2002) Mucoactive medications and airway disease. *Paediatr Respir Rev* 3: 104–9
47 Rubin BK (2002) The pharmacologic approach to airway clearance: mucoactive agents. *Respir Care* 47: 818–22
48 Kaneko Y, Yanagihara K, Seki M, Kuroki M, Miyazaki Y, Hirakata Y, Mukae H, Tomono K, Kadota J, Kohno S (2003) Clarithromycin inhibits overproduction of muc5ac core protein in murine model of diffuse panbronchiolitis. *Am J Physiol Lung Cell Mol Physiol* 285: L847–53
49 Shimizu T, Shimizu S, Hattori R, Gabazza EC, Majima Y (2003) *in vivo* and *in vitro* effects of macrolide antibiotics on mucus secretion in airway epithelial cells. *Am J Respir Crit Care Med* 168: 581–7
50 Parnham MJ, Orešković K (2004) Antibiotics for asthma? *Ped Pulmonol* (Suppl) 26: 52
51 Carević O, Djokić S (1988) Comparative studies on the effects of erythromycin A and azithromycin upon extracellular release of lysosomal enzymes in inflammatory processes. *Agents Actions* 25: 124–31
52 Oyama K, Sakuta T, Matsushita K, Maruyama I, Nagaoka S, Torii M (2000) Effects of roxithromycin on tumor necrosis factor-alpha-induced vascular endothelial growth fac-

tor expression in human periodontal ligament cells in culture. *J Periodontol* 71: 1546–53
53 Chabot-Fletcher M (2000) Cellular signalling to NFκB: Role in inflammation and therapeutic promise. In: LG Letts, DW Morgan (eds): *Inflammatory processes: Molecular mechanisms and therapeutic opportunities*. Birkhäuser Verlag, Basel, 23–7
54 Manning AM (2000) Small molecule regulators of AP-1 and NFκB. In: LG Letts, DW Morgan (eds): *Inflammatory processes: Molecular mechanisms and therapeutic opportunities*. Birkhäuser Verlag, Basel, 23–7
55 Tak PP, Firestein GS (2001) NFκB: a key role in inflammatory diseases. *J Clin Invest* 107: 7–11
56 Seymour RA, Heasman PA (1995) Tetracyclines in the management of periodontal diseases: A review. *J Clin Periodontol* 22: 22–35
57 Stone M, Fortin PR, Pacheco-Tena C, Inman RD (2003) Should tetracycline treatment be used more extensively for rheumatoid arthritis? Meta-analysis demonstrates clinical benefit with reduction in disease activity. *J Rheumatol* 30: 2112–22
58 Nieman GF, Zerler BR (2001) A role for the anti-inflammatory properties of tetracyclines in the prevention of acute lung injury. *Curr Med Chem* 8: 317–25
59 Zernicke RF, Wohl GR, Greenwald RA, Moak SA, Leng W, Golub LM (1997) Administration of systemic matrix metalloproteinase inhibitors maintains bone mechanical integrity in adjuvant arthritis. *J Rheumatol* 24: 1324–31
60 Sewell KL, Breedveld F, Furrie E, O'Brien J, Brinckerhoff C, Dynesius-Trentham R, Nosaka Y, Trentham DE (1996) The effect of minocycline in rat models of inflammatory arthritis: correlation of arthritis suppression with enhanced T cell calcium flux. *Cell Immunol* 167: 195–204
61 Kuzin II, Snyder JE, Ugine GD, Wu D, Lee S, Bushnell T, Insel RA, Young FM, Bottaro A (2001) Tetracyclines inhibit activated B cell function. *Int Immunol* 13: 921–31
62 Yrjänheikki J, Tikka T, Keinänen R, Goldsteins G, Chan PH, Koistinaho J (1999) A tetracycline derivative, minocycline, reduces inflammation and protects against focal cerebral ischemia with a wide therapeutic window. *Proc Natl Acad Sci USA* 96: 13496–500
63 Tikka T, Usenius T, Tenhunen M, Keinanen R, Koistinaho J (2001) Tetracycline derivatives and ceftriaxone, a cephalosporin antibiotic, protect neurons against apoptosis induced by ionizing radiation. *J Neurochem* 78: 1409–14
64 Szczypka M, Obminska-Mrukowicz B (2003) Comparative effects of fluoroquinolones on subsets of T lymphocytes in normothermic and hyperthermic mice. *J Vet Pharmacol Ther* 26: 253–58
65 Khan AA, Slifer TR, Araujo FG, Suzuki Y, Remington JS (2000) Protection against lipopolysaccharide-induced death by fluoroquinolones. *Antimicrob Agents Chemother* 44: 3169–173
65a Breban M, Fournier C, Gougerot-Pocidalo MA, Muffat-Joly M, Pocidalo JJ (1992) Protective effects of ciprofloxacin against type II collagen induced arthritis in rats. *J Rheumatol* 19: 216–22

66 Honda J, Okubo Y, Kusaba M, Kumagai M, Saruwatari N, Oizumi K (1998) Fosfomycin (FOM: 1R-2S-epoxypropylphosphonic acid) suppresses the production of IL-8 from monocytes *via* the suppression of neutrophil function. *Immunopharmacol* 39: 149–55
67 Yoneshima Y, Ichiyama T, Ayukawa H, Matsubara T, Furukawa S (2003). Fosfomycin inhibits NF-_B activation in U-937 and Jurkat cells. *Int J Antimicrob Agents* 21: 589–92
68 Morikawa K, Zhang J, Nonaka M, Morikawa S (2002) Modulatory effect of macrolide antibiotics on the Th1- and Th2-type cytokine production. *Int J Antimicrobial Agents* 19: 53–9
69 Labro MT (2002) Antibiotics as anti-inflammatory agents. *Curr Opinion Invstig Drugs* 3: 61–8
70 Kishi K, Hirai K, Hiramatsu K, Yamasaki T, Nasu M (1999) Clindamycin suppresses endotoxin released by ceftazidime-treated Escherichia coli O55:B5 and subsequent production of tumor necrosis factor alpha and interleukin-1β. *Antimicrob Agents Chemother* 43: 616–22
71 Hirata N, Hiramatsu K, Kishi K, Yamasaki T, Ichimaya T, Nasu M (2001) Pretreatment of mice with clindamycin improves survival of endotoxic shock by modulating the release of inflammatory cytokines. *Antimicrob Agents Chemother* 45: 2638–42
72 Di Marco R, Khademi M, Wallstrom E, Muhallab S, Nicoletti F, Olsson T (1999) Amelioration of experimental allergic neuritis by sodium fusidate (fusidin): suppression of IFN-gamma and IFN-alpha and enhancement of IL-10. *J Autoimmun* 13: 187–95
73 Di Marco R, Puglisi G, Papaccio G, Nicoletti A, Patti F, Reggio A, Bendtzen K, Nicoletti F (2001) Sodium fusidate (fusidin) ameliorates the course of monophasic experimental allergic encephalomyelitis in the Lewis rat. *Mult Scler* 7: 101–4
74 Van Zyl JM, Basson K, Kriegler A, van der Walt BJ (1991) Mechanisms by which clofazimine and dapsone inhibit the myeloperoxidase system. A possible correlation with their anti-inflammatory properties. *Biochem Pharmacol* 42: 599–608
75 Dhondt A, Vanholder R, Waterloos MA, Glorieux G, De Smet R, Ringoir S (1998) *In vitro* effect of cefodizime, imipenemcilastin and co-trimoxazole on dexamethasone and cyclosporine A depressed phagocytosis. *Infection* 26: 120–5
76 Sandborn WJ, Feagan BG (2003) Review article: mild to moderate Crohn's disease – defining the basis for a new treatment algorithm. *Aliment Pharmacol Ther* 18: 263–77
77 Geletka R, St. Clair EW (2003) Treatment of early rheumatoid arthritis. *Best Practice Res Clin Rheumatol* 17: 791–809
78 Dijkstra G, Moshage H, Jansen PL (2002) Blockade of NF-kappaB activation and donation of nitric oxide: new treatment options in inflammatory bowel disease? *Scand J Gastroenterol* (Suppl) 236: 37–41
79 Ward C, Dransfield I Chilvers ER, Haslett C, Rossi AG (1999) Pharmacological manipulation of granulocyte apoptosis: potential therapeutic targets. *Trends Pharmacol Sci* 20: 503–9
80 Chilvers ER, Rossi AG, Murray J, Haslett C (1998) Regulation of granulocyte apoptosis and implication for anti-inflammatory therapy. *Thorax* 53: 533–4

81 Athens JW, Mauer AM, Aschenbrucker H, Cartwright GE, Wintrobe MM (1961) Leukokinetic studies III: The distribution of granulocyte in the blood of normal subjects. *J Clin Invest* 40: 159–61
82 Athens JW, Raab OP, Raab SO, Mauer AM, Aschenbrucker H, Cartwright GE, Wintrobe MM (1961) Leukokinetic studies IV: The total blood, circulating and marginal granulocyte pools and the granulocyte turnover rate in normal subjects. *J Clin Invest* 40: 989–97
83 Savill J (1992) Macrophage recognition of senescent neutrophils. *Clin Sci* 83: 649–55
84 Aoshiba K, Nafai A, Konno K (1995) Erythromycin shortens neutrophil survival by accelerating apoptosis. *Antimicrob Agents Chemother* 39: 872–7
85 Adachi T, Motojima S, Hirata A, Fukuda T, Kihara N, Kosaku A, Ohtake H, Makino S (1996) Eosinophil apoptosis caused by theophylline, glucocorticoids, and macrolides after stimulation with IL-5. *J Allergy Clin Immunol* 98: S207–S215
86 Inamura K, Ohta N, Fukase S, Kasajima N, Aoyagi M (2000) The effect of erythromycin on human peripheral neutrophil apoptosis. *Rhinology* 38: 124–9
87 Koch CC, Esteban DJ, Chin AC, Olson ME, Read RR, Ceri H, Morck DW, Buret AG (2000) Apoptosis, oxidative metabolism and interleukin-8 production on human neutrophils exposed to azithromycin: effects of *Streptococcus pneumoniae*. *J Antimicrob Chemother* 46: 19–26
88 Healy DP, Silverman P, Neely AN, Holder I.A, Babcock GF (2002) Effects of antibiotics on human polymorphonuclear neutrophil apoptosis. *Pharmacotherapy* 22: 578–85
89 Chin AC, Lee WD, Murrin KA, Morck DW, Merrill JK, Dick P, Buret AG (2000) Tilmicosin induces apoptosis in bovine peripheral neutrophils in the presence or in absence of *Pasteurella hemolytica* and promotes neutrophil phagocytosis by macrophages. *Antimicrob Agents Chemother* 44: 2465–70
90 Grdiša M, Lopatar N, Pavelić K (1998) Effects of a 17-membered azalide on tumor cell growth. *Chemotherapy* 44: 331–6
91 Lawrence T, Gilroy DW, Colville-Nash PR, Willoughby DA (2001) Possible new role for NFκB in the resolution of inflammation. *Nature Med* 7: 1291–7
92 Yamaryo T, Oishi K, Yoshimine H, Tsuchihashi Y, Matsushima K, Nagatake T (2003) Fourteen-member macrolides promote the phosphatidylserine receptor-dependent phagocytosis of apoptic neutrophils by alveolar macrophages. *Antimicrob Agents Chemother* 47: 48–53
93 Yerramasetti R, Gollapudi S, Gupta S (2002) Rifampicin inhibits CD95-mediated apoptosis of Jurkat T cells *via* glucocorticoid receptors by modifying the expression of molecules regulating apoptosis. *J Clin Immunol* 22: 37–47
94 Gollapudi S, Jaidka S, Gupta S (2003) Molecular basis of rifampicin-induced inhibition of anti CD95-induced apoptosis of peripheral blood T lymphocytes: the role of CD95 ligand and FLIPs. *J Clin Immunol* 23: 11–22
95 Azuma Y, Ohura K (2003) Alteration of constitutive apoptosis in neutrophils by quinolones. *Inflammation* 27: 115–22
96 Eichenfeld AH (1999) Minocycline and autoimmunity. *Curr Opin Pediatr* 11: 447–56

The cytoprotective interactions of antibiotics with human ciliated airway epithelium

Charles Feldman[1] *and Ronald Anderson*[2]

[1]Division of Pulmonology, Department of Medicine, Faculty of Health Sciences, University of the Witwatersrand Medical School, 7 York Road, Parktown 2193, Johannesburg, South Africa; [2]MRC Unit for Inflammation and Immunity, Department of Immunology, University of Pretoria, Pretoria, and Tshwane Academic Division of the National Health Laboratory Service, South Africa

Introduction

The human airway is lined by a specialized epithelium, consisting of a number of different cells, of which the ciliated columnar epithelial cell is particularly important. The action of the cilia in concert with the mucus secreted from the specialized goblet and mucous cells in the epithelium constitutes the mucociliary transport mechanism. This mechanical clearance mechanism of the airway protects the epithelium and is the first line defense of the lower respiratory tract against the harmful effects of inhaled bacteria, bacterial products, dusts and several other endogenous and exogenous mediators and toxins [1]. Primarily as a consequence of the actions of this mucociliary clearance mechanism the lower respiratory tract is normally sterile. However, there are a number of chronic airway disorders in which this defence mechanism is perturbed, by either primary or secondary mechanisms [2]. The consequences of this attenuation of function may include persistent airway colonization by bacteria, chronic infection or inflammation, mucosal injury and even bacterial invasion, which occur in diverse chronic airway disorders such as asthma, cystic fibrosis, diffuse panbronchiolitis, chronic obstructive pulmonary disease and bronchiectasis.

Injury to airway epithelium

A number of bacteria that colonize the respiratory tract or cause respiratory tract infections have the ability to perturb the structure and function of the ciliated epithelium, primarily through the production of toxic virulence factors [2–4]. A particularly well-studied bacterial virulence factor is pneumolysin, a thiol-activated protein toxin that is produced by all clinically relevant strains of *Streptococcus pneumoniae* (pneumococcus) [4]. This toxin has been shown to slow ciliary beating, which may aid in colonization of the epithelial surface by the pneumococcus [4].

Antibiotics as Anti-Inflammatory and Immunomodulatory Agents, edited by Bruce K. Rubin and Jun Tamaoki

Subsequent damage to the epithelium induced by this toxin may assist the organism to penetrate the mucosa, with subsequent systemic invasion. Other microorganisms that produce factors that may affect the ciliated epithelium include *Haemophilus influenzae*, *Staphylococcus aureus*, *Klebsiella pneumoniae* and *Pseudomonas aeruginosa* [2, 3].

In addition, there are a number of endogenous and exogenous chemicals or toxins that may also injure the respiratory epithelium and may act as mediators or contri butory factors to airway disorders [2]. One example is the bioactive phospholipids, including platelet-activating factor (PAF), lyso-PAF (LPAF) and lysophosphopha tidylcholine (LPC) [2]. These endogenous substances are thought to be important possible mediators of airway disorders such as asthma, and PAF in particular is the only mediator that has been shown to be able to mimic all the important manifestations of asthma. Other examples of endogenous mediators that affect function of the mucociliary mechanism include reactive oxidants, protease enzymes, prostaglandins, cytokines, leukotrienes, among many others [2].

Cytoprotective effects of macrolides, azalides and ketolides

The macrolides, azalides and ketolides are a group of antibiotics that are also cytoprotective of human ciliated epithelium, which may serve to protect the epithelium from bacterial and chemical mediator-induced injury [2, 5]. In the case of bacterial infection these actions may interfere with colonization of the respiratory epithelium by bacteria and/or protect the epithelium from damage induced by colonization and the host response to this process. In both bacterial and chemical mediator induced injury, these antimicrobial agents protect against effects on ciliary function as well as injury to the structural integrity of the epithelium, and the cytoprotective effects of these antibiotics are mediated both directly and indirectly [2, 5].

Effects on bacterial adherence and epithelial injury

For some time it has been known that the macrolide antibiotics on their own have a positive affect on both ciliary function and mucus secretion of airway epithelium. Erythromycin has been shown to stimulate ciliary beat frequency (CBF) of rabbit tracheal epithelial cells [6]. Roxithromycin has been shown to stimulate the mucociliary activity of the Eustachian tube of guinea pigs and to enhance ciliary activity and mucociliary transport velocity of rabbit tracheal epithelial cells [7, 8].

Conversely, many bacteria and several endogenous mediators antagonize these effects. *H. influenzae* infection of nasal epithelial respiratory mucosa has been shown to cause ciliary slowing and damage to the respiratory epithelium [9, 10]. However, in one study, incubation of the nasal epithelial tissue with sub-MIC concentrations of dirithromycin significantly reduced the slowing of CBF and the epithelial disruption caused by culture filtrates of this organism [9]. The effects on structural integrity were found to be more complete than the effects on CBF and were confirmed by transmission electron microscopy (TEM). The authors suggested that these effects were likely to be direct, rather than indirect, due to activity against inflammatory cells, since the strips of epithelium used in the studies were obtained from healthy volunteers and contained very few inflammatory cells [9]. They further suggested that these direct effects could possibly be associated with elevations in cyclic AMP [9], which they had previously shown may protect epithelial cells against damage from bacterial toxins [11] and other investigators have confirmed the ability of the macrolide antibiotic, roxithromycin, to elevate cyclic AMP [12].

Using an organ culture model of human adenoid tissue the same investigators showed that *H. influenzae* caused significant mucosal damage [9]. Culturing of *H. influenzae* with sub-MIC concentrations of dirithromycin prior to infection of the organ culture had no effect on the structural integrity of human respiratory mucosa [9]. In contrast, incubation of adenoid tissue with sub-MIC concentrations of dirithromycin prior to assembly of the organ culture reduced the mucosal damage by as much as 50%, in association with a decrease in the amount of adherent bacteria, probably as a consequence of a decrease in the amount of damaged epithelium to which the bacterium could adhere [9]. In that study it appeared that dirithromycin, in concentrations that are achievable *in vivo*, protected respiratory mucosa by a direct cytoprotective effect [9]. In another similar study, protection against *H. influenzae*-induced injury of human respiratory mucosa was also demonstrated for sub-MIC concentrations of other antibiotics, including amoxicillin, loracarbef and ciprofloxacin [10].

Sub-inhibitory concentrations of erythromycin have been shown in an *in vitro* cell culture to inhibit the adherence of *Streptococcus pneumoniae* to human respiratory epithelial cells [13]. In that study there was a small, non-significant, decrease in the number of pneumococci recovered when comparing the control and test preparations, suggesting that there was no effect of the antibiotic on the viability of the microorganism. Similarly disruption of epithelial integrity (as measured by a decrease in transepithelial resistance) was delayed in the presence of erythromycin [13]. The authors suggested that the mechanism might be related to interference with pneumolysin release, since in a pneumococcal suspension in cell culture medium without respiratory epithelium these investigators demonstrated that the addition of erythromycin almost completely prevented the release of pneumolysin [13]. Other investigators have demonstrated the ability of macrolides and similar agents to inhibit pneumolysin production [14].

Similar studies have documented that culture filtrates of *Pseudomonas aeruginosa* cause ciliary slowing and damage to the structural integrity of human nasal ciliated epithelium, particularly in the presence of neutrophils [15]. Addition of erythromycin to the culture filtrates had no effect on these injurious actions. Filtrates of *Pseudomonas aeruginosa* cultured in erythromycin (which had no effect on growth of the microorganism) caused less slowing of CBF and less disruption of structural integrity in both the absence and presence of neutrophils [15]. This study suggested that sub-MIC concentrations of erythromycin suppress the production of toxins by *P. aeruginosa* that damage the epithelium directly, as well as protect the epithelium indirectly as a consequence of inhibition of neutrophil-associated cytotoxicity [15].

Similar cytoprotection of the respiratory mucosa against *Haemophilus influenzae* and *Pseudomonas aeruginosa* infection has been demonstrated for chemotherapeutic agents other than these antibiotics, including salmeterol [11, 16, 17], a long-acting β2-agonist. Rolipram, a type IV phosphodiesterase inhibitor, also has the ability to prevent *P. aeruginosa*-induced epithelial damage [18], and to a greater extent than salmeterol. These studies suggested that the mechanism of airway protection may be due to elevation of intracellular levels of cAMP as rolipram was more effective than salmeterol at elevating cAMP [18]. In addition fluticasone propionate has also been shown to reduce mucosal damage caused by *P. aeruginosa* in an organ culture model and to preserve the ciliated cells [17]. In that experimental system, fluticasone propionate acted synergistically with salmeterol in the preservation of ciliated cells [17].

Effects on mediator injury

Endogenous mediators of inflammation, such as the bioactive phospholipids (PL), PAF, LPAF and LPC cause dose-dependent slowing of ciliary beating and damage to the structural integrity of human ciliated epithelium at concentrations >1 μg/ml [19]. These effects are both direct, probably as a consequence of their nonspecific membrane-disruptive, detergent-like activity (but not due to oxidant injury), as well as indirect, through their activation of human neutrophils [19]. We have shown that the macrolide antibiotics, roxithromycin, clarithromycin and erythromycin, the azalide agent azithromycin and the ketolide agents HMR 3004 and HMR 3647 (now called telithromycin) protect against these effects, both directly, as well as indirectly through the inhibition of polymorphonuclear leukocyte-mediated injury (Tab. 1) [19, 20]. The greater protective activity of the ketolides, particularly HMR 3004, is almost certainly related to their superior level of intracellular accumulation, associated with both direct protection and enhanced indirect protection due to inhibition of reactive oxidant release by activated neutrophils [21].

The cytoprotective effects of the macrolides/azalides/ketolides on the epithelium are mirrored by their membrane stabilizing ability (measured using a hemolytic assay), and are further associated with inhibition of neutrophil superoxide production (measured using lucigenin-enhanced chemiluminescence) due to inhibition of NADPH oxidase activation consequent on stabilization of the neutrophil mem-

Table 1 - Percentage slowing of CBF and induction of ED in ciliated respiratory epithelium exposed to PAF- and LPC- treated PMNL in the presence and absence of azithromycin, clarithromycin, and roxithromycin

	Test preparation[a]		Preincubation with epithelial strips[b]		Preincubation with PMNL[c]	
	% Slowing CBF	% ED	% Slowing CBF	% ED	% Slowing CBF	% ED
PAF						
Azithromycin experiments	39%	55%	22%	25%	6%	10%
Clarithromycin experiments	32%	50%	19%	25%	12%	20%
Roxithromycin experiments	30%	55%	3%	20%	2%	15%
LPC						
Azithromycin experiments	26%	40%	9%	20%	7%	10%
Clarithromycin experiments	28%	50%	9%	30%	0.5%	0%
Roxithromycin experiments	36%	45%	8%	30%	0%	0%

The mean (± SEM) CBF for the control systems was 11.2 ± 0.4 Hz. There was no epithelial damage in any of the control systems (reproduced with permission from [19]).

[a]*Epithelial strips were exposed to neutrophils and LPC or PAF in the absence of the antimicrobial agents.*

[b]*In this system the epithelial strips were pretreated with the macrolides prior to addition of PMNL and PAF or LPC.*

[c]*In this system the PMNL were pretreated with the macrolides prior to exposure to PAF or LPC followed by addition to epithelial strips.*

brane [20, 21]. All these effects are found, to a greater or lesser extent, with the 14-member macrolides (erythromycin, clarithromycin and roxithromycin), the 15-member azalide (azithromycin), and the ketolides, but are not seen with the 16-member macrolides, such as spiramycin or josamycin [22]. Spiramycin does not decrease reactive oxidant production in N-formyl-L-methionyl-L-leucyl-L-phenylalanine (FMLP)-activated human neutrophils, and has only very weak membrane-stabilizing activity, as compared with clarithromycin [22].

Other agents that we have demonstrated may protect against PL-induced injury to ciliated epithelium include vitamin E, which also antagonized reactive oxidant production by PL-activated human neutrophils and which had membrane stabilizing activity, inhibiting PL-induced hemolysis of sheep erythrocytes [23].

Effects on other epithelial cells

Studies have confirmed that clarithromycin is cytoprotective of gastric epithelium against damage induced by ethanol in rats, most probably as a consequence of increased fluid volume and the mucus volume retained in the gastric lumen, the latter possibly related to α_2-adrenoceptor effects. The effects were not mediated *via* endogenous prostaglandins, sulfhydryl compounds of the gastric mucosa or changes in the gastric contractile patterns [24].

Cytoprotective properties of macrolides which are secondary to anti-inflammatory activity

Notwithstanding the direct cytoprotective effects of macrolides on respiratory epithelium, described above, these agents also maintain ciliated epithelial cell structure and function by protecting these cells against inflammation-mediated damage and dysfunction. These anti-inflammatory properties are achieved by two mechanisms. Firstly, by modulation of the activities of various types of inflammatory cells, particularly neutrophils, a property common to 14- and 15-, but not 16-member macrolides. Secondly, by interference with the synthesis of bacterial derived mediators of inflammation, a property that is presumably common to all macrolides.

Anti-inflammatory effects of macrolides on neutrophils

Fourteen-member macrolides, as well as azalides and ketolides possess a range of anti-inflammatory activities, which enable these cells not only to control the influx of inflammatory cells, particularly neutrophils, into the airways, but also to suppress the generation of toxic reactive oxidants (ROS) by phagocytes and to decrease the reactivity of neutrophil elastase. Moreover, these agents shorten the lifespan of neutrophils, which may also contribute to the control of neutrophil-mediated tissue damage. Although these activities could be construed as being potentially negative in the context of compromising host defenses against microbial pathogens, this must be offset against the adverse consequences of over-exuberant inflammatory responses, which pose the risk of excessive production of reactive oxidants, and inflammation-mediated tissue damage [25, 26]. Ciliated respiratory epithelial cells are especially vulnerable to the cytotoxic actions of neutrophil-derived ROS and proteases [27, 28].

Macrolides and neutrophil migration

The inhibitory effects of macrolides on neutrophil migration [29–31] are achieved in part by interference with the generation of neutrophil-selective chemoattractants, particularly IL-8, by neutrophils themselves [32], eosinophils [33], monocytes [34], bronchial epithelial cells [35, 36] and fibroblasts [37]. This is a particularly significant anti-inflammatory property of macrolides because not only does IL-8 amplify neutrophil influx but this chemokine also confers resistance to corticosteroid-induced apoptosis of these cells [38].

In addition to inhibition of synthesis of IL-8 by a variety of inflammatory cell types, macrolides also interfere with the synthesis and expression of the adhesion molecules, ICAM-1 and VCAM-1 on vascular endothelium [32, 39], as well as with upregulated expression of β2-integrins on activated neutrophils [32]. With the exception of β2-integrin expression, these inhibitory effects of macrolides on key events in the transendothelial migration of neutrophils, as well as eosinophils and monocytes, appear to be related to interference with the nuclear translocation of the transcription factors AP-1 and NFκB [31, 34, 36, 37].

Anti-oxidative interactions of macrolides with neutrophils

It is well accepted that 14 member macrolides, as well as azithromycin and telithromycin, but not 16-member macrolides such as josamycin and spiramycin, inhibit the generation of reactive oxidants by neutrophils and monocytes/macrophages, targeting both NADPH oxidase and nitric oxide synthase [21, 32, 40–45]. In the case of nitric oxide synthase, the macrolides appear to interfere with the induction of type II nitric oxide synthase mRNA as opposed to inhibiting the production of nitric oxide [31, 46]. However, in the case of NADPH oxidase, macrolides appear to inhibit the activity of the fully assembled oxidase, without affecting transductional events involved in its activation, or by scavenging of superoxide [40, 43].

Although the exact molecular/biochemical mechanisms of macrolide-mediated inhibition of NADPH oxidase remain to be conclusively established, we believe that these anti-oxidative, cytoprotective activities of the macrolides are attributable, at least in part, to the membrane-stabilizing activities described above. Efficient functioning of NADPH oxidase is dependent on membrane fluidity and lateral mobility, facilitating juxtaposition of the components of the election transporter [47].

The proposed relationship between macrolide-mediated membrane stabilization and inhibition of NADPH oxidase is supported by our observations that the inhibitory actions of the macrolides on oxidase activity are reversed by the membrane destabilizing agents LPC, LPAF and PAF [21, 44, 45]. Conversely, macrolides antagonize the sensitizing actions of LPC, LPAF and LPC on neutrophil NADPH oxidase activity, which, given the extremely high concentrations of these bioactive phospholipids, particularly LPC in inflamed airways [48], represents a potentially important cytoprotective, anti-inflammatory activity of these antimicrobial agents. Moreover, macrolides have also been reported to antagonize the pro-oxidative interactions of the *Pseudomonas aeruginosa*-derived pigments, pyocyanin and 1-hydroxyphenazine [49].

Macrolide-mediated antagonism of neutrophil elastase

Elastase is cytotoxic for airway epithelium [28]. Importantly, macrolides possess anti-elastolytic properties, which are achieved by several different mechanisms. Notwithstanding the inhibitory effects of these antimicrobial agents on the influx of neutrophils into inflamed airways, with a concomitant reduction in the elastase load [29], erythromycin and flurithromycin have been reported to function as direct inhibitors of this protease [50]. In addition, and albeit somewhat speculatively, macrolide-mediated attenuation of phagocyte NADPH oxidase activity, if operative *in vivo*, may protect α-1-proteinase inhibitor (API), the primary inhibitor of neutrophil elastase in the airways, against phagocyte-mediated oxidative inactivation. This is of considerable potential significance given firstly that API possesses anti-inflammatory and possibly antimicrobial properties [51–53], and secondly that loss of anti-protease activity is associated with unfavorable outcome in patients with severe sepsis, including community-acquired pneumonia [54, 55].

Pro-apoptotic properties of macrolides

Macrolides have been proposed to accelerate resolution of inflammation by promoting neutrophil apoptosis both directly [56], as well as indirectly by inhibiting the synthesis of anti-apoptotic IL-8 by various inflammatory cell types, including neutrophils themselves [32–37]. Moreover, these antimicrobial agents have also been reported to increase the efficiency of removal of apoptotic neutrophils by alveolar macrophages, in the setting of minimal release of mediators of inflammation and tissue damage, including elastase [57].

Neutrophil-directed anti-inflammatory activities of macrolides are summarized in Table 2.

Table 2 - Neutrophil-directed anti-inflammatory activities of macrolides

Anti-inflammatory activity	Mechanism	Refs.
Inhibition of neutrophil accumulation	Interference with:	
	i) synthesis and release of IL-8	[32–37]
	ii) synthesis of ICAM-1 and VCAM 1	[32, 39]
	iii) upregulation of β2 integrins	[32]
Inhibition of generation of ROS	Interference with:	
	i) the activity of NADPH oxidase	[21, 32, 40–45]
	ii) the synthesis of type II nitric oxide synthase (macrophages)	[31, 46]
Anti-elastase	Interference with:	
	i) elastolytic activity	[50]
	ii) decreased elastase load secondary to inhibition of neutrophil influx	[29]
	iii) possible protection of API against oxidative inactivation	unproven
Pro-apoptotic	I) inhibition of synthesis of IL-8	[32–37]
	ii) induction of apoptosis	[56]
	iii) accelerated clearance of apoptotic neutrophils by macrophages	[57]

Anti-inflammatory activities of macrolides secondary to antimicrobial activity

Macrolides, by virtue of their primary inhibitory actions on polypeptide synthesis, interfere with the production of microbial, proinflammatory virulence factors, some of which may initiate and sustain a cascade of futile inflammatory events, which result in damage to host tissues. For example, in spite of the fact that *P. aeruginosa* is insensitive to the antimicrobial effects of erythromycin, exposure of this microbial pathogen to the macrolide is accompanied by decreased synthesis of proinflammatory, virulence factors, including *Pseudomonas* protease and hemolysin [58]. Erythromycin and roxithromycin also inhibit the synthesis of bacterial-derived neutrophil chemoattractants [59], as well as the production and release of proinflammatory, pore-forming cytotoxins such as pneumolysin [13, 14]. Interference with the synthesis of pneumolysin may be of particular importance because this toxin is a potent activator of the proinflammatory activity of both neutrophils and macrophages which, somewhat counter-intuitively, favors extra-pulmonary dissemination of the pneumococcus, probably as a consequence

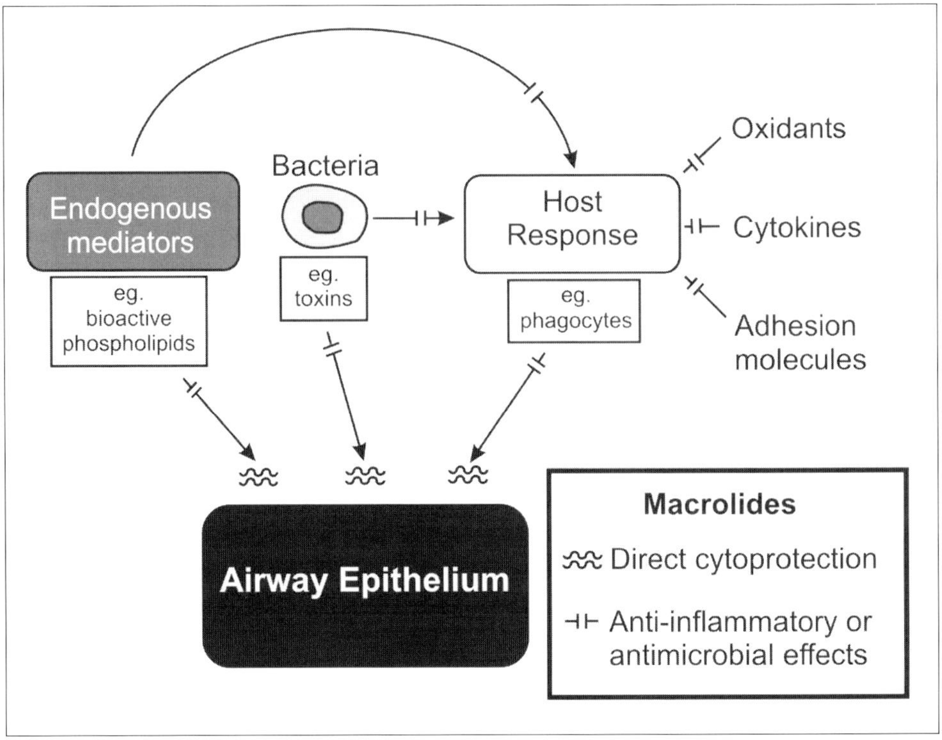

Figure 1
The cytoprotective and anti-inflammatory activities of macrolides and their relationship to protection of airway epithelium
The cytoprotective activity enables the epithelium to resist the direct damaging actions of microbial- and host-derived cytotoxins (≈). The indirect effects are targeted at inhibiting one or more of synthesis, release or activity of bacterial toxins or mediators of host inflammatory responses (≈).

of inflammation-mediated damage to epithelium [60, 61]. The cytoprotective and anti-inflammatory activities of macrolides in relation to protection of respiratory epithelium are summarized in Figure 1.

References

1 Feldman C (2000) Nonspecific host defenses: Mucociliary clearance and cough. In: MS Niederman, GA Sarosi, J Glassroth (eds): *Respiratory infections*. Williams and Wilkins, Philadelphia, 13–25

2 Feldman C, Anderson R, Rutman A, Cole PJ, Wilson R (1998) Human ciliated epithelium *in vitro* – mechanisms of injury and protection. In: GL Baum, Z Priel, Y Roth, N Liron, EJ Ostfeld (eds): *Cilia, mucus, and mucociliary interactions.* Marcel Dekker, Inc., New York, 461–71

3 Wilson R, Dowling RB, Jackson AD (1996) The biology of bacterial colonization and invasion of the respiratory mucosa. *Eur Respir J* 9: 1523–30

4 Feldman C, Read R, Rutman A, Jeffery PK, Brain A, Lund V, Mitchell TJ, Andrew PW, Boulnois GJ, Todd HC et al (1992) The interaction of *Streptococcus pneumoniae* with intact human respiratory mucosa *in vitro. Eur Respir J* 5: 576–83

5 Feldman C, Anderson R, Theron AJ, Cole P, Wilson R (2001) The cytoprotective effects of macrolides, azalides, and ketolides on human ciliated epithelium *in vitro.* In: M Salathe (ed): *Cilia and mucus. From development to respiratory defense.* Marcel Dekker, Inc., New York, 145–53

6 Tamaoki J, Chiyotani A, Sakai N, Takeyama K, Takizawa T (1992) Effect of erythromycin on ciliary motility in rabbit airway epithelium *in vitro. J Antimicrob Chemother* 29: 173–8

7 Sugiura Y, Ohashi Y, Nakai Y (1997) Roxythromycin stimulates the mucociliary activity of the Eustachian tube and modulates neutrophil activity in the healthy guinea pig. *Acta Otolaryngol (Stockh)* (Suppl) 531: 34–8

8 Nakano T, Ohashi Y, Tanaka A, Kakinoki Y, Washio Y, Nakai Y (1998) Roxythromycin reinforces epithelial defence function in rabbit trachea. *Acta Otolaryngol (Stockh)* (Suppl) 538: 233–8

9 Rutman A, Dowling R, Wills P, Feldman C, Cole PJ, Wilson R (1998) Effect of dirithromycin on *Haemophilus influenzae* infection of the respiratory mucosa. *Antimicrob Agents Chemother* 42: 772–8

10 Tsang KW, Rutman A, Kanthakumar K, Belcher J, Lund V, Roberts DE, Read RC, Cole PJ, Wilson R (1993) *Haemophilus influenzae* infection of human respiratory mucosa in low concentrations of antibiotics. *Am Rev Respir Dis* 148: 201–7

11 Dowling RB, Rayner CFJ, Rutman A, Jackson AD, Kanthakumar K, Dewar A, Taylor GW, Cole PJ, Johnson M, Wilson R (1997) Effect of salmeterol on *Pseudomonas aeruginosa* infection of respiratory mucosa. *Am J Respir Crit Care Med* 155: 327–36

12 Takeyama K, Tamaoki J, Chiyotani A, Tagaya E, Konno K (1993) Effect of macrolide antibiotics on ciliary motility in rabbit airway epithelium *in vitro. J Pharm Pharmacol* 45: 756–8

13 Lagrou K, Peetermans WE, Jorissen M, Verhaegen J, Van Damme J, Van Eldere J (2000) Sub-inhibitory concentrations of erythromycin reduce pneumococcal adherence to respiratory epithelial cells *in vitro. J Antimicrob Chemother* 46: 717–23

14 Spreer A, Kerstan H, Bottcher T, Gerber J, Siemer A, Zysk G, Mitchell TJ, Eiffert H, Nau R (2003) Reduced release of pneumolysin by *Streptococcus pneumoniae in vitro* and *in vivo* after treatment with nonbacteriolytic antibiotics in comparison to ceftriaxone. *Antimicrob Agents Chemother* 47: 2649–54

15 Tanaka E, Kanthakumar K, Cundell DR, Tsang KWT, Taylor GW, Kuze F, Cole PJ, Wil-

son R (1994) The effect of erythromycin on *Pseudomonas aeruginosa* and neutrophil mediated epithelial damage. *J Antimicrob Chemother* 33: 765–75

16 Dowling RB, Johnson M, Cole PJ, Wilson R (1998) Effect of salmeterol on *Haemophilus influenzae* infection of respiratory mucosa *in vitro*. *Eur Respir J* 11: 86–90

17 Dowling RB, Johnson M, Cole PJ, Wilson R (1999) Effect of fluticasone propionate and salmeterol on *Pseudomonas aeruginosa* infection of the respiratory mucosa *in vitro*. *Eur Respir J* 14: 363–9

18 Dowling RB, Johnson M, Cole PJ, Wilson R (1999) The effect of rolipram, a type IV phosphodiesterase inhibitor, on *Pseudomonas aeruginosa* infection of respiratory mucosa. *J Pharmacol Exp Ther* 282: 1565–71

19 Feldman C, Anderson R, Theron AJ, Ramafi G, Cole PJ, Wilson R (1997) Roxithromycin, clarithromycin, and azithromycin attenuate the injurious effects bioactive phospholipids, on human respiratory epithelium *in vitro*. *Inflammation* 21: 655–65

20 Feldman C, Anderson R, Theron A, Mokgobu I, Cole PJ, Wilson R (1999) The effect of ketolides on bioactive phospholipid-induced injury to human ciliated epithelium *in vitro*. *Eur Respir J* 13: 1022–8

21 Mokgobu I, Theron AJ, Anderson R, Feldman C (1999) The ketolide antimicrobial agent HMR-3004 inhibits neutrophil superoxide production by a membrane-stabilizing mechanism. *International J Immunopharmacology* 21: 365–77

22 Theron AJ, Feldman C, Anderson R (2000) Investigation of the anti-inflammatory and membrane-stabilizing potential of spiramycin *in vitro*. *J Antimicrob Chemother* 46: 269–71

23 Feldman C, Anderson R, Theron AJ, Steel HC, van Rensburg CEJ, Cole PJ, Wilson (2001) Vitamin E attenuates the injurious effects of bioactive phospholipids on human ciliated epithelium *in vitro*. *Eur Respir J* 18: 122–9

24 Gutierrez-Cabano CA, Raynald AC (1999) Gastroprotective effect of intragastric clarithromycin again damage induced by ethanol in rats. *Dig Dis Sci* 44: 1721–31

25 Dallegri F, Ottonello L (1997) Tissue injury in neutrophilic inflammation. *Inflamm Res* 46: 382–91

26 Witko-Sarsat V, Rieu P, Descamps-Latscha B, Lesavre P, Halbwachs-Mecarelli L (2000) Neutrophils: molecules and pathophysiological aspects. *Lab Invest* 80: 617–53

27 Feldman C, Anderson R, Kanthakumar K, Vargas A, Cole PJ, Wilson R (1994) Oxidant-mediated ciliary dysfunction in human respiratory epithelium. *Free Rad Biol Med* 17: 1–10

28 Lewis S, Berg JR, Kleine TJ (1995) Modulation of epithelial permeability by extracellular molecules. *Physiol Rev* 75: 561–89

29 Ichikawa Y, Ninomiya H, Koga H, Tanaka M, Kinoshita M, Tokuna N, Yano T, Oizumi K (1992) Erythromycin reduces neutrophils and neutrophil-derived elastolytic-like activity in the lower respiratory tract of bronchiolitis patients. *Am Rev Resp Dis* 146: 196–203

30 Oda H, Kadota J, Kohno S, Hara K (1994) Erythromycin inhibits neutrophil chemotaxis in bronchoalveoli of diffuse panbronchiolitis. *Chest* 106: 1116–23

31 Ianaro A, Ialenti A, Maffia P, Sautebin L, Rombolo L, Carnuccio R, Iuvone T, D'Acquisto F, Di Rosa M (2000) Anti-inflammatory activity of macrolide antibiotics. *J Pharmacol Exp Ther* 292: 156–63

32 Lin HC, Wang CH, Liu CY, Yu CT, Kuo HP (2000) Erythromycin inhibits beta 2-integrins (CD11b/CD18) expression, interleukin-8 release and intracellular oxidative metabolism in neutrophils. *Respir Med* 94: 654–60

33 Kohyama T, Takizawa H, Kawasaki S, Akiyama N, Sato M, Ito K (1999) Fourteen-member macrolides inhibit interleukin-8 release by human eosinophils from atopic donors. *Antimicrob Agents Chemother* 43: 907–11

34 Kikuchi T, Hagiwara K, Honda Y, Gomi K, Kobayashi T, Takahashi H, Tokue Y, Watanabe A, Nukiwa T (2002) Clarithromycin suppresses lipolysaccharide-induced interleukin-8 production by human monocytes through AP-1 and NF-kappa B transcription factors. *J Antimicrob Chemother* 49: 745–55

35 Kawasaki S, Takizawa H, Ohtoshi T, Takeuchi N, Kohyama T, Nakamura H, Kasama T, Kobayashi K, Nakahara K, Morita Y et al (1998) Roxithromycin inhibits cytokine production by and neutrophil attachment to bronchial epithelial cells *in vitro*. *Antimicrob Agents Chemother* 42: 1499–1502

36 Abe S, Nakamura H, Inoue S, Takeda H, Saito H, Kato S, Mukaida N, Matsushima K, Tomoike H (2000) Interleukin-8 gene repression by clarithromycin is mediated by the activator protein-1 binding site in human bronchial epithelial cells. *Am J Resp Cell Mol Biol* 22: 51–60

37 Takaki M, Ushikai M, Deguchi K, Nishimoto K, Matsune S, Kurono Y (2003) The role of nuclear factor-kappa B in interleukin-8 expression by human adenoidal fibroblasts. *Laryngoscope* 113: 1378–85

38 Strickland I, Kisich H, Hauk PJ, Vottero A, Chrousos GP, Klemm DJ, Leung DYM (2001) High constitutive glucocorticoid receptor β in human neutrophils enables them to reduce their spontaneous rate of cell death in response to corticosteroids. *J Exp Med* 193: 585–93

39 Li YJ, Azuma A, Takahashi S, Usuki J, Matsuda K, Aoyama A, Kudo S (2002) Fourteen-membered ring macrolides inhibit vascular cell adhesion molecule 1 messenger RNA induction and leukocyte migration: role in preventing lung injury and fibrosis in bleomycin-challenged mice. *Chest* 122: 2137–45

40 Anderson R (1989) Erythromycin and roxithromycin potentiate human neutrophil locomotion by inhibition of leukoattractant-activated superoxide generation and autooxidation. *J Infect Dis* 159: 966–73

41 Hand WL, Hand DL, King-Thomson NL (1990) Antibiotic inhibition of the respiratory burst in human polymorphonuclear leukocytes. *Antimicrob Agents Chemother* 34: 863–70

42 Perry DK, Hand WL, Edmondson DE, Lambeth JD (1992) Role of phospholipase D-derived diarydyl-glycerol in the activation of the human neutrophil respiratory burst oxidase. *J Immunol* 149: 2749–58

43 Umeki S (1993) Anti-inflammatory action of erythromycin: its inhibitory effect on neutrophil NADPH-oxidase activity. *Chest* 104: 1191–3

44 Anderson R, Theron AJ, Feldman C (1996) Membrane-stabilizing, anti-inflammatory interactions of macrolides with neutrophils. *Inflammation* 20: 693–705

45 Abdelghaffar H, Vazifeh D, Labro MT (1997) Erythromycin A-derived macrolides modify the functional activities of human neutrophils by altering the phospholipase D-phosphatidate phosphohydrolase transduction pathway. *J Immunol* 159: 3995–4005

46 Kohri K, Tamaoki J, Kondo M, Aoshiba K, Tagaya E, Nagai A (2000) Macrolide antibiotics inhibit nitric oxide generation by rat pulmonary alveolar macrophages. *Eur Respir J* 15: 62–7

47 Shao DM, Segal AW, Dekker LV (2003) Lipid rafts determine efficiency of NADPH oxidase activation in neutrophils. *Febs Lett* 550: 101–6

48 Chilton FH, Averill FJ, Hubbard WC, Fonteh AN, Triggiana M, Liu MC (1996) Antigen-induced generation of lysophospholipids in human airways. *J Exp Med* 183: 2235–45

49 Ras GJ, Anderson R, Taylor GW, Savage JE, van Niekerk E, Joone G, Koornhof HJ, Saunders J, Wilson R, Cole PJ (1992) Clindamycin, erythromycin, and roxithromycin inhibit the proinflammatory interactions of *Pseudomonas aeruginosa* pigments with human neutrophils *in vitro*. *Antimicrob Agents Chemother* 36: 1236–40

50 Gorrini M, Lupi A, Viglio S, Pamparana F, Cetta G, Iadarola P, Powers JC, Luisetti M (2001) Inhibition of human neutrophil elastase by erythromycin and flurythromycin, two macrolide antibiotics. *Am J Resp Cell Mol Biol* 25: 492–9

51 Woolhouse IS, Bayley DL, Stockley RA (2002) Sputum chemotactic activity in chronic obstructive pulmonary disease: effect of alpha (1)-antitrypsin deficiency and the role of leukotriene B4 and interleukin 8. *Thorax* 57: 709–14

52 Lieberman J (2000) Augmentation therapy reduces frequency of lung infections in antitrypsin deficiency – A new hypothesis with supporting data. *Chest* 118: 1480–5

53 Hiemstra PS (2002) Novel roles of protease inhibitors in infection and inflammation. *Biochem Soc Transact* 30: 116–20

54 Greene C, Taggart C, Lowe G, Gallagher P, McElvaney N, O'Neill S (2003) Local impairment of anti-neutrophil elastase activity in community-acquired pneumonia. *J Infect Dis* 188: 769–76

55 Lim YP, Bendelja K, Opal SM, Siryaporn E, Hixson DC, Palardy JE (2003) Correlation between mortality and the levels of inter-alpha inhibitors in the plasma of patients with severe sepsis. *J Infect Dis* 188: 919–26

56 Aoshiba K, Nayai A, Konno K (1995) Erythromycin shortens neutrophil survival by accelerating apoptosis. *Antimicrob Agents Chemother* 39: 872–7

57 Yamaryo T, Oishi K, Yoshimine H, Tsuchihashi Y, Matsushima K, Nagatake T (2003) Fourteen-member macrolides promote the phosphatidylserine receptor-dependent phagocytosis of apoptotic neutrophils by alveolar macrophages. *Antimicrob Agents Chemother* 47: 48–53

58 Sofer D, Gilboa-Garber N, Belz A, Garber NC (1999) "Subinhibitory" erythromycin

represses production of *Pseudomonas aeruginosa* lectins, autoinducer and virulence factors. *Chemother* 45: 335–41
59 Jain A, Sangal L, Basal E, Kaushal GP, Agarwal SK (2002) Anti-inflammatory effects of erythromycin and tetracycline on *Propionobacterium acnes* induced production of chemotactic factors and reactive oxygen species by human neutrophils. *Dermatol Online J* 8: 2
60 Musher DM, Phan HB, Baughn R (2001) Protection against bacteremic pneumococcal infection by antibody to pneumolysin. *J Infect Dis* 183: 827–30
61 Jounblat R, Kadioglu A, Mitchell TJ, Andrew PW (2003) Pneumococcal behavior and host responses during bronchopneumonia are affected differently by the cytolytic and complement-activating activities of pneumolysin. *Infect Immun* 71: 1813–9

Chemotaxis

Jun-ichi Kadota

Division of Pathogenesis and Disease Control, Department of Infectious Diseases, Oita University Faculty of Medicine, 1-1 Hasama, Oita 879-5593, Japan

Introduction

The immunomodulatory properties of antimicrobial agents and their clinical impact have been the focus of worldwide interest in recent years [1–3]. Macrolides, in particular, modulate some inflammatory parameters *in vivo*, including neutrophilia and the release of inflammatory mediators into bronchoalveolar lavage (BAL) of patients with diffuse panbronchiolitis (DPB) [4]. In this review, we focus on the immunomodulatory effects of macrolides with respect to their ability to inhibit neutrophil migration into the airway in DPB and to modulate adhesion molecules, chemotactic factors and neutrophil chemotaxis into sites of inflammation.

Effects of macrolides on chemotactic factors that mediate neutrophil migration into the airway of DPB

One of the key cellular features of the inflammatory process in DPB is the excessive accumulation of neutrophils into the airways, which is demonstrated by marked neutrophilia in BAL fluid. Long-term erythromycin treatment significantly reduces the percentage of neutrophils in BAL fluid [4]. This reduction occurs irrespective of the outcome of sputum bacterial cultures, suggesting that the antibacterial activity is not the only determinant of the efficacy of erythromycin [5]. Based on these findings, studies on the immunomodulatory effect of macrolide antibiotics has become focused on their ability to inhibit neutrophil transmigration from the blood to the site of inflammation in the lung.

BAL fluid from DPB patients has a high level of neutrophil chemotactic activity (NCA). After treatment with erythromycin, the NCA of the BAL fluid is reduced in parallel with improvement in clinical parameters and BAL fluid neutrophilia [4] (Fig. 1). Gel-filtration chromatography of BAL fluid results in four NCA peaks, including molecular weights of approximately 8,000 Daltons, which closely corresponds to the molecular weight of the major neutrophil chemoattractant, interleukin (IL)-8 [6]. Analysis of the cytokine profile reveals that IL-8 in the BAL fluid

Antibiotics as Anti-Inflammatory and Immunomodulatory Agents, edited by Bruce K. Rubin and Jun Tamaoki

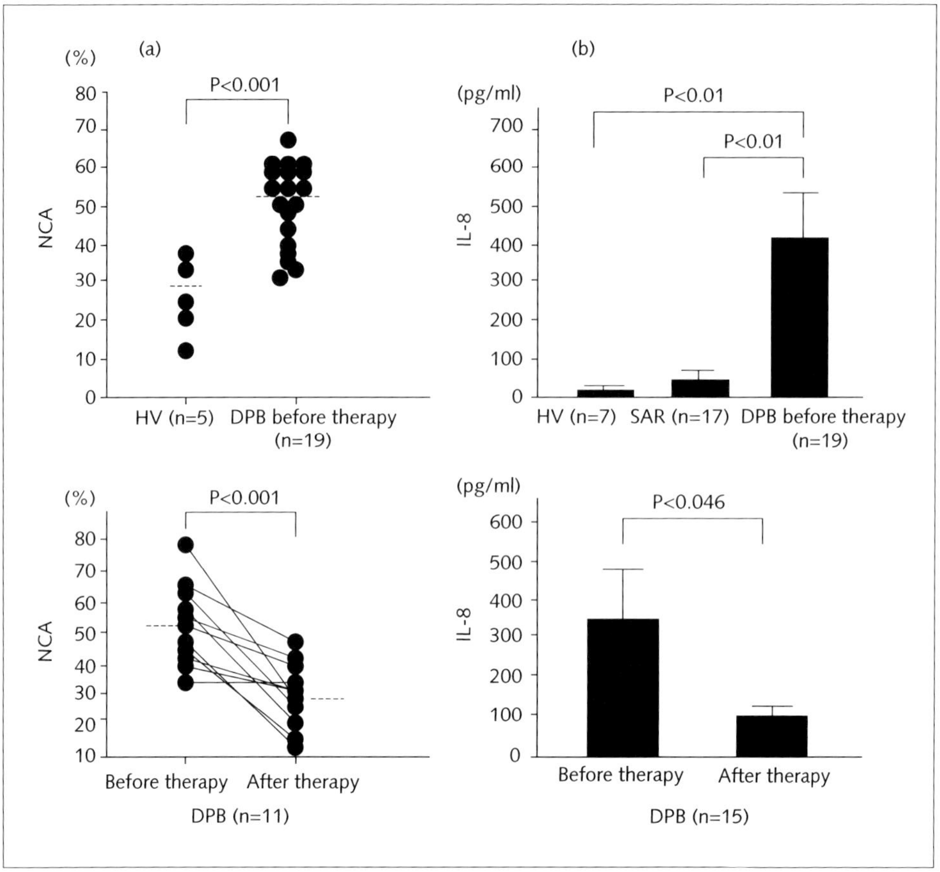

Figure 1
NCA (a) and IL-8 levels (b) in BAL fluid before and after macrolide therapy
HV, healthy volunteers; DPB, diffuse panbronchiolitis; SAR, sarcoidosis. From [4] and [7] with permission of the American Lung Association and S. Karger AG, Basel, respectively.

from DPB patients is significantly elevated and that erythromycin treatment reduces these levels (Fig. 1) as well as the number of neutrophils in the BAL fluid [7]. *In situ* hybridization shows positive staining for IL-8 mRNA in alveolar macrophages, bronchiolar epithelial cells, and endothelial cells in open lung biopsy specimens of patients with DPB [unpublished observations], indicating that these cells are an important cellular source for IL-8 in the lung.

Due to these findings, several *in vitro* and *ex vivo* studies have been conducted on the immunomodulatory effects of macrolides. A 4-week administration of oral

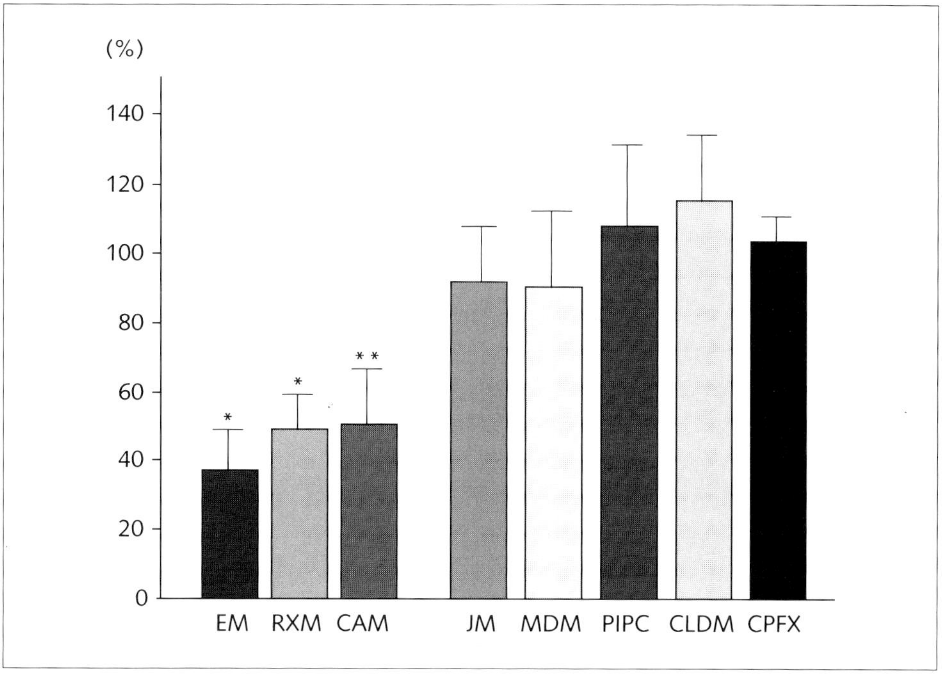

Figure 2
*Effect of various antibiotics on IL-8 by 1α,25-dihydroxyvitamin D3-stimulated THP-1 cells. THP-1 cells were stimulated with 10 ng/ml of lipopolysaccharide and 1% normal human serum and were simultaneously incubated with 10 μg/ml of the test antibiotic. IL-8 concentration in the culture supernatant was measured, and results are expressed as the percentage of IL-8 production compared to untreated control cells. Data represent the mean ± SEM. *P<0.01, **P<0.05 compared to control.*
EM, erythromycin; RXM, roxithromycin; CAM, clarithromycin; JM, josamycin; MDM, midecamycin acetate; PIPC, piperacillin sodium; CLDM, clindamycin; CPFX, ciprofloxacin hydrochloride. From [8] with permission of the American Society for Microbiology.

erythromycin in healthy individuals results in a decreased production of IL-8 by alveolar macrophages [7]. Similarly, macrolides inhibit IL-8 production in 1α, 25-dihydroxyvitamin D3-stimulated THP-1 cells, a human macrophage-lineage cell line [8]. In these studies, erythromycin, roxithromycin, and clarithromycin, each at a concentration of 10 μg/ml, significantly reduces the production of IL-8 in response to stimulation by lipopolysaccharide (10 ng/ml) and 1% normal human serum. Macrolides other than those with a 14-membered macrocyclic ring structure and non-macrolide drugs, such as ciprofloxacin hydrochloride, piperacillin sodium, and

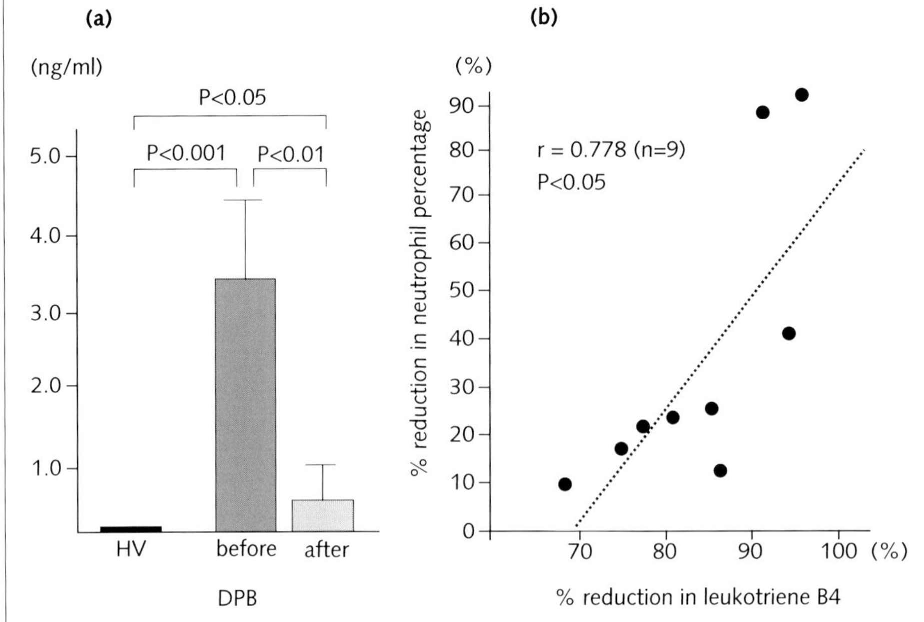

Figure 3
LTB4 levels in BAL fluid obtained from healthy volunteers (HV) and DPB before and after macrolide therapy (a) and correlation between the percent reduction in the neutrophil percentages and LTB4 levels in BAL fluid of DPB patients after macrolide therapy (b).
From [11] with permission of the American College of Chest Physicians.

clindamycin, fail to influence IL-8 production [8] (Fig. 2), supporting the specific clinical effects of 14-membered macrolides on DPB. Other *in vitro* studies have also shown that erythromycin dose-dependently reduces *Pseudomonas*-induced production of IL-8 by neutrophils [9] and inhibits IL-8 release from bronchial epithelial cells stimulated with endotoxin from *H. influenzae* [10]. Post-treatment reduction of leukotriene B4 (LTB4), another neutrophil chemotactic factor, in the BAL fluid of DPB patients also correlates with a reduction in the neutrophil count [11] (Fig. 3). Together, these observations suggest that neutrophil chemotactic factors produced at the site of inflammation during DPB, including IL-8 and LTB4, contribute to excessive neutrophil accumulation in the airways. Furthermore, these results indicate that the therapeutic effects of macrolides may be mediated through suppression of these factors. In fact, the immunomodulatory activity of macrolides on neutrophil migration has been found even in some animal models of acute non-infectious inflammation, such as intratracheal lipopolysaccharide, IL-8 challenge [4], carrageenan-induced paw edema [12] and pleurisy [13], and bleomycin-induced acute lung injury [14]. Collectively, these results suggest that macrolides reduces neutrophil chemotaxis into the inflammatory sites, such as the bronchoalveoli, through inhibition of the cytokine network and, thus, prevent excessive neutrophil accumulation.

Effects of macrolides on adhesion molecules

Interaction between neutrophil adhesion molecules, such as L-selectin or Mac-1, and P-selectin, E-selectin or intercellular adhesion molecule (ICAM)-1 on endothelial cells is also important for neutrophil migration into sites of inflammation. Flow cytometry, using an anti-CD11b antibody, demonstrates that Mac-1 expression on resting neutrophils is higher in the peripheral blood and BAL fluid from DPB patients than from healthy volunteers. The level of Mac-1 on neutrophils from the DPB patients is similar to that found on N-formyl-methionyl-leucyl-phenylalanine (FMLP)-activated neutrophils from healthy volunteers, and, furthermore, there is no difference between Mac-1 expression on neutrophils from peripheral blood and BAL fluid. This study also showed that long-term macrolide treatment causes a decrease in the expression of Mac-1 on resting peripheral neutrophils in parallel with improvement in clinical findings, although a decrease in Mac-1 expression does not occur in non-responders [15] (Fig. 4). In addition, serum levels of other soluble adhesion molecules, such as L-, E-, P-selectin, ICAM-1, and vascular cell adhesion molecule (VCAM)-1, are all significantly elevated in DPB patients, and macrolide treatment significantly reduces these levels as well as the BAL fluid levels of IL-1β and IL-8. Furthermore, in DPB patients, there is a significant correlation between soluble E-selectin in the serum and IL-1β in BAL fluid as well as between soluble L-selectin in the serum and IL-8 in BAL fluid. Incubation of neutrophils with macrolides *in vitro* does not directly affect L-selectin shedding from neutrophils following stimulation with IL-8 or the level of Mac-1 expression on peripheral neutrophils from patients with DPB. Based on these results, the downregulation of these adhesion molecules may be secondary to the inhibition of inflammatory cytokines release. In contrast, therapeutic concentrations of roxithromycin inhibit neutrophil adhesion to cultured human bronchial epithelial cells (Bet-1A cells) and directly decrease the expression of ICAM-1 on interferon (IFN)-γ treated epithelial cells [16]. Finally, a recent report shows that 14-membered ring macrolides directly inhibit VCAM-1 mRNA induction and leukocyte migration into the lung in a bleomycin-induced pulmonary fibrosis mouse model [17]. These studies suggest that 14-membered ring macrolides either directly or indirectly downregulate adhesion molecules, resulting in the inhibition of neutrophil migration into sites of inflammation.

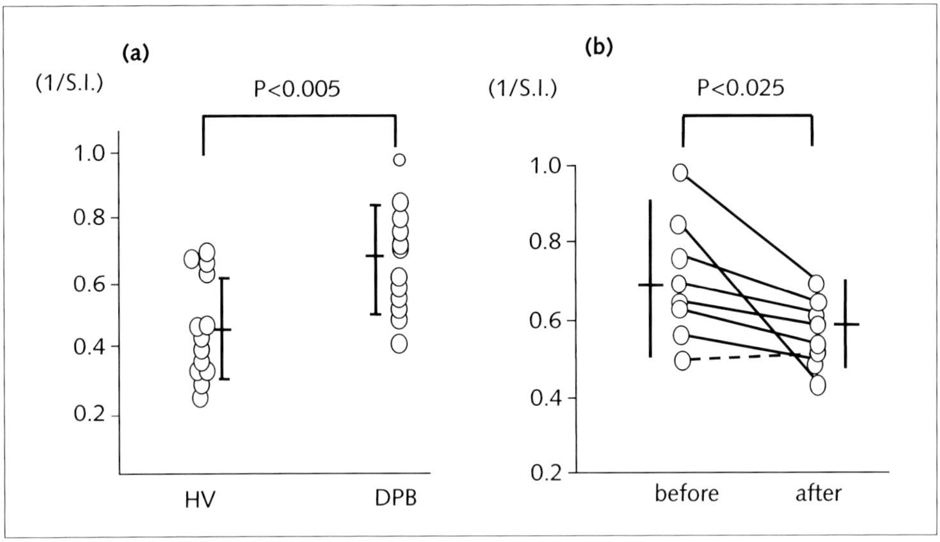

Figure 4
Mac-1 expression on peripheral neutrophils from patients with DPB
Comparison between healthy volunteers (HV) and DPB (a), before and after macrolide therapy (b). The dotted line shows the change in Mac-1 expression in a non-responder before and after therapy. S.I., stimulation index. 1/S.I. was calculated as the ratio of the mean channel fluorescence without stimulation to that with FMLP stimulation. Bars represent the mean ± SD. From [15] with permission of S. Karger AG, Basel.

Direct effects of macrolides on neutrophil chemotaxis

Macrolides can directly affect neutrophil chemotaxis *in vitro* and *in vivo*, although some of the data are conflicting (Tab. 1). Higher concentrations of macrolides (around 20 μg/ml) enhance [18], inhibit [4] or do not change [19, 20] neutrophil chemotaxis induced by various stimuli. Most of these reports demonstrate that the lower serum concentrations (< 1–2 μg/ml) that can be clinically achieved lack direct effects. Brennan et al. [21] also demonstrated that *in vitro* treatment of neutrophils from children with cystic fibrosis or normal individuals with 1–100 μg/ml erythromycin has no effect on IL-8-stimulated migration. *Ex vivo* effects of macrolides on neutrophil chemotaxis also vary. For example, Torre et al. [22] reported that FMLP-stimulated neutrophil chemotaxis is inhibited after administrating 2,250 mg erythromycin per day to healthy subjects for 4 days, while a one-time dose of 500 mg to healthy subjects enhances endotoxin activated serum-stimulated neutrophil chemotaxis when measured 90 min after dosing [23]. In addition, three-time

Table 1 - Direct effect of macrolides on neutrophil chemotaxis

	Subjects	Drug/dose		Stimulant	Result	Ref.
in vitro						
		EM, RXM;	2.5–5 μg/ml	EAS	no change	[18]
			10–20 μg/ml	EAS	enhance	
			2.5–20 μg/ml	FMLP	enhance	
	healthy volunteers	EM;	2 μg/ml	IL-8	no change	[4]
			20 μg/ml	IL-8	inhibit	
		RXM;	1–100 μg/ml	FMLP, ZAS	no change	[19]
		EM;	1–100 μg/ml	FMLP	no change	[20]
	CF patients	EM;	1–100 μg/ml	IL-8	no change	[21]
ex vivo						
	healthy volunteers	EM; 2,250 mg/day, 4 days		FMLP	inhibit	[22]
		EM; 500 mg, once, 90 min. later		EAS	enhance	[23]
	CF patients	EM; 750 mg/day, 4 weeks		IL-8	slight decrease	[21]

EM, erythromycin; RXM, roxithromycin; EAS, endotoxin activated serum; FMLP, N-formyl-methionyl-leucyl-phenylalanine; ZAS, zymosan activated serum; IL-8, interleukin-8; CF, cystic fibrosis

daily treatment of children with cystic fibrosis using 250 mg erythromycin for 4 weeks causes a slight, but insignificant, decrease in the responsiveness of neutrophils to IL-8 [21]. Thus, the direct effects of macrolides on neutrophil chemotaxis are still controversial because the results vary with the experimental conditions. Further studies are necessary to determine whether immune parameters modified by macrolides *in vitro* are clinically relevant.

Although the effects of other non-macrolide antibiotics on neutrophil chemotaxis have also been evaluated, the results vary with the experimental condition as well as those of macrolides. Burgaleta et al. demonstrated the effects of four beta-lactams (cefotaxime, cefoxitin, ceftazidime and latamoxef) using agarose migration and a Boyden chamber method. Cefoxitin (25–200 μg/ml) and cefotaxime (25–200 μg/ml) but not ceftazidime and latamoxef reduced agarose migration, while the Boyden chamber method showed no significant inhibition of chemotaxis by any of beta-lactam antibiotics [24]. Additionally, chemotaxis was not altered by cefotaxime even at concentration as high as 1,000 μg/ml [25]. Van Rensburg et al. also reported the inhibitory effect of cefotaxime on *in vitro* neutrophil migration towards endotoxin-activated serum and towards FMLP in contrast to *in vivo* studies before and after intramuscular injection of therapeutic doses of cefotaxime (1 g) that showed no changes in neutrophil functions [26]. Similarly, exposure to 40 µg/ml of ceftriaxone resulted in the marked inhibition of *in vitro* chemotaxis, while the *in vivo* effects of ceftriaxone before and 30 min after intravenous injection at a dose of 2 g showed no change in any neutrophil function [27]. On the other hand, cefpirome at therapeutic concentrations of 10 and 50 µg/ml significantly enhanced chemotaxis *in vitro* [28]. Other beta-lactam antibiotics, including cefodizime, cefixime, and cefdinir, in the range of their attainable therapeutic concentrations exhibited no significant effects on neutrophil chemotaxis *in vitro* [25, 29]. Carbapenem antibiotics, such as meropenem and imipenem/cilastatin, reduce chemotaxis only at very high concentrations (2,000 and 4,000 µg/ml) [30]. Aminoglycosides, penicillins, glycopeptides, and fluoroquinolone, generally, do not influence neutrophil chemotaxis *in vitro* [31–34]. Collectively, these findings seem to indicate that, so far, no definite conclusion can be drawn on the *in vivo* significance of *in vitro* findings regarding non-macrolide antibiotics, and further studies are required, especially in the clinical setting to fully exploit the potential of the immunomodulatory effect of these drugs during, for example, immunosuppression, chronic airway inflammatory diseases, and acute inflammatory diseases.

Clinical effects of macrolides on neutrophil-mediated inflammatory diseases

Macrolides are likely to be useful for not only DPB but also other chronic neutrophil-induced airway inflammatory diseases, such as cystic fibrosis (CF) [35], sinusitis [36] and chronic obstructive pulmonary disease (COPD) [37] as discussed in other Chapters. Furthermore, new potentials for application of macrolide were recently reported in the field of dermatology and gynecology. Preliminary results show 14-membered ring macrolides (erythromycin 600 mg/day or clarithromycin 200 mg/day for 3 months) improved the clinical symptoms of 20 patients with pustulosis palmaris et plantaris, which is usually involved in hand and/or foot with aseptic pustules and neutrophil-associated inflammation, and it may be explained by the inhibitory action of clarithromycin on TNF-α or staphylococcal enterotoxin B plus IFN-γ-stimulated IL-8 secretion from epidermal keratinocytes [38]. Clarithromycin (200 mg/day for 4 months) also improved the clinical status of five patients suffering from pyometra, a chronic intrauterine infection, in parallel with a decrease in the neutrophil percentages in the lavage fluid of the uterine endometrial cavity and in the level of IL-8 [39]. However, a large-scale clinical study would be necessary to clarify the effect of macrolides on neutrophil-associated inflammatory diseases in the near future.

Cytokines

Hajime Takizawa

Department of Respiratory Medicine, University of Tokyo, Graduate School of Medicine, 7-3-1 Hongo, Bunkyo-ku, Tokyo 113-8655, Japan

Introduction

Erythromycin (EM) is a macrolide antibiotic that is widely used for the treatment of upper and lower respiratory tract infections. Recent reports have further showed that EM and other fourteen-membered ring macrolides, such as clarithromycin (CAM) and roxithromycin (RXM), are effective for the treatment of chronic airway diseases such as diffuse panbronchiolitis (DPB), bronchial asthma and chronic sinusitis [1–3]. This effectiveness is considered to be aside from their antimicrobial actions, because they are effective at half the recommended dosage as antibiotics, and even in cases without concomitant infection. It has been recently shown that azithromycin (AZM), a 15-membered ring macrolide, has a beneficial effect on the clinical course of patients with cystic fibrosis [4, 5], which is a serious hereditary disorder among Caucasian people. Their precise mechanisms, however, remain unclear. Several cytokines including IL-1, TNF-α and IL-8 have been reported to be elevated in bronchoalveolar lavage fluids (BALF) from patients with such airway inflammatory diseases, and to be decreased after appropriate therapy, suggesting important roles in airway inflammatory processes [2, 6]. Kadota and his associates [2] demonstrated an increase of neutrophil chemotactic activity (NCA) in BALF, which showed a clear correlation with neutrophil numbers. They further showed that inflammatory cytokines such as IL-8, IL-1β and TNF-α were also increased in BALF from patients with chronic airway inflammatory diseases such as DPB and bronchiectasis [6]. The treatment with 14-ring member macrolide antibiotics such as EM induced a marked decrease in both neutrophil number and these inflammatory cytokines and chemokines. These cytokines are potent activators of neutrophils, among which IL-8 is one of the most potent chemotactic factors in the airways. Therefore, it is probable that EM attenuates airway inflammatory responses by decreasing the local cytokine/chemokine levels and thus decreasing the recruitment of inflammatory cells such as neutrophils. Airway epithelial cells are one of the potent sources of cytokines and chemokines [7], and their anatomical location suggests their pivotal role in the regulation of cell recruitment into the airways. There

Antibiotics as Anti-Inflammatory and Immunomodulatory Agents, edited by Bruce K. Rubin and Jun Tamaoki

is increasing evidence that macrolide antibiotics show modulating effects on cytokine expression in clinical and experimental settings. *In vitro* studies further indicated that these drugs have inhibitory actions on cytokine production and/or expression in various cells. This review will focus on the effects of the macrolides on cytokine/chemokine production and its potential molecular mechanisms.

Studies on airway epithelium from patients with chronic airway inflammatory disease

We studied whether or not macrolides had any effect on cytokine expression and production by human bronchial epithelial cells. We evaluated the changes in IL-8 mRNA levels and IL-8 protein release by airway epithelial cells before and after macrolide therapy [8]. Patients with chronic airway diseases (DPB, chronic bronchitis, and diffuse bronchiectasis) received oral EM or CAM therapy for more than 3 months with no side effects. In accordance to the clinical changes, IL-8 mRNA levels corrected by β-actin transcripts were decreased in patients who responded to macrolide therapy when assessed by reverse transcription and polymerase chain reaction (RT-PCR). Spontaneous IL-8 release from epithelial cells was also decreased by macrolide therapy.

Inhibitory actions of macrolides on cytokine/chemokine production by various types of cells: *In vitro* findings

Airway epithelial cells

It is well documented that normal human bronchial epithelial cells release a variety of cytokines and chemokines, and proinflammatory cytokines such as IL-1α , IL-1β and TNF-α stimulate their production *in vitro* (Fig. 1) [9]. We cultured human normal and transformed bronchial epithelial cells, and studied the effect of EM, CAM and RXM on IL-6 [10], IL-8 [8, 11] and GM-CSF [12] production. Among the antimicrobes tested, only 14-member macrolides EM, CAM and RXM showed an inhibitory action on IL-6 and IL-8 release by unstimulated and cytokine-stimulated human bronchial epithelial cells, whereas a 16-ring member macrolide, josamycin (JM), failed to show such effects. LDH release assay, trypan blue dye exclusion test as well as a colorimetric MTT assay showed that this effect was not due to cytotoxicity. To assess the effect of macrolide antibiotics on IL-8 production by inflamed airway epithelium, bronchial epithelial cells were obtained from patients with chronic airway disease including DPB, sinobronchial syndrome, and diffuse bronchiectasis under fiber optic bronchoscope. Spontaneous IL-8 release by airway epithelial cells from inflamed airways were significantly inhibited with the addition

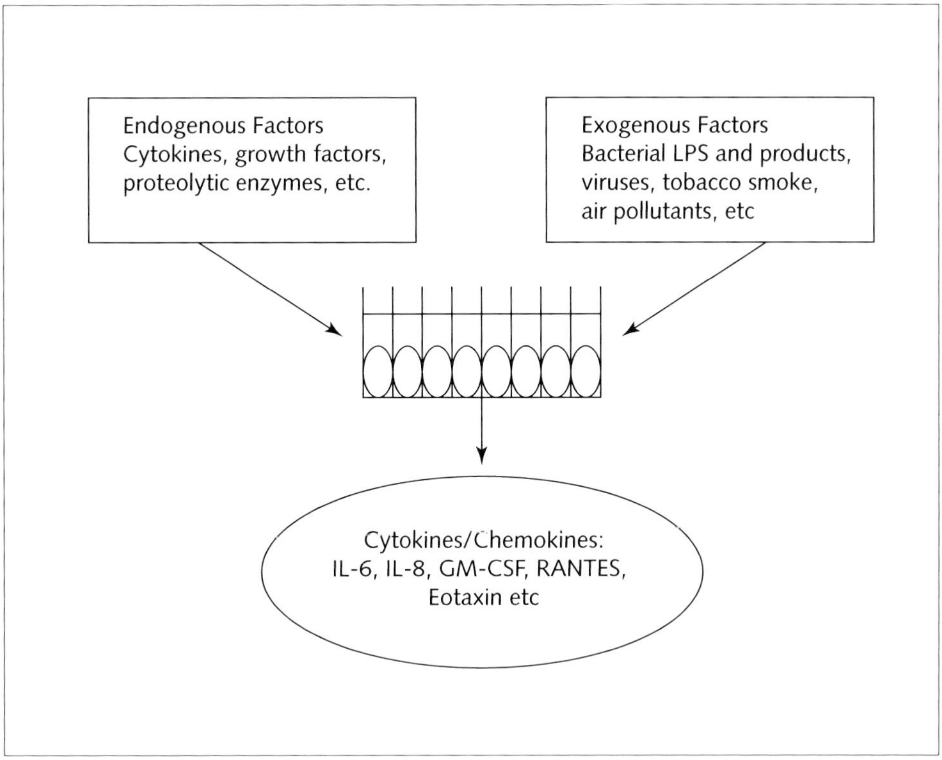

Figure 1
Airway epithelial cells as sources of cytokines and chemokines in the airways
Airway epithelial cells express and release a variety of cytokines/chemokines, adhesion molecules and lipid mediators, and thereby participate in the regulation of inflammatory responses in the airways.

of EM and CAM, but not with ABPC *in vitro* [8]. Khair et al. [13] reported that EM inhibited release of IL-8 as well as of IL-6 from *H. influenzae* endotoxin-stimulated normal bronchial epithelial cells.

Alveolar macrophages and neutrophils

Alveolar macrophages are another important source of cytokines in the lung. Iino and co-workers [14] reported that EM suppressed IL-1β and TNF-α production by human peripheral blood monocytes. Fujii and co-workers [15] demonstrated that 14-ring member macrolides uniquely inhibited IL-8 production by a vitamin D-differentiated macrophage cell line THP-1 cells. Similarly, CAM and AZM inhibit the production of IL-1α, IL-1β, IL-6, IL-10, TNF-α, and granulocyte and macrophage colony stimulating factor [16].

Sugiyama and associates showed that chronic administration of EM in rats induced an inhibitory changes in cytokine production such as GRO/CINC-1 and CINC-2α, homologues of human IL-8 and MIP-2, respectively [17]. Oishi and his co-workers [18] studied the production of IL-8 by neutrophils which were stimulated with inactivated *Pseudomonas aeruginosa* bacilli, and they found that IL-8 release from the neutrophils were inhibited by EM treatment.

Effects on other kind of cells

EM has been described in literature as an alternative therapy for intractable bronchial asthma [19]. It was reported that EM has a corticosteroid-sparing effect; however, EM alone decreased airway hyper responsiveness and asthma severity [20]. Konno et al. [21] found that RXM inhibited production of IL-2 and IL-4 by peripheral lymphocytes. Nakahara and associates [22] reported that production of Th2-derived cytokines IL-4 and IL-5 were significantly suppressed by EM, whereas that of Th1-derived cytokines such as IFN-γ rather increased. Therefore, it is probable that macrolides exert inhibitory effects on Th2 cytokines in asthma patients. Kohyama et al. [23] showed that EM significantly suppressed IL-8 release from human peripheral blood eosinophils from atopic donors.

Molecular mechanisms of anti-inflammatory actions of macrolides

We evaluated the effects of macrolides on steady state levels of IL-6 and IL-8 mRNA by Northern blot analysis [8, 11]. Human bronchial epithelial cells expressed constitutive IL-6 and IL-8 mRNA, which were significantly upregulated by the cytokines such as IL-1α, β and TNF-α. EM, CAM and RXM inhibited steady state levels of IL-6 and IL-8 expression in normal and immortalized bronchial epithelial cells. This action appeared to be unique, because other antibiotics, including a 16-member macrolide JM, did not show any effect. Therefore, it is probably one mechanism of the clinical beneficial effect of these macrolide antibiotics.

It is well known that the transcriptional rates of IL-8 are regulated by several transcription factors such as NF-κB and AP-1. Abe and co-workers [24] demonstrated that CAM repressed TNF-α-induced AP-1 activation in human bronchial epithelial cells. We studied the effect of EM and CAM on the phorbol myristate acetate (PMA)-induced activation of NF-κB and AP-1. Pretreatment of EM and CAM at therapeutic concentration before the PMA treatment showed an inhibitory effect on both of the transcription factors as assessed by electrophoretic mobil-

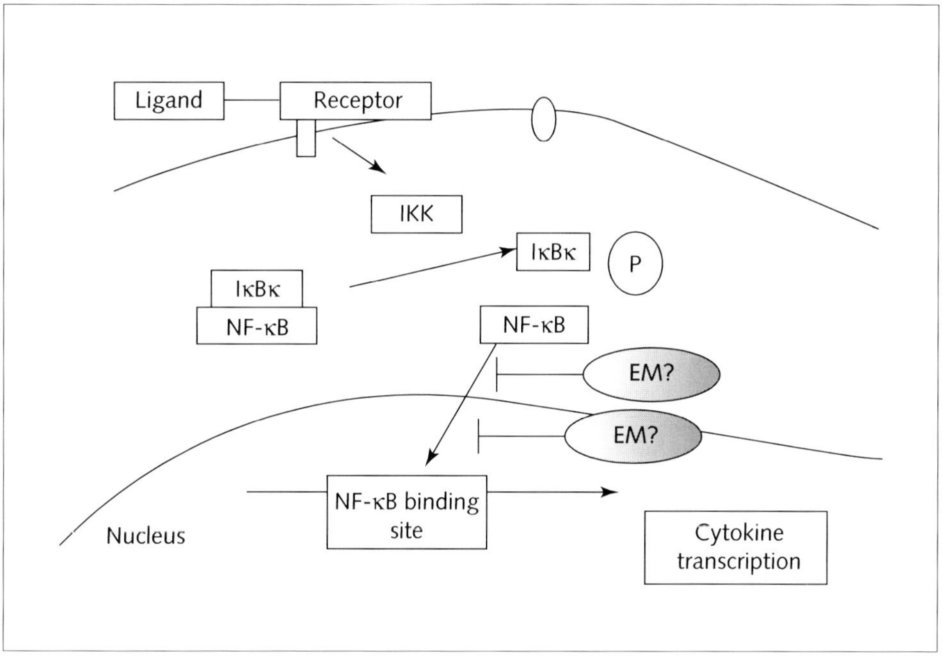

Figure 2
Potential mechanisms of macrolides on the regulation of transcription factors
The NF-κB is present in latent form in the cytoplasm by binding to the inhibitor protein IκB. Cytokine-induced signal transduction results in selective IκB phosphorylation, which is in turn ubiquitinated and degraded by proteasome pathway. Free NF-κB migrates to the nucleus by several localization signals. Binding of NF-κB to its specific site of genes induces transcription of several NF-κB-dependent genes. NF-κB is then inactivated by newly synthesized IκB both in cytoplasm and nucleus. IKK: IκB kinase

ity shift assay (EMSA) [25]. Such findings have been reported in nasal epithelial cells and fibroblasts, monocytes and macrophage cell line [26–28]. In contrast, the macrolides showed no effect on the activation of cyclic AMP-responsive element binding protein (CREB), suggesting that the suppressive effect on some transcription factors is somewhat specific [25]. We further evaluated the effect of EM on the phosphorylation of inhibitor of NF-κB (IκB), which is a crucial step for transactivation of NF-κB. EM did not influence the phosphorylation processes *in vitro* [29]. These data suggest that EM act at the process of nuclear translocation of NF-κB, or at the stages of DNA binding within the nucleus (Fig. 2). Further studies are necessary to elucidate the molecular events important for their potentials.

Anti-tumor effect of macrolides: Their potential as a biological response modifier

Oral administration of erythromycin increased survival in tumor-bearing mice [30]. The tumorcidal activity of macrophages increased as serum IL-4 levels elevated. They further showed that anti-IL-4 antibody abolished the effect of erythromycin. Recent *in vitro* studies showed that the macrolides induce IL-4 production by splenic cells [31]. Although there are only few reports to show that macrolides induce cytokines and other biological response peptides *in vitro* and *in vivo*, it needs further investigation to clarify the anti-tumor activity of these drugs.

Immunomodulatory effects of other classes of antimicrobial agents

The above observations strongly suggest that macrolide antibiotics exert their clinically beneficial effects, at least in part, by their anti-inflammatory or immunomodulatory effects. However, such effects have also been reported in other kinds of antimicrobial agents. Most fluoroquinolone derivatives induce IL-2 synthesis, but inhibit synthesis of IL-1 and TNF-α [32]. They also enhance the synthesis of colony-stimulating factor (CSF). The potential molecular mechanisms that are not fully elucidated include effects on intracellular cyclic AMP and phosphodiesterases, effects on transcription factors such as NF-κB, AP-1, NF-AT and NF-IL-6. However, the reported effects are very diverse, and different results have been reported in different cells, stimuli and study methods. It should also be noted that the drugs show their modulating effects only at high concentrations. *In vivo* experiments using LPS-injected animals showed that quinolones protect the animals by decreasing TNF-α and IL-12 and by increasing IL-10 [33]. A few clinical trials have been conducted to show the attenuating effects on neutropenia, with controversial results [34].

Like macrolides, the immunomodulatory effects of quinolones may contribute to their clinical efficacy in chronic infections. However, this possibility has not yet been exploited. Tetracycline derivatives have also been reported to show immunomodulatory effects. Doxycycline reduces mortality to lethal endotoxemia by reducing nitric oxide synthesis *via* an IL-10-independent mechanism [35]. This drug also inhibits matrix metalloproteinase (MMP) activity to attenuate periodontal bone loss in rat models [36].

Anti-inflammatory or immunomodulatory effects of quinolones and tetracyclines seem to have partially common features with those of macrolides. Comparative studies among these drugs may facilitate researches for better clinical applications.

Future directions

EM also has a motilin-like stimulating activity on gastrointestinal smooth muscles [37]. Therefore, inhibitory effect on cytokine expression in human cells, as summarized here, may be a third bioactivity of the macrolide antibiotic. We found that some of the derivatives with no antimicrobial activity have an inhibitory effect on IL-8 production by human airway epithelial cells. These analogues also showed inhibitory action on the activation of NF-κB and AP-1 assessed by EMSA [29]. Characterization of the chemical structure responsible for its potential would be important to pursue, and further investigation for the molecular mechanism would be necessary for a possible new type anti-inflammatory agent.

Conclusion

The above data suggest that the anti-inflammatory or immunomodulatory properties of macrolides are, at least in part, by inhibitory effects on cytokine gene expression through actions on transcription factors. The effects of the macrolides reported so far are generally normalizing the activated states, but not suppressing the basal levels. Although the mechanisms for their anti-inflammatory actions are being partially elucidated, it still remains unclear which macrolide structure is critical for their effects, which has the best efficacy and fewest adverse effects, duration of the anti-inflammatory effect with long-term macrolide therapy, and the long-term impact of continuous antimicrobial coverage.

Acknowledgements
This work is supported in part by The Diffuse Lung Disease Research Committee, Japan Ministry of Welfare and Labor, Japan.

References

1 Kudoh S, Uetake T, Hagiwara K, Hirayama M, Hus L-H, Kimura H, Sugiyama Y (1987) Clinical effect of low-dose, long-term erythromycin chemotherapy on diffuse panbronchiolitis. *Jpn J Thorac Dis* 25: 632–42
2 Kadota J, Sakito O, Kohno S, Sawa H, Mukae H, Oda H, Kawakami K, Fukushima K, Hiratani K, Hara K (1993) A mechanism of erythromycin treatment in patients with diffuse panbronchiolitis. *Am Rev Respir Dis* 147: 153–9
3 Spector S, Katz F, Farr R (1974) Troleandomycin: Effectiveness in steroid-dependent asthma and bronchitis. *J Allergy Clin Immunol* 54: 367–79
4 Equi A, Balfour-Lynn IM, Bush A, Rosenthal M (2002) Long term azithromycin in chil-

dren with cystic fibrosis: a randomized, placebo-controlled crossover trial. *Lancet* 360 (9338): 978–84

5 Saiman L, Marshall BC, Mayer-Hamblett N, Burns JL, Quittner AL, Cibene DA, Coquillette S, Fieberg AY, Accurso FJ, Campbell PW 3rd; Macrolide Study Group (2003) Azithromycin in patients with cystic fibrosis chronically infected with Pseudomonas aeruginosa: a randomized controlled trial. *JAMA* 290 (13): 1749–56

6 Sakito O, Kadota J, Kohno S, Kabe K, Shirai R, Hara K (1994) Interleukin 1 beta, tumor necrosis factor alpha, and interleukin 8 in bronchoalveolar lavage fluid of patients with diffuse panbronchiolitis: A potential mechanism of macrolide therapy. *Respiration* 63: 42–8

7 Takizawa H (1998) Airway epithelial cells as regulators of airway inflammation. *Int J Mol Med* 1: 367–78

8 Takizawa H, Ohtoshi T, Kawasaki S, Kohyama T, Sato M, Tanaka M, Kasama T, Kobayashi K, Nakajima K, Ito K (1997) Erythromycin modulates IL-8 expression in human bronchial epithelial cells: Studies with normal and inflamed airway epithelium. *Am J Respir Crit Care Med* 156: 266–71

9 Nakamura H, Yoshimura K, Jaffe HA, Crystal RG (1991) Interleukin-8 gene expression in human bronchial epithelial cells. J Biol Chem 266: 19611–17

10 Takizawa H, Desaki M, Ohtoshi T, Kikutani T, Okazaki H, Sato M, Akiyama N, Shoji S, Hiramatsu K, Ito K (1995) Erythromycin suppresses interleukin 6 expression by human bronchial epithelial cells. *Biochem Biophys Res Commun* 210: 781–6

11 Kawasaki S, Takizawa H, Ohtoshi T, Takeuchi N, Kohyama T, Nakamura H, Kasama T, Kobayashi K, Nakahara K, Morita Y, Yamamoto K (1998) Roxithromycin inhibits cytokine production and neutrophil attachment with human bronchial epithelial cells *in vitro*. *Antimicrob Agents Chemother* 42: 1499–502

12 Takizawa H, Ohtoshi T, Takeuchi N, Ito K (1996) Effect of macrolide antibiotics on the expression and release of inflammatory cytokines in human bronchial epithelial cells. *J Jpn Bronchoesophagol Soc* 47: 185–8 (in Japanese)

13 Khair OA, Devalia JL, Abdelaziz MM, Sapsford RJ, Davis RJ (1995) Effect of erythromycin on *Hemophilus influenzae* endotoxin-induced release of IL-6, IL-8 and sICAM-1 by cultured human bronchial epithelial cells. *Eur Respir J* 8: 1451–7

14 Iino Y, Toriyama M, Kudo K, Natori Y, You A (199). Erythromycin inhibition of lipopolysaccharide-stimulated tumour necrosis factor-alpha production by human monocytes *in vitro*. *Ann Otol Rhinol Laryngol* 101: 16–20

15 Fujii T, Kadota J, Morikawa T, Matsubara Y, Kawakami K, Iida K, Shirai R, Taniguchi H, Kaseda M, Kawamoto S, Kohno S (1996) Inhibitory effect of erythromycin on interleukin 8 production by 1alpha,25-dihydroxyvitamin D3-stimulated THP-1 cells. *Antimicrob Agents Chemother* 40: 1548–51

16 Khan AA, Slifer TR, Araujo FG, Remington JS (1999) Effect of clarithromycin and azithromycin on production of cytokines by human monocytes. *Int J Antimicrob Agents* 11 (2): 121–32

17 Sugiyama Y, Yanagisawa K, Tominaga S-I, Kitamura S (1999) Effects of long-term

administration of erythromycin on cytokine production in rat alveolar macrophages. *Eur Respir J* 14: 1113–16
18 Oishi K, Sonoda F, Kobayashi S, Iwagaki A, Nagatake T, Matsushima K, Matsumoto K (1994) Role of interleukin-8 (IL-8) and an inhibitory effect of erythromycin on IL-8 release in the airways of patients with chronic airway diseases. *Infect Immunity* 62: 4145–52
19 Kamada AK, Hill MR, Ikle DN, Brenner AM, Szefler SJ (1993) Efficacy and safety of low-dose troleandomycin therapy in children with severe, steroid-requiring asthma. *J Allergy Clin Immunol* 91(4): 873–82
20 Miyatake H, Taki F, Taniguchi H, Suzuki R, Takagi K, Satake T (1991) Erythromycin reduces the severity of bronchial hyperresponsiveness in asthma. *Chest* 99: 670–3
21 Konno S, Asano K, Kurokawa M, Ikeda K, Okamoto K, Adachi M (1994) Anti-asthmatic activity of a macrolide antibiotic, roxithromycin. Inter Archv Allergy Immunol 105: 308–16
22 Nakahara H, Higashida A, Nogami J, Iwanaga K, Ueshima H, Sawaguchi H, Haraguchi R, Muraki M, Kubo Y, Nakajima S (1997) Effect of roxithromycin on cytokine production by peripheral monocytes derived from patients with bronchial asthma. *Jpn J Antibiotics* 50 (Suppl): 113–15 (in Japanese)
23 Kohyama T, Takizawa H, Kawasaki S, Akiyama N, Sato M, Ito K (1999) Fourteen-member macrolides inhibit IL-8 release by human eosinophils from atopic donors. *Antimicrob Agents Chemother* 43: 907–11
24 Abe S, Nakamura H, Inoue S, Takeda H, Saito H, Kato S, Mukaida N, Matsushima K, Tomoike H (2000) Interleukin-8 gene repression by clarithromycin is mediated by the activator protein-1 binding site in human bronchial epithelial cells. *Am J Respir Cell Mol Biol* 22: 51–60
25 Desaki M, Takizawa H, Ohtoshi T, Kasama T, Kobayashi K, Sunazuka T, Omura S, Yamamoto K, Ito K (2000) Erythromycin suppresses nuclear factor-kappa B and activator protein-1 activation in human bronchial epithelial cells. *Biochem Biophys Res Commun* 267: 124–8
26 Miyanohara T, Ushikai M, Matsune S, Ueno K, Katahira S, Kurono Y (2000) Effects of clarithromycin on cultured human nasal epithelial cells and fibroblasts. *Laryngoscope* 110 (1): 126–31
27 Kikuchi T, Hagiwara K, Honda Y, Gomi K, Kobayashi T, Takahashi H, Tokue Y, Watanabe A, Nukiwa T (2002) Clarithromycin suppresses lipopolysaccharide-induced interleukin-8 production by human monocytes through AP-1 and NF-kappa B transcription factors. *J Antimicrob Chemother* 49 (5): 745–55
28 Ichiyama T, Nishikawa M, Yoshitomi T, Hasegawa S, Matsubara T, Hayashi T, Furukawa S (2001) Clarithromycin inhibits NF-kappaB activation in human peripheral blood mononuclear cells and pulmonary epithelial cells. *Antimicrob Agents Chemother* 45 (1): 44–7
29 Desaki M, Okazaki H, Sunazuka T, Omura S, Yamamoto K, Takizawa H (2004) Molecular mechanisms of anti-inflammatory action of erythromycin in human bronchial

epithelial cells: possible role in the signaling pathway that regulates nuclear factor-kappaB activation. *Antimicrob Agents Chemother* 48 (5): 1581–5

30 Hamada K, Kita E, Sawaki M, Mikasa K, Narita N (1995) Antitumor effect of erythromycin in mice. *Chemotherapy* 41 (1): 59–69

31 Ortega E, Escobar MA, Gaforio JJ, Algarra I, Alvarez De Cienfuegos G (2004) Modification of phagocytosis and cytokine production in peritoneal and splenic murine cells by erythromycin A, azithromycin and josamycin. *J Antimicrob Chemother* 53 (2): 367–70

32 Dalhoff A, Shalit I (2003) Immunomodulatory effects of quinolones. *Lancet Infect Dis* 3 (6): 359–71

33 Khan AA, Slifer TR, Araujo FG, Suzuki Y, Remington JS (2000) Protection against lipopolysaccharide-induced death by fluoroquinolones. *Antimicrob Agents Chemother* 44 (11): 3169–73

34 Broide E, Douer D, Shaked N, Yellin A, Lieberman Y, Rosen N, Segev S, Rubinstein E (1992) Effect of short-term therapy with ciprofloxacin, ceftriaxone and placebo on human peripheral WBC and marrow-derived granulocyte-macrophage progenitor cells (CFU-GM) *Eur J Haematol* 48(5): 276–7

35 D'Agostino P, La Rosa M, Barbera C, Arcoleo F, Di Bella G, Milano S, Cillari E (1998) Doxycycline reduces mortality to lethal endotoxemia by reducing nitric oxide synthesis *via* an interleukin-10-independent mechanism. *J Infect Dis* 177 (2): 489–92

36 Ramamurthy NS, Rifkin BR, Greenwald RA, Xu JW, Liu Y, Turner G, Golub LM, Vernillo AT (2002) Inhibition of matrix metalloproteinase-mediated periodontal bone loss in rats: a comparison of 6 chemically modified tetracyclines. *J Periodontol* 73(7): 726–34

37 Kondo Y, Torii K, Omura S, Itoh Z (1988) Erythromycin and its derivatives with motilin-like biological activities inhibit the specific binding of ^{125}I-motilin to duodenal muscle. *Biochem Biophys Res Commun* 150: 877–82

Antibacterial agents and the oxidative burst

Marie-Thérèse Labro

INSERM U479, CHU X. Bichat, 16 rue Henri Huchard, 75018 Paris, France

Introduction

In 1883, the term "phagocytes" was coined by the Russian zoologist Elie Metchnikoff, following his observation of specialized cells ingesting bacteria, and from there phagocytosis was recognized as a major defence mechanism in multicellular organisms. Polymorphonuclear neutrophils (PMN) and monocytes/macrophages are the professional phagocytes in mammals. PMN have a prominent role against microbial pathogens. In general, they are the first host cell to arrive at sites of microbial invasion, and they have an innate capacity to ingest and kill a wide range of microorganisms. Phagocytes ingest microorganisms into intracellular compartments called phagosomes, where they direct an arsenal of digestive and antibacterial agents (oxygen-independent antibacterial system). In 1933, Baldridge and Gerard discovered that phagocytosing neutrophils undergo explosive oxygen consumption (50- to 100-fold increase) – the "oxidative burst" – unrelated to mitochondrial respiration, which reflects the activity of the NADPH oxidase system, a multicomponent enzyme that assembles at the phagosomal membrane. The oxidants generated by this enzyme are used to destroy ingested pathogens, but when released in the extracellular medium, they can cause "collateral damage" to host cells and tissues, and so can be involved in the pathophysiological process of inflammatory reactions and various inflammatory diseases. Recent studies have evidenced the existence of phagocyte-type NADPH oxidases in many non-phagocytic cells (fibroblasts, vascular smooth muscle cells, endothelial cells, renal mesangial cells and tubular cells), the Nox/Duox family of NADP oxidases; ROS production by these oxidases may serve a signaling role or lead to oxidative damage. Modulation of oxidant production remains a therapeutic target to dampen an excessive inflammatory response. The immunomodulatory (anti-inflammatory) properties of some antimicrobial agents has been reviewed recently [1, 2]. This chapter completes a previous review on the interference of antibacterial drugs with the phagocyte oxidative burst [3]. After briefly presenting the structure of the NADPH oxidase, the biochemistry of the oxidative burst, its beneficial and detrimental roles, its regulation and the laborato-

Antibiotics as Anti-Inflammatory and Immunomodulatory Agents, edited by Bruce K. Rubin and Jun Tamaoki

ry methods of analysis, the main characteristics of the *in vitro* interference of antibacterial agents with ROS production and activity will be detailed, before concluding on the potential therapeutic consequences of these effects.

The oxidative burst

The phagocyte NADPH oxidase

Studies in many laboratories over a number of years have established the identity of the phagocyte NADPH oxidase as a multiprotein enzyme whose catalytic and regulatory subunits are partitioned between the cytosol and plasma (and granule) membrane in resting cells and assemble at the cytosolic face of the plasma membrane after activation [4–7]. The core enzyme consists of five subunits: in unstimulated cells, three of these, $p40^{phox}$, $p47^{phox}$, $p67^{phox}$ (phox for phagocyte oxidase) form a cytosolic complex of undefined stoichiometry that can be purified by gel filtration chromatography with an apparent molecular mass of 250–300 kDa; $p22^{phox}$ and $gp91^{phox}$ form a heterodimeric, membrane-bound flavocytochrome, known as cytochrome b_{558} according to its infrared absorbance (or cyto b_{-245} from its oxidation–reduction mid-point potential). In resting cells, approximately 85% of the cytochrome is located on the membrane of peroxidase-negative (specific and secretory) granules, and the rest is found on the cytoplasmic membrane. Interactions among the various oxidase components occur through a number of specific regions, including SH3 domains and proline-rich motifs. Upon exposure to appropriate stimuli, multiple phosphorylation events in the cytosolic components take place, which induce rearrangements in a number of protein–protein interactions, ultimately leading to translocation of the cytoplasmic complex to the membrane and association with cytochrome b_{558}. Activation requires also the participation of two low molecular weight guanine nucleotide-binding proteins: Rac2, which in the resting cells form a complex with Rho-GDI (guanine nucleotide dissociation inhibitor), and Rap1A which is found on membranes and can be co-purified with the cytochrome. During activation, Rac2 binds GTP, dissociates from its inhibitor and migrates to the membrane. Activation also triggers the fusion of the secretory vesicle, and later specific granule, membranes with the plasma membrane where the active enzyme complex is finally assembled. The knowledge of the enzyme derives from studies of the human genetic disorder chronic granulomatous disease (CGD) [8]. Phagocytes of CGD patients are missing, or have an abnormal form of, one or another of the protein components of the respiratory burst oxidase. Thus, various genetic defects can lead to a failure of the respiratory burst and the associated microbicidal defect is responsible for the clinical features of CGD. The genetic heterogeneity was appreciated early on, as the disease was transmitted in an X-linked fashion in some kindred and as an autosomal recessive trait in others.

Biochemistry of the oxidative burst

The phagocyte oxidase catalyzes the one-electron reduction of oxygen into superoxide anion at the expense of NADPH, and from there a vast assortment of reactive oxidants (reactive oxygen species [ROS], reactive oxygen intermediates [ROI]) are generated [9]. Much is known about the reactive oxygen species released into the extracellular surroundings when PMN respond to soluble stimuli. However, the enzymatic and chemical reactions involved in oxidant production are dependent on environmental conditions, which may vary markedly between the phagosome and the extracellular medium. Knowledge of the biochemistry within the phagosome is limited by its inaccessibility to standard detectors and scavengers. Consequently, the oxidant species directly responsible for killing bacteria are still open to speculation. Much, if not all, of the extra oxygen consumed in the respiratory burst is converted to the superoxide anion ($O_2^{\bullet-}$) by the one-electron reduction of oxygen using NADPH (provided by hexose monophosphate shunt [HMPS] activity) as the electron donor: $2\ O_2 + NADPH \rightarrow 2\ O_2^{\bullet-} + NADP^+ + H^+$. In the acidic vacuolar medium (or in the presence of superoxide dismutase) superoxide anion is further dismutated into hydrogen peroxide (H_2O_2): $2\ O_2^{\bullet-} + 2\ H^+ \rightarrow H_2O_2 + O_2$. Myeloperoxidase (MPO), which is released from azurophilic (primary) granules of PMN and monocytes by a degranulation process, reacts with H_2O_2 to form a complex that can oxidize a large variety of substances [10]. Among the latter is chloride, which is oxidized initially to hypochlorous acid (HOCl), with the subsequent formation of highly reactive chloramines (R-NHCl). Other toxic species include hydroxyl radical ($OH^\bullet$), and singlet oxygen ($^1O_2$): in the presence of metal, (Haber–Weiss reaction), $H_2O_2 + O_2^{\bullet-} \rightarrow OH^\bullet + OH^- + {}^1O_2$. Hydroxyl radical can also be produced by the reaction between $O_2^{\bullet-}$ and HOCl, and singlet oxygen may be formed in a reaction between HOCl and hydrogen peroxide. A schematic presentation of the oxidative burst is given in Figure 1.

Regulation of the oxidative burst

Activation of the NADPH oxidase can be obtained *via* a number of transduction pathways following phagocytosis (particulate stimuli) or stimulation with various humoral mediators, through selective recognition *via* membrane receptors, including the receptors for opsonins (Fcγ-Rs but not CR3, although this remains controversial [11]) and chemoattractant receptors (for bacterial formulated peptides, C5a, or leukotrienes). Some ligands do not directly stimulate a functional response but increase oxidase activity after a second stimulus. This is referred to as " priming" and is observed with some cytokines, endotoxin and suboptimal concentrations of directly activating stimuli. *In vitro* assays also use direct modulators of signaling pathway, for instance phorbol esters (PMA) to activate protein kinase C, calcium

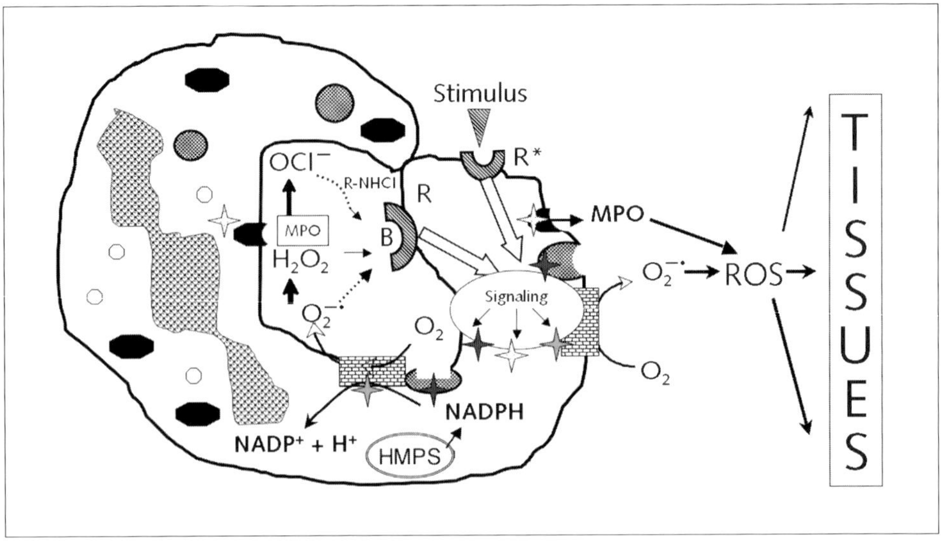

Figure 1
In and out the phagolysosome: oxidant production by phagocytes
See text for details and abbreviations.

ionophores to increase intracellular calcium, etc. Binding of a ligand to its receptor triggers a sequence of events known as a biochemical signaling pathway. The first, proximal event, related to the structure of the receptor, directs the main signaling pathways. Various receptor subgroups are defined according to the primary signal, including pertussis toxin insensitive heterotrimeric G-protein-linked receptors and tyrosine-kinase receptors. The knowledge of the transductional pathways involved in NADPH oxidase activation is of particular importance to propose new therapeutic targets. The mechanism for the activation of phagocytic NADPH oxidase has not been fully elucidated. No selective pathway for NADPH oxidase activation has been described and redundancy of effectors (Phospholipases C, D, A_2, protein kinases C, A, MAP kinases) and of second messengers (Ca^{2+}, diacylglycerol, arachidonic acid, etc.) is the rule for other phagocytic functions. It has been proposed that the stimulation of neutrophils by receptor-binding ligands results in an intracellular signaling cascade, including the activation of phospholipase C, which releases IP3 and diacylglycerol, which, in turn, increase intracellular Ca^{2+} concentration and activate protein kinase C (PKC), respectively. The two pathways function synergistically for $O_2^{\bullet-}$ generation. Activations of phospholipase D, mitogen-activated protein (MAP) kinase, phosphoinositide 3-kinase, and probably phospholipase A_2 are also functionally linked to $O_2^{\bullet-}$ generation.

The oxidative burst in health and diseases

The oxidative burst is essential for killing a number of microorganisms, as shown by the susceptibility to infections of individuals with CGD [12]. Patients with CGD experience, usually from early childhood, recurrent and often life-threatening bacterial and fungal infections, as well as a granulomatous response in affected tissues. The biochemical basis for this severe clinical phenotype is an absence of the respiratory burst in neutrophils and other phagocytic cells. Thus, invading microorganisms are ingested normally but remain viable within phagocytic vacuoles, since no ROS are generated. One exception to this rule is the normal killing of microorganisms that produce significant quantities of H_2O_2 (e.g., pneumococci), thereby supplying a missing ingredient that the CGD neutrophil can use to reconstitute the activity of the MPO–H_2O_2–halide antimicrobial system and avoid infections caused by these catalase-negative bacteria. However, the actual role of ROS in the bactericidal defect of PMN from patients with CGD, has been questioned recently following various observations: first, MPO deficiency is common, but seldom leads to microbicidal defects; also, early variation in the pH of the phagosome (rapid increase followed by a progressive acidification) is not observed in PMN from CGD patients (drop below pH 7 after a delay of 30 min); lastly, double knock-out mice for elastase and cathepsin G have a defective bactericidal defect similar to that of mice with NADPH oxidase defect. NADPH oxidase not only transfers electrons, but also protons, from the cytosol to the phagosome for compensation of the charge. However, some studies have shown that part of charge compensation is due to the transfer of K^+ instead of H^+ (which explains the early alkalinization of the medium), as this ion is necessary for the liberation of proteases such elastase and cathepsin G from the acid proteoglycan matrix. The debate now centers on the question whether ROS do have a role by themselves or not [13]. In normal PMN, the synergistic interaction of oxygen-dependent (ROS) and independent (granular bactericidal peptides and proteins) microbicidal systems results in pathogen killing. There is limited information on the resistance of pathogens, such as *Anaplasma phagocytophilum* to PMN functions [14]. By contrast, bacterial subversion of macrophage metabolism is widely acknowledged, and owing to a less potent oxidative burst and the absence of MPO, these phagocytes represent safe harbours for many intracellular pathogens. This defective bactericidal function can be boosted by cytokine stimulation. In particular, proinflammatory cytokines, interferon, bacteria and their products synergistically induce NO synthase, which may be the major pathway of macrophage bactericidal activity.

The products of the MPO–H_2O_2–chloride system are powerful oxidants [10] and, when released in the extracellular medium, a reaction with chloride can induce damage to adjacent tissue and apoptosis in other immune reactive cells. As a protection against excessive oxidation, there exists a complex set of interactive antioxidant systems. Over-activity of phagocytic NADPH oxidase can provoke functional impairment of T lymphocytes, cytotoxicity against endothelial cells, direct DNA damage in bystander cells, and metabolism of drugs to cytotoxic, genotoxic and immunogenic metabolites. Neutrophil priming by agents such as tumour necrosis factor-alpha (TNF-α), granulocyte/macrophage colony-stimulating factor (GM-CSF) and lipopolysaccharide causes a dramatic increase in the response of these cells to an activating agent; this process has been shown to be critical for neutrophil-mediated tissue injury both *in vitro* and *in vivo*. Increase in ROS production has been associated with, and may be causally related to, a variety of acute and chronic inflammatory states, e.g., bacterial sepsis, adult respiratory distress syndrome, inflammatory bowel disease, rheumatic diseases, vasculitis as well as cancer, auto-immunity and aging. It has been suggested that pulmonary injury, renal glomerular damage, and the initiation of atherosclerotic lesions may be caused by the MPO system. The involvement of ROS in the bactericidal function of macrophages has been assumed for many years, but it is now clear that the H_2O_2 produced by the respiratory burst, functions as a second messenger and activates major signaling pathways in these cells [15]. Both the nuclear factor-κB and activator protein-1 transcription factors are activated by H_2O_2 produced by the respiratory burst, and control the inducible expression of genes whose products are part of the inflammatory response; this may be a critical link between the respiratory burst and other inflammatory responses.

Methods of analysis

The development of methods to measure the generation/release of phagocyte respiratory burst products is of great importance for the clinical diagnosis and prognosis of various diseases, and a number of different techniques are currently in use for this purpose [16]. These techniques are valuable tools in basic as well as more clinically oriented research dealing with phagocyte function. Diseases within virtually every subspecialty of medicine have been studied in this respect, but most investigations have focused on infectious and autoimmune conditions. The analysis of drug-mediated modulation of the oxidative burst involves either global assays such as oxygen consumption, iodination and luminol-amplified chemiluminescence, or measurement of specific oxygen species, mainly superoxide anion (cytochrome C reduction and lucigenin- amplified chemiluminescence). Stimulation of phagocyte functions is obtained with agents that mimic bacterial chemotaxins (formylated peptides such as N-formyl methionyl-leucylphenylalanine [fMLP]), or directly activate protein kinase C (phorbol esters such as PMA), or increase Ca^{2+} flux (calcium ionophores). Phagocyte activity can be boosted by priming agents before stimulation. Fluorescence-activated cell sorter (FACS) analysis gives information on many phagocyte functions and membrane antigens, and permits rapid evaluation of individual phagocyte responses. The main problems encountered *in vitro* are due to non-standardization of techniques in different laboratories and artifacts introduced by the technique itself. Various tools used in the research setting are available for in-depth analysis of all transductional pathways underlying the drug-induced immunomodulatory effects. Phagocytic cell lines derived from human or animal cells (HL-60, PLB-985) are commonly used *in vitro* after differentiation into more mature forms with functional oxidase activity. These standardized cell lines theoretically avoid the problem of intra/interspecies variability and heterogeneity. Cell-free systems (xanthine–xanthine oxidase reaction, orthodianizine reduction, etc.) are also used to explore possible oxidant scavenging or oxidative properties of some drugs. *Ex vivo* assays explore the functional capabilities of phagocytes after drug administration to animals, healthy volunteers or patients. Isolation and separation of phagocytes from their context, along with intra/interspecies differences and chronobiology, often generate unsubstantiated extrapolation of results.

Antibacterial agents and the oxidative burst

The interference of antibacterial agents with the oxidative burst can occur in different ways (Tab. 1). Indirect effects concern all the events related either to phagocyte differentiation (myelopoiesis) (1), or receptor functionality (2), or modulation of host cell responses with release of priming or inhibitory factors (e.g., cytokines) (3), or alteration of bacterial structure and metabolism (4) leading to release of by-products (e.g., endotoxin), increased stimulating effects or increased susceptibility to ROS. These indirect aspects will not be explored here. This Chapter will deal with the direct effects that concern the interference with the signaling pathways (5), or the modulation of enzyme (oxidase, MPO) activities (6), or the scavenging of ROS (7). Conversely, alteration of antibacterial agents by ROS themselves (8) can lead to changes in antibacterial activity or increased toxicity to host cells.

Modification of antibacterial agents by ROS

The possibility that antibiotics are inactivated by phagocytes, their products, or the intracellular medium, has rarely been investigated. There are no data clearly demonstrating a loss of activity due to intraphagosomal or extracellular oxidization. Conversely, the modification of antibacterial agents leading to increased activity inside the phagosome or in the vicinity of phagocytes has not been investigated either, although there are reports that various quinolones (ciprofloxacin, fleroxacin, pefloxacin) as well as macrolides (erythromycin A, roxithromycin), clindamycin and amoxycillin, display optimal intracellular efficacy when PMN have an intact oxidant-generating system [3]. Antibacterial synergy between josamycin and acellular oxidant-producing systems has also been observed [3]. However, the possibility that oxidant-induced alteration of bacteria was responsible for the observed synergy

Table 1- Interactions of antibacterial agents (ABA) with the phagocyte oxidative burst

Target	**Antibacterial agents**	**Effects** **1-Indirect**	**Consequences**
(1) Progenitors	β-lactams, chloramphenicol, dapsone, etc.	e.g., toxicity	↓ number of PMN
(2) Phagocytes	?	receptor functionality	↓/↑ oxidative response
(3) Bacteria	all ABA (sub/supra MIC)	alteration structure metabolism	release pro-oxidative products, ↑ opsonization, stimulation, susceptibility
(4) Host cells	Macrolides, quinolones fosfomycin	production pro-/ anti-inflammatory CK	↓/↑ oxidative burst
Target	**Antibacterial agents**	**Effects** **2-Direct**	**Consequences**
(5) Phagocytes	(see text)	alteration of signaling pathways	↓/↑ oxidative burst
(6) NADPH oxidase	?	alteration of activity	↓/↑ oxidative burst?
MPO	cefdinir, dapsone INH	↓ activity	↓ ROS
(7) ROS	cyclines, rifampicin clofazimine, penicillin, some cephalosporins	scavenging	↓ ROS
(8) ABA	INH, quinolones (?) macrolides (?)	alteration by ROS	↑ activity (?)
	quinolones, dapsone INH, chloramphenicol		↑ toxicity

could not be excluded. Recent studies have shown that superoxide anion increases the antimicrobial effect of isoniazid (INH), in fact a prodrug requiring oxidative activation by the catalase-peroxidase hemoprotein of *Mycobacterium tuberculosis* [17]. Neutrophils and monocytes can metabolize drugs to reactive metabolites, especially those drugs that have nitrogen or sulfur in a low oxidation state [3, 18]. The major system involved in this oxidation is the combination of NADPH oxidase and myeloperoxidase, which generates HOCl. Reactive metabolites, by their very nature, have short half-lives, and most of their effects will be exerted on the cells that formed them. Therefore, they are likely to be important for adverse reactions that involve leukocytes, such as agranulocytosis and immune-mediated reactions. However, the mechanism of these reactions is unknown and evidence for the association of leukocyte-derived reactive metabolites with such reactions is circumstantial at present. The oxidation of drugs by leukocytes requires activation of the cells; therefore, infection or other inflammatory conditions that activate leukocytes may represent one of the risk factors for idiosyncratic drug reactions. Increased toxicity by antibiotics oxidized by PMN oxidants has been observed, for instance with INH. The hematologic toxicity of dapsone and chloramphenicol could rely on their oxidative metabolism. The cellular phototoxicity of various quinolones (e.g., lomefloxacin, ciprofloxacin, norfloxacin) might also be linked to an oxygen-dependent mechanism [19].

Artefactual effects and inhibition of enzyme activity

Scavenging of oxidant species or interference with the detection methods may lead to false appreciations of the actual effect of an antibiotic on phagocyte activity, and appropriate controls using cell-free oxidant-generating systems are required to validate the results. Various antibiotics (danofloxacin, ceftiofur, oleandomycin, oxytetracycline, doxycycline, lincomycin, etc.) decrease the chemiluminescence of bovine PMN by interacting with the detection system (absorption of the blue light emitted by luminol) or scavenging of oxidative species [20, 21]. Rifampicin also has light-absorbing property and quenches superoxide anion; cyclines scavenge hypochlorous acid as do clofazimine, sulfapyridine and various aminothiazolyl cephalosporins. Penicillin G and ampicillin inhibit the chemiluminescence of PMN and cell-free systems by scavenging HOCl and hydrogen peroxide, whereas chloramphenicol increases it. Dapsone and INH directly inhibit MPO activity and impair the production of HOCl by the MPO–H_2O_2–halide system. Dapsone irreversibly inhibits MPO, by converting the enzyme into its inactive (ferryl) form. Cefdinir, a hydroxyimino-aminothiazolyl cephalosporin, impairs MPO activity in the external medium but not in the phagolysosome, likely because it does not enter neutrophils. Direct inhibition of oxidase activity has not yet been reported. The major question arising from these results is the impact on bactericidal function and the tissue-destructive potential of neutrophils. Various *in vitro* experiments suggest, for instance, that some β-lactams may have a cytoprotective role or prevent antiprotease inactivation by activated neutrophils. The anti-inflammatory potential of dapsone, INH, clofazimine and cyclines may be due to their impact on HOCl generation either directly or indirectly (since this oxidant is a potent activator of latent collagenase activity). However, superoxide limits the potency of the drugs that inhibit MPO reversibly, by trapping it as its inactive redox intermediate and reducing it to the active enzyme. Furthermore, under conditions where the activity of MPO exceeds that of the hydrogen peroxide-generating system, which is most likely to occur in phagosomes, partial inhibition of MPO does not affect hypochlorous acid production [22].

Direct modulation of the oxidative burst *in vitro*

Drug-induced modulation of the oxidative burst is generally accompanied by modifications of other phagocyte functions, suggesting either an effect on a transductional target (or a second messenger) involved in several activation pathways. A simplified summary of the main effects of antibacterial agents on the oxidative burst is given in Table 2. The inhibitory effect of aminoglycosides (at therapeutic concentrations) on PMN oxidative metabolism is conflicting. Interestingly, amikacin enhances the PMN oxidative burst at low concentrations *in vitro*, whereas concentrations higher than 1 g/l inhibit this phenomenon, likely as a result of oxidant scavenging [23]. Gentamicin, netilmicin, and tobramycin are ineffective in a wide range of concentrations. Ansamycins impair various PMN functions, including the oxidative burst (although artefactual effects have been noted [see above]). The most active new compounds are derivatives carrying an acidic substituent at C3 [24, 25]. PMN from patients with rheumatoid arthritis (RA) are more susceptible to the depressive effect of rifamycin SV than are cells from healthy subjects. β-lactams have been largely investigated in this context, but no class or subgroup effect has been demonstrated [26]. Particular behaviors have been linked to certain chemical features. Cefotaxime enhances the oxidative burst of PMN by priming the cell response to a second stimulation with complement-opsonized particles; the presence of an acetoxy at position 3 of the cephem ring is crucial for this effect. High concentrations of meropenem decrease superoxide anion production by PMN whereas faropenem enhances it, probably because of its interference at a site where Ca^{2+} regulates NADPH oxidase activation. Three chemically unrelated β-lactams (cefmetazole, imipenem and cefoxitin) have similar stimulatory effects on various PMN functions including the oxidative burst. These antibiotics also significantly stimulate protein carboxy methylation, increase intracellular cyclic GMP levels, and decrease ascorbate content. Clofazimine increases superoxide anion production by stimulated neutrophils, and TNF-α potentiates this enhancement. The pro-oxidative effect of clofazimine analogs is largely dependent on the nature of the alkylimino group at position 2 of the phenazine nucleus and, to a lesser extent, on halogenation. The mechanism underlying this pro-oxidative effect seems to involve stimulation of phospholipase A_2 (PLA_2) activity, with subsequent accumulation of arachidonic acid and lysophospholipids, which act as second messengers to activate the oxidase. At therapeutic concentrations, the gyrase B inhibitor coumermycin impairs chemotaxis, superoxide anion production and intracellular killing by PMN. Cyclines have been widely studied in this context, most reports confirming an inhibitory action on various phagocyte functions (including the oxidative burst) at therapeutic concentrations. Outside their scaveng-

Table 2 - In vitro effects of antibacterial agents on the oxidative burst

Antibacterial agents	Effects	Mechanisms
Aminoglycosides	e.g., No effect. Amikacin increases (low concentrations)	
Ansamycins	e.g., Decrease	Scavenging + ?
β-Lactams	Variable effects (see text)	
Clindamycin	Increase or no effect[1]	
Clofazimine	Increase	stimulation of PLA_2
Coumermycin	Decrease	
Cyclines	Decrease	Ca^{2+} chelation + scavenging
Dapsone	Decrease	
Fosfomycin	Decrease	PKC?
Fusidic acid	No effect[2] . Decrease[3]	
Macrolides	Josamycin (increase);	
	Erythromycin A derivatives (decrease)	PLD-PPH, PKA?
Quinolones	Variable effects (see text)	
Sulfonamides	Decrease	Ca^{2+}?
TMP	Decrease	PPH

[1]depends on concentrations; [2]in monocytes; [3]in PMN

ing effect, various mechanisms have been forwarded to explain the inhibitory action of cyclines (Ca^{2+} chelation, binding of intracellular Mg^{2+} or cellular toxicity). Structure-activity studies indicate a parallel increase in lipid solubility (and possibly cellular accumulation) and inhibitory properties, although others stress the different chemical reactivity of the various molecules to UV light. Dapsone inhibits neutrophil functions such as chemotaxis and oxidant production. Twelve analogues of dapsone showed comparable or greater effects on zymosan mediated oxidative burst [27]. The most effective compounds were the 2-nitro-4-amino, 2-hydroxy-4 aminopropyl, and 2-methoxy-4-aminoethyl derivatives. In general, potency was inversely associated with lipophilicity. *In vitro*, fosfomycin increases basal extracellular oxidant production by PMN (the effect was non-significant for PMA-stimulated cells) and intracellular Ca^{2+} concentrations [28]. By contrast, other authors have noted an inhibitory effect of fosfomycin on PMA-stimulated oxidant production by PMN, suggesting an effect on PKC-dependent activation pathways [29].

Fusidic acid decreases PMN but not monocyte oxidative burst. Macrolides are the subject of worldwide studies for their immunomodulatory potential [30–33]. Whereas josamycin increases oxidant production by PMN and monocytes, erythromycin A derivatives time- and concentration-dependently impair the oxidative burst of PMN. Similar results are obtained with the phagocytic cell line PLB-985, after differentiation into PMN [34], eosinophils [35] and the monocytic cell line THP-1 [36]. In this latter study, it is interesting to note that short-term treatment with clarithromycin potentiates the release of NO, H_2O_2, IL-1 and TNF, whereas longer treatment (2–4 h) reverses the process and decreases both the release of mediators and the activity of hydrolytic enzymes. Structure-activity studies have shown that only erythromycin A derivatives, including the azalide azithromycin, impair the phagocyte oxidative burst [37, 38]. The chemical entity responsible for these effects is the L cladinose at position 3 of the lactone ring, but other structures may also interfere with phagocytic transductional targets [39]: various ketolides (RU 64004 [HMR 3004], HMR 3647 [telithromycin], and ABT-773 [cethromycin]), which are deprived of cladinose, also impair oxidant production by neutrophils [40–42]. The structure involved in the inhibitory effect of HMR 3004 and ABT-773 seems to be the quinoline linked by a butyl chain to the C11–C12 carbazate. The inhibitory structure in HMR 3647 has not yet been identified. The transductional pathway by which erythromycin A derivatives interfere with neutrophils seems to be the phospholipase D–phosphatidate phosphohydrolase (PLD–PPH) pathway [37]. In resting PMN these drugs directly stimulate PLD activity, which results in the accumulation of phosphatidic acid (PA), a messenger important for triggering exocytosis, while in stimulated neutrophils, these drugs impair PPH activity, resulting in a decrease in diradylglycerol (the natural PKC activator) production. The cellular target of macrolides is unknown. Preliminary results from our group have shown that roxithromycin and HMR 3004 impair the activity of PKA (a protein kinase shown to downregulate PLD activity) which could be a possible target for these two erythromycin A derivatives. It must be noted that PKA inhibitors decrease the inhibitory effect of macrolides but they also impair macrolide uptake and *in vitro* conditions which modify cellular uptake can interfere with the inhibitory effect of these drugs. For instance, pentoxifylline and its derivatives increase the uptake of roxithromycin and dirithromycin and the combination inhibits oxidant generation to a larger extent than either drug alone [43]. TNF-α and GM-CSF reduce the inhibitory effect of HMR 3647 on oxidant production by neutrophils, while these cytokines do not modify (or increase) the effect of roxithromycin and HMR 3004 [41]. Similarly, Kadota et al. [44] have observed marked suppression of superoxide anion generation by G-CSF-primed neutrophils exposed to therapeutic concentrations of erythromycin A, but the uptake of this drug was not investigated. Recently, Abeyama et al. [45] have provided a link between macrolide-induced impairment of the oxidative response and modulation of cytokine production: they observed that roxithromycin, erythromycin A and clarithromycin, by decreasing oxidants produced by LPS-stimulated leukocytes and THP-1 monocytes, preferentially inhibited ROS-mediated "proinflammatory events". NF-κB activation mediated by ROS, and subsequent proinflammatory cytokine production were decreased, whereas macrolides could rapidly increase intracellular cAMP levels and CREB (cAMP-responsive element-binding protein) activities by LPS-primed leukocytes. Although an understanding of the mechanism of redox signaling is in its infancy, it is becoming clear that the ROS produced by the respiratory burst have a profound effect on intracellular signaling pathways and ultimately in modulating gene expression. Quinolones have also been widely studied for their immunomodulatory properties [46, 47]. At therapeutic concentrations, quinolones differently affect phagocytosis, adhesion, and oxidant production by rat peritoneal macrophages and human PMN; their effect (increase, decrease, no effect) on oxidant production appears to depend on the animal species and the quinolone structure. The ofloxacin-induced increase in the PMN oxidative response is due to the enhancement of PKC activity, whereas norfloxacin increases oxidant production by mouse macrophages through enhanced mobilization of NADPH oxidase subunits [48]. A similar transient potentiation of the oxidative burst of rat macrophages has been reported with ofloxacin, fleroxacin, sparfloxacin and levofloxacin. Lower concentrations were more effective than higher concentrations. Moxifloxacin [49] and gatifloxacin [50] do not alter oxidant production by phagocytes. Interestingly, grepafloxacin [51] exerts a priming effect on the PMN oxidative burst triggered by fMLP, LTB4 (not PMA), through translocation of $p47^{phox}$ and $p67^{phox}$; GTP-binding protein, tyrosine phosphorylation and PKC activity were not involved in the priming effect. A non-antibiotic quinolone derivative, 2-phenyl-4-quinolone (YT-1), which possesses cytotoxicity against several human cancer cell lines inhibits the respiratory burst of rat neutrophils in response to fMLP but not to PMA; the inhibition of phosphodiesterase (PDE), probably PDE4 isoenzyme, rather than the activation of adenylate cyclase by YT-1 contributes to an increase in the cellular cyclic AMP level, which, in turn, activates PKA and inhibits the respiratory burst in fMLP-activated neutrophils [52]. In most studies, trimethoprim (TMP), alone or in combination, has an inhibitory effect on PMN functions. Interestingly, TMP-SMX (sulfamethoxazole) increases nitric oxide (NO) production by PMN from patients with chronic granulomatous disease [53]. Brodimoprim, in which the methoxy group in position 4 of the TMP benzyl ring is replaced by a bromine atom, displays greater lipophilicity and cellular uptake than TMP, and no inhibitory effects on PMN functions. At high (therapeutically irrelevant) concentrations, TMP impairs PPH activity. In general, sulfonamides inhibit phagocyte functions, and many agents in this class have been switched from infections to anti-inflammatory indications. The mechanisms underlying these immunomodulatory effects are unclear. Inhibition of the elevation of intracellular Ca^{2+} after stimulation has been reported with sulfasalazine and sulfapyridine.

Consequences of modulation of the oxidative burst: *In vivo/ex vivo* effects

The therapeutic relevance of the immunomodulatory actions of antibacterial agents is controversial, and there is no general agreement on whether these effects must be taken into account when choosing an antibacterial treatment [1, 2, 54]. With regards to their interference with the phagocyte oxidative burst, and owing to the ambivalent role of ROS, two possible consequences must be envisaged: alteration of the bactericidal function and/or modulation of oxidant-triggered inflammatory reactions.

Consequences on the bactericidal activity of phagocytes

The possibility that antibiotic-induced stimulation of oxidant production by phagocytes results in increased bactericidal activity has rarely been investigated: for instance, the enhancement of bacterial killing by cefotaxime-treated PMN *in vitro* was linked to the pro-oxidative effect of this drug [3]. The cephalosporin cefodizime does not modify the phagocyte oxidative burst *in vitro*, but restores various phagocytic functions in immunocompromised patients *ex vivo*: in particular, in chronic hemodialysis patients, with a depressed oxidative response, as assessed by hexose-monophosphate shunt activity, cefodizime given for 10 days, significantly increased this phagocytic response compared to cotrimoxazole and placebo treatment [55]. Similarly, 30 patients with severe bacterial infections were treated with cefodizime or ceftriaxone and the effect of cefodizime and ceftriaxone on the phagocytic capacity and generation of reactive oxygen intermediates after phagocytosis by granulocytes was assessed prior to, during, and after therapy. Prior to therapy, patients in both groups exhibited a decreased capacity to phagocytize *Escherichia coli* and to generate reactive oxygen intermediates. Granulocyte function increased after the initiation of therapy and normalized within 7 days for the ceftriaxone-treated patients and within 3 days for the cefodizime group [56]. The only apparent clinical advantage was earlier defervescence in the cefodizime group. No reports are available on the consequences of prophylactic administration of cefodizime in patients at risk of infections. However, in the field of infectious diseases, the clinical benefit of the immunostimulating/restoring effects of antibacterial agents is considered minimal compared to their direct antibacterial activity. Conversely, antibacterial-induced inhibition of the oxidative burst does not seem to result in a decrease in bacterial killing, likely because of other bactericidal (oxygen-independent) mechanisms [57] or because this inhibition does not occur in the bacteria-containing phagolysosome [58].

Consequences on the inflammatory reaction

Owing to the detrimental role of ROS in various pathological settings, modulation of oxidant production by phagocytes remains a critical target. Some *ex vivo* studies have confirmed the results obtained *in vitro*. For instance, decreased oxidant production by PMN from patients with acute myocardial infarction and treated with doxycycline has been observed *ex vivo* [59]. The effect of rifampicin on oxidant production *in vivo* has not been assessed, but recently, in a rabbit model of *Streptococcus pneumoniae* meningitis, it was shown that leukocytes (neutrophils and monocytes) from rifampicin-treated rabbits produced smaller amounts of reactive oxygen species than leukocytes from ceftriaxone-treated animals [60]. *Ex vivo* modulation of the oxidative burst by macrolides has also been reported: in general early analysis after administration demonstrates an increased response, while delayed analysis or prolonged administration result in progressive inhibition of the oxidative potential [61, 62]. By contrast, *in vitro* enhancement of the oxidative burst by antibacterial agents does not seem to increase the inflammatory response. However, it has been hypothesized that quinolone-related arthropathy was linked to the stimulation of the respiratory burst of immature articular chondrocytes [63]. The beneficial consequences of antibacterial agent-induced decrease in oxidant production are difficult to ascertain in patients suffering from inflammatory diseases, since, in general, these drugs alter also various proinflammatory components such as cytokine production. The recent study of Abeyama et al. [45] may provide a unifying hypothesis for macrolide-induced decrease in oxidant generation and subsequent modulation of cytokine production. Various antibacterial agents that impair oxidant production are showing promise in inflammatory diseases [2]. Three classes have stimulated widespread interest in the context of inflammatory diseases, namely cyclines, ansamycins and macrolides. Tetracyclines, particularly minocycline, are used in rheumatoid arthritis and their potential benefit has been studied in ischemia-reperfusion injury [64–66]; ansamycins may be useful in Crohn's disease, and macrolides display the largest panorama of potential non-antibiotic use from chronic inflammatory sinopulmonary diseases, including cystic fibrosis, up to Crohn's disease, psoriasis, asthma, and coronary artery disease [67–71].

Conclusions

Attempts to modify inflammatory reactions by either enzymes that metabolize ROS (superoxide dismutase and catalase) or by scavengers (low-molecular weight reducing compounds like mannitol, N-acetyl cysteine) have proved effective *in vitro* and in various *in vivo* models, but failed as clinically useful drugs. Specific and potent inhibitors of NADPH oxidase are not yet available. The NADPH oxidase is not restricted to the myelomonocytic lineage and B lymphocytes of mammals. It is also found in fish, insects and appears to be a part of the host defence apparatus of plants. In addition, related oxidases are present in a variety of host cells (fibroblasts, endothelial cells, thyroid cells, vascular smooth cells, etc.) and can produce small amounts of superoxide anion that play a role in physiology and act as biological signals. Pathological activation of vascular NAD(P)H oxidases are common in cardiovascular diseases: ROS production following angiotensin II-mediated stimulation of these oxidases lead to events such as inflammation, hypertrophy, remodeling and angiogenesis and contribute to cardiovascular diseases including atherosclerosis and hypertension. No studies have yet been published on the effects of antibacterial agents on Nox/Duox enzymes, but there is no doubt that such investigations could bring further light on the therapeutic potential of some antibacterial agents (e.g., macrolides) in cardiovascular diseases.

References

1 Labro MT (2000) Interference of antibacterial agents with phagocyte functions: immunomodulation or "immuno-fairy" tales. *Clin Microbiol Rev* 13: 615–50
2 Labro MT (2002) Antibiotics as anti-inflammatory drugs. *Curr Op Investig Drugs* 3: 61–8
3 Labro MT, El Benna J (1993) Interaction of antibiotics with the phagocyte oxidative burst. In: Faist E, Meakins JL, Schildberg FW (eds): *Host defense dysfunction in trauma, shock and sepsis*. Springer-Verlag, Berlin, Heidelberg, 953–64
4 Clark RA (1999) Activation of the neutrophil respiratory burst oxidase. *J Infect Dis* 179 (Suppl 2): S309–317
5 Babior BMC (1999) NADPH oxidase: an update. *Blood* 93: 1464–76
6 Vignais PV (2002) The superoxide-generating NADPH oxidase: structural aspects and activation mechanism. *Cell Mol Life Sci* 59: 1428–59
7 Bokoch GM, Diebold BA (2002) Current molecular models for NADPH oxidase regulation by Rac GTPase. *Blood* 100: 2692–6
8 Heyworth PG, Cross AR, Curnutte JT (2003) Chronic granulomatous disease. *Curr Opin Immunol* 15: 578–84
9 Hampton MB, Kettle AJ, Winterbourn CC (1998) Inside the neutrophil phagosome: oxidants, myeloperoxidase, and bacterial killing. *Blood* 92: 3007–17
10 Klebanoff SJ (1999) Myeloperoxidase. *Proc Assoc Am Physicians* 111 (5): 383–9
11 Karlsson A, Dahlgren C (2002) Assembly and activation of the neutrophil NADPH oxidase in granule membranes. *Antioxid Redox Signal* 9: 49–60
12 Roos D, van Bruggen R, Meischl C (2003) Oxidative killing of microbes by neutrophils. *Microbes Infect* 5: 1307–15
13 Geiszt M, Kapus A, Ligeti E (2001) Chronic granulomatous disease: more than the lack of superoxide? *J Leukoc Biol* 69: 191–6
14 Rosen H (2004) Bacterial responses to neutrophil phagocytosis. *Curr Opin Hematol* 11: 1–6
15 Iles KE, Forman HJ (2002) Macrophage signaling and respiratory burst. *Immunol Res* 26: 95–105
16 Dahlgren C, Karlsson A (1999) Respiratory burst in human neutrophils. *J Immunol Methods* 232: 3–14
17 Bulatovic VM, Wengenack NL, Uhl JR, Hall L, Roberts GD, Cockerill III FR, Rusnak

F (2002) Oxidative stress increases susceptibility of Mycobacterium tuberculosis to isoniazid. *Antimicrob Agents Chemother* 46: 2765–71

18 Uetrecht JP (1994) Metabolism of drugs by leukocytes. *Drug Metabol Drug Interact* 11: 259–82

19 Saniabadi AR, Wada K, Umemura K, Nakashima M (1996) Impairment of phagocytic cell respiratory burst by UVA in the presence of fluoroquinolones: an oxygen-dependent phototoxic damage to cell surface microvilli. *Photochem Photobiol* 33: 137–42

20 Hoeben D, Dosogne H, Heyneman R, Burvenich C (1997) Effect of antibiotics on the phagocytotic and respiratory burst activity of bovine granulocytes. *Eur J Pharmacol* 332: 289–97

21 Hoeben D, Burvenich C, Heyneman R (1998) Antibiotic commonly used to treat mastitis and respiratory burst of bovine polymorphonuclear leukocytes. *J Dairy Sci* 81: 403–10

22 Kettle AJ, Gedye CA, Winterbourn CC (1993) Superoxide is an antagonist of antiinflammatory drugs that inhibit hypochlorous acid production by myeloperoxidase. *Biochem Pharmacol* 45: 2003–10

23 Gressier B, Brunet C, Dine T, Luycks M, Ballester L, Cazin M, Cazin JC (1998) *In vitro* activity of aminoglycosides on the respiratory burst response in human polymorphonuclear neutrophils. *Methods Find Exp Clin Pharmacol* 20: 819–24

24 Spisani S, Traniello S, Martuccio C, Rizzuti O, Cellai L (1997). Rifamycins inhibit human neutrophil functions: new derivatives with potential antiinflammatory activity. *Inflammation* 21: 391–400

25 Spisani S, Traniello S, Onori AM, Rizzuti O, Martuccio C, Cellai L (1998) 3-(Carboxyalkylthio) rifamycin S and SV derivatives inhibit human neutrophil functions. *Inflammation* 22: 459–69

26 Labro MT (1995) Resistance to and immunomodulation effect of cephalosporin antibiotics. *Clin Drug Invest* 9 (Suppl 3): 31–44

27 Coleman MD, Smith JK, Perris AD, Buck NS, Seydi JK (1997) Studies on the inhibitory effects of analogues of dapsone on neutrophil functions *in vitro*. *J Pharm Pharmacol* 49: 53–7

28 Krause R, Patruta S, Daxböck F, Fladerer P, Wenisch C (2001) The effect of fosfomycin on neutrophil function. *J Antimicrob Chemother* 47: 141–6

29 Hamada M, Honda J, Yoshimoto T, Fumimori T, Okamoto M, Aizawa H (2002) Fosfomycin inhibits neutrophil function *via* a protein kinase C-dependent signaling pathway. *Int Immunopharmacol* 2: 511–18

30 Labro MT (1997) Effects of macrolides on leukocytes and inflammation. In: Zinner SH, Young LS, Acar JF, Neu HC (eds): *Expanding indications for the new macrolides, azalides and streptogramins*. Marcel Dekker, New York, 101–16

31 Labro MT (1998) Antiinflammatory activity of macrolides: a new therapeutic potential? *J Antimicrob Chemother* 41 (Suppl. B): 37–46

32 Labro MT (1998) Immunological effects of macrolides. Curr Op Infect Dis 11: 681–8

33 Culic O, Erakovic V, Parnham MJ (2001) Anti-inflammatory effects of macrolide antibiotics. *Eur J Pharmacol* 429: 209–29
34 Abdelghaffar H, Soukri A, Babin-Chevaye C, Labro MT (2003) Interactions of macrolides and ketolides with the phagocytic cell line PLB-985. *J Chemother* 28: 350–6
35 Cui CH, Honda K, Saito N, Yamada Y, Sannobe S, Ueki S, Hamada H, Yamaguchi K, Kobayashi Y, Adachi T, Kayaba H, Chihara J (2001) Effect of roxithromycin on eotaxin-primed reactive oxygen species from eosinophils. *Int Arch Allergy Immunol* 125 (Suppl 1): 38–41
36 Ives TJ, Schwab UE, Ward ES, Butts JD, Hall IH (2001) Disposition and functions of clarithromycin in human THP-1 monocytes during stimulated and unstimulated conditions. *Res Commun Mol Pathol Pharmacol* 110: 183–208
37 Abdelghaffar H, Vazifeh D, Labro MT (1997) Erythromycin A-derived macrolides modify the functional activities of human neutrophils by altering the phospholipase D-phosphatidate phosphohydrolase transduction pathway. *J Immunol* 159: 3995–4005
38 Theron AJ, Feldman C, Anderson R (2000) Investigation of the antiinflammatory and membrane-stabilizing potential of spiramycin *in vitro*. *J Antimicrob Chemother* 46: 269–71
39 Abdelghaffar H, Kirst H, Soukri A, Babin-Chevaye C, Labro MT (2002) Structure-activity relationships among 9-N-alkyl derivatives of erythromycylamine and their effect on the oxidative burst of human neutrophils *in vitro*. *J Chemother* 14: 132–9
40 Vazifeh D, Preira A, Bryskier A, Labro MT (1998) Interactions between HMR 3647, a new ketolide, and human polymorphonuclear neutrophils. *Antimicrob Agents Chemother* 42: 1944–51
41 Vazifeh D, Bryskier A, Labro MT (2000) Effect of proinflammatory cytokines on the interplay between roxithromycin, HMR 3647, or HMR 3004 and human polymorphonuclear neutrophils. *Antimicrob Agents Chemother* 44: 511–21
42 Abdelghaffar H, Babin-Chevaye C, Labro MT (2004) Interaction between the new ketolide, ABT-773 (cethromycin) and human polymorphonuclear neutrophils and the phagocytic cell line PLB-985 *in vitro*. *Antimicrob Agents Chemother* 48: 1096–1104
43 Hand WL, Hand DL (1995) Influence of pentoxifylline and its derivatives on antibiotic uptake and superoxide generation by human phagocytic cells. *Antimicrob Agents Chemother* 39: 1574–9
44 Kadota JI, Iwashita T, Matsubara Y, Ishimatsu Y, Yoshinaga M, Abe K, Kohno S (1998) Inhibitory effect of erythromycin on superoxide anion production by human neutrophils primed with granulocyte-colony stimulating factor. *Antimicrob Agents Chemother* 42: 1866–7
45 Abeyama K, Kawahara K-I, Iino S, Hamada T, Arimura S-I, Matsushita T, Nakajima T, Maruyama I (2003) Antibiotic cyclic AMP signaling by "primed" leukocytes confers anti-inflammatory cytoprotection. *J Leukoc Biol* 74: 908–15
46 Riesbeck K (2002) Immunomodulating activity of quinolones: Review. *J Chemother* 14: 3–12

47 Dalhoff A, Shalit I (2003) Immunomodulatory effects of quinolones. *Lancet Infect Dis* 3: 359–71
48 El Bekay R, Alvarez M, Carballo M, Martin-Nieto J, Monteseirin J, Pintado E, Bedoya FJ, Sobrino F (2002) Activation of phagocytic cell NADPH oxidase by norfloxacin: a potential mechanism to explain its bactericidal action. *J Leukoc Biol* 71: 255–61
49 Fischer S, Adam D (2001) Effects of moxifloxacin on neutrophil phagocytosis, burst production, and killing as determined by a whole-blood cytofluorometric method. *Antimicrob Agents Chemother* 45: 2668–9
50 Braga PC, Dal Sasso M, Bovio C, Zavaroni E, Fonti E (2002) Effects of gatifloxacin on phagocytosis, intracellular killing and oxidant production by human polymorphonuclear neutrophils. *Int J Antimicrob Agents* I19: 183–7
51 Niwa M, Kanamori Y, Hotta K, Matsuno H, Kozawa O, Fujimoto S, Uematsu T (2002) Priming by grepafloxacin on respiratory burst of human neutrophils: its possible mechanism. *J Antimicrob Chemother* 50: 469–78
52 Wang JP, Raung SL, Huang LJ, Kuo SC (1998) Involvement of cyclic AMP generation in the inhibition of respiratory burst by 2-phenyl-4-quinolone (YT-1) in rat neutrophils. *Biochem Pharmacol* 56: 505–14
53 Tsuji S, Taniuchi S, Hasui M, Yamamoto A, Kobayashi Y (2002). Increased nitric oxide production by neutrophils from patients with chronic granulomatous disease on trimethoprim-sulfamethoxazole. *Nitric Oxide* 7: 283–8
54 Labro MT (1993) Immunomodulation by antibacterial agents. Is it clinically relevant? *Drugs* 45: 319–28
55 Vanholder R, Dagrosa EE, Van Landschoot N, Waterloos MA, Ringoir SM (1993) Antibiotics and energy delivery to the phagocytosis-associated respiratory burst in chronic hemodialysis patients: a comparison of cefodizime and cotrimoxazole. *Nephron* 63: 65–72
56 Wenisch C, Parshalk B, Hasenhundl M, Wiesinger E, Graninger W (1995) Effects of cefodizime and ceftriaxone on phagocytic functions in patients with severe infections. *Antimicrob Agents Chemother* 39: 672–6
57 Abdelghaffar H, Vazifeh D, Labro MT (2002) Effect of telithromycin (HMR 3647) on polymorphonuclear neutrophil killing of Staphylococcus aureus in comparison with roxithromycin. *Antimicrob Agents Chemother* 46: 1364–74
58 Labro MT, El Benna J, Charlier N, Abdelghaffar H, Hakim J (1994) Cefdinir (CI-983), a new oral amino-2-thiazolyl cephalosporin, inhibits human neutrophil myeloperoxidase in the extracellular medium but not the phagolysosome. *J Immunol* 152: 2447–2455
59 Takeshita S, Ono Y, Kozuma K, Suzuki M, Kawamura Y, Yokoyama N, Furukawa S, Isshiki T (2002) Modulation of oxidative burst of neutrophils by doxycycline in patients with acute myocardial infarction. *J Antimicrob Chemother* 49: 411–13
60 Bottcher T, Gerber J, Wellmer A, Smirnov AV, Fakhrjanali F, Mix E, Pilz J, Zettl UK, Nau R (2000) Rifampin reduces production of reactive oxygen species of cerebrospinal

fluid phagocytes and hippocampal neuronal apoptosis in experimental *Streptococcus pneumoniae* meningitis. J Infect Dis 181: 2095–8

61 Culic O, Erakovic V, Cepelak I, Barisic K, Brajsa K, Ferencic Z, Galovic R, Glojnaric I, Manojlovic Z, Munic V et al (2002) Azithromycin modulates neutrophil function and circulating inflammatory mediators in healthy human subjects. *Eur J Pharmacol* 450: 277–89

62 Labro MT, Bryskier A, Babin-Chevaye C, Hakim J (1988) Interaction de la roxithromycine avec les polynucléaires neutrophiles humains *in vitro* et *ex vivo*. *Pathol Biol* 36: 711–14

63 Hayem G, Petit PX, Levacher M, Gaudin C, Kahn MF, Pocidalo JJ (1994) Cytofluorometric analysis of chondrotoxicity of fluoroquinolone antimicrobial agents. *Antimicrob Agents Chemother* 38: 243–7

64 Alarcon GS (2000) Tetracyclines for the treatment of rheumatoid arthritis. *Expert Opin Investig Drugs* 9: 1491–8

65 Reasoner DK, Hindman BJ, Dexter F, Subieta A, Cutkomp J, Smith T (1997) Doxycycline reduces early neurologic impairment after cerebral arterial air embolism in the rabbit. *Anesthesiol* 87: 569–76

66 Smith JR, Gabler WL (1995) Protective effects of doxycycline in mesenteric ischemia and reperfusion. *Res Commun Mol Pathol Pharmacol* 88: 303–15

67 Jaffé A, Bush A (2001) Anti-inflammatory effects of macrolides in lung disease. *Pediatr Pulmonol* 31: 464–73

68 Gaylor AS, Reilly JC (2002) Therapy with macrolides in patients with cystic fibrosis. *Pharmacother* 22: 327–35

69 Carey KW, Alwami A, Danziger LH, Rubinstein I (2003) Tissue reparative effects of macrolide antibiotics in chronic inflammatory sinopulmonary diseases. *Chest* 123: 261–5

70 Cazzola M, Salzillo A, Diamant F (2000) Potential role of macrolides in the treatment of asthma. *Monaldi Arch Chest Dis* 55: 231–6

71 Leiper K, Morris AI, Rhodes JM (2000) Open label trial of oral clarithromycin in active Crohn's disease. *Aliment Pharmacol Ther* 14: 801–6

Immune system

Jun-ichi Kadota

Division of Pathogenesis and Disease Control, Department of Infectious Diseases, Oita University Faculty of Medicine, 1-1 Hasama, Oita 879-5593, Japan

Introduction

Lung lymphocytes are important for pulmonary immunoregulation, and their elimination is critical for terminating the immune response of the murine lung to intratracheal particulate antigen exposure [1]. The first subset of T helper lymphocytes (Th1 cells) produces various cytokines, including interferon (IFN)-γ, interleukin (IL)-2, tumor necrosis factor (TNF)-α and IL-1β, while Th2 cells produce IL-4 and IL-5. The Th1 cytokines play a prominent role in the enhancement of cell-mediated immunity [2] and are central in host defense mechanisms against various pathogenic microorganisms [3]. IL-12, produced by macrophages, plays a central role in the development of Th1 cells from naive T cells [4]. In contrast, the Th2 cytokines inhibit the production and biological activities of Th1 cytokines, thus attenuating host defense mechanisms against pathogenic organisms. The Th2 immune response induces a chronic, fatal disease such as chronic graft *versus* host disease, systemic sclerosis, and atopic disorder characterized by production of IL-3, IL-4, IL-5, IL-6 and IL-10, which act together to promote humoral immunity [2]. Thus, the commitment of specific Th cells to differentiation into Th1 or Th2 cells may determine the susceptibility of the host to particular pathogenic microorganisms.

It is well known that low-dose and long-term macrolide treatment is strikingly effective in the clinical setting for diffuse panbronchiolitis (DPB). DPB is pathologically characterized by chronic inflammation localized predominantly in the respiratory bronchiole with excessive infiltration of mononuclear cells, including lymphocytes, plasma cells and macrophages [5]. In this context, it is of interest to determine whether cytokines mediate the effects of macrolide antibiotics on the interaction between mononuclear cells during the immune response, and whether the drugs directly affect mononuclear cells function.

Effects of macrolides on lymphocyte accumulation in the lung

The efficacy of macrolides on lymphocyte accumulation in the lung was first described in a human study [6]. Lymphocyte accumulation around respiratory bron-

Antibiotics as Anti-Inflammatory and Immunomodulatory Agents, edited by Bruce K. Rubin and Jun Tamaoki

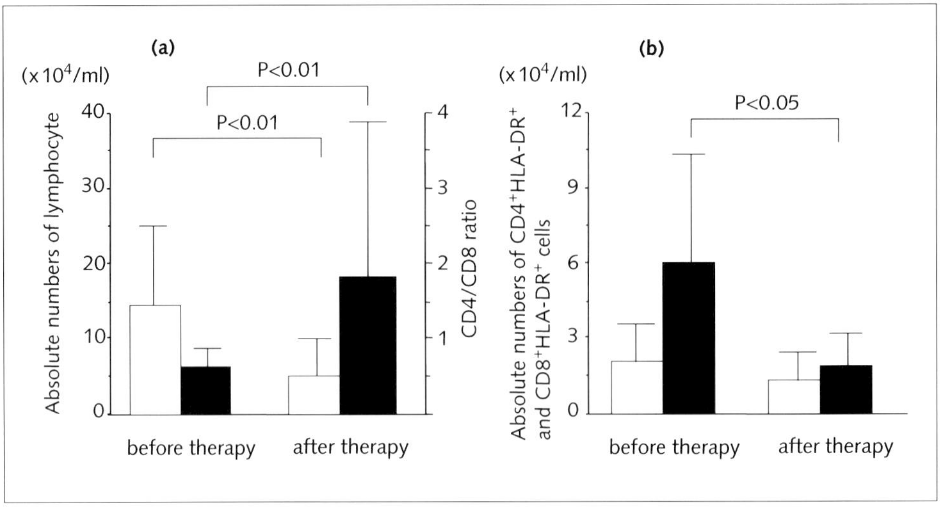

Figure 1
Absolute numbers of lymphocytes (open column) and the CD4/CD8 ratio (closed column) (a) and absolute numbers of CD4⁺ HLA-DR⁺ (open column) and CD8⁺ HLA-DR⁺ cells (closed column) (b) in bronchoalveolar lavage (BAL) fluid from patients with diffuse panbronchiolitis (DPB) before and after macrolide therapy.
From [6] with permission of the American Lung Association.

chioles is a striking pathological feature of DPB. In addition, there is typically an elevated number of lymphocytes and a reduced CD4/CD8 ratio as well as activation of $CD8^+$ cells in the airway lumen [6]. These findings suggest that lymphocytes are important cellular components of chronic bronchial inflammation in DPB.

Long-term (2–6 months) macrolide therapy of DPB patients with erythromycin, clarithromycin or roxithromycin causes a significant reduction in the number of lymphocytes and activated $CD8^+$ cells and increases the CD4/CD8 ratio in the bronchoalveolar lavage (BAL) fluid (Fig. 1) [6]. Another study reported that the activation of $CD8^+$ cells, particularly $CD8^+CD11b^-$ cytotoxic T cells, in the airway lumen of patients with DPB may contribute to chronic bronchial inflammation, possibly through upregulation of adhesion molecules. Treatment with macrolides reduces the number of these $CD8^+$ cells. On the other hand, there is an increase in the number of $CD4^+$ cells, the majority of which are $CD4^+CD29^+$ memory T cells, that is unaffected by treatment with macrolides (Fig. 2) [7]. Although the number of lymphocytes, $CD8^+CD11b^-$ cells, $CD8^+$HLA-DR^+ cells, and $CD4^+CD29^+$ cells is higher in DPB patients with bacterial infections than in those without bacterial infection, macrolide therapy reduces the number of these cells irrespective of whether or not bacterial infection can be identified (Tab. 1) [7]. The authors of this study conclud-

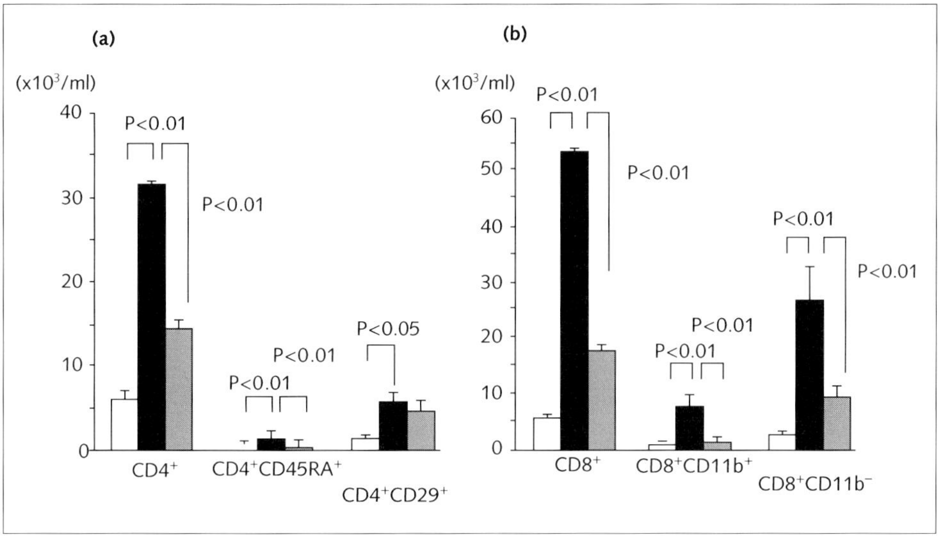

Figure 2
Absolute numbers of CD4⁺ (a) and CD8⁺ (b) cell subsets in bronchoalveolar lavage (BAL) fluid samples from healthy volunteers (open column) and patients with diffuse panbronchiolitis (DPB) before (closed column) and after macrolide therapy (shaded column).
From [7] with permission of Blackwell Publishing Ltd.

ed that macrolide antibiotics bring about an improvement in the clinical condition by reducing inflammation *via* a direct or indirect suppression of cytotoxic T cells [7].

Anti-inflammatory properties of macrolide antibiotics observed in human DPB have also been found in a murine model of DPB [8]. In this model, which is based on chronic respiratory infection with *P. aeruginosa*, there is an increase in the total number of pulmonary lymphocytes and a steady fall in the lung CD4/CD8 ratio that commences on day 7 and persists until day 120. Following a 10-day course of oral clarithromycin (10 mg/kg/day), the number of pulmonary lymphocytes and the CD4/CD8 ratio is brought back to the normal baseline without any change in the number of *P. aeruginosa* in the lungs. In contrast, ofloxacin reduces the number of bacteria but does not influence the number of lymphocytes or the CD4/CD8 ratio. A similar result has also been found in acute inflammation. In mice with influenza virus-induced pneumonia, erythromycin treatment reduces the mortality rate by suppressing IFN-γ production through inhibition of lymphocyte infiltration into the lungs, an effect that is independent of the lung viral load [9]. Thus, macrolides cause a definite decrease in bio-inflammation, despite an inability to eradicate the microorganisms. Additionally, high doses of 14- or 16-membered ring macrolides

Table 1 - Effects of macrolide therapy on lymphocyte subsets in bronchoalveolar lavage fluid of DPB patients with and without bacterial infection

			With bacterial infection (n)		Without bacterial infection (n)	
Total cells	($\times 10^5$/ml)	Before	16.14 ± 3.86 (23)		8.27 ± 1.07 (6)	
		After	3.82 ± 1.80 (10)	*	2.38 ± 0.39 (15)	*
Lymphocytes	($\times 10^5$/ml)	Before	1.24 ± 0.23 (23)		0.70 ± 0.29 (6)	
		After	0.26 ± 0.06 (10)	*	0.43 ± 0.06 (15)	
$CD4^+CD29^+$ cells	($\times 10^3$/ml)	Before	7.52 ± 2.54 (10)		4.75 ± 1.37 (4)	
		After	2.78 ± 0.98 (7)		5.26 ± 1.57 (9)	
$CD8^+CD11b^-$ cells	($\times 10^3$/ml)	Before	25.74 ± 5.77 (17)		9.74 ± 1.84 (3)	
		After	4.37 ± 1.26 (10)	**	6.89 ± 1.55 (13)	
$CD4^+$HLA-DR^+ cells	($\times 10^3$/ml)	Before	2.90 ± 0.41 (15)		2.65 ± 0.97 (3)	
		After	2.17 ± 0.98 (8)		2.34 ± 0.92 (10)	
$CD8^+$HLA-DR^+ cells	($\times 10^3$/ml)	Before	22.82 ± 4.56 (15)		11.73 ± 5.24 (3)	
		After	2.15 ± 1.04 (8)	*	4.35 ± 1.75 (10)	

** $P < 0.001$ vs. before therapy; ** $P < 0.05$ vs. before therapy. (From [7] with permission of the Blackwell Publishing Ltd.)*

suppress the proliferation of mitogen-activated human peripheral T-lymphocytes, apparently by inhibiting IL-2 production *in vitro* [10]. However, it is not known if the anti-lymphocytic activity of macrolides observed *in vitro* explains its effects on lymphocytes *in vivo*.

Effect of macrolides on inflammatory cytokines related to lymphocyte accumulation into the lung

TNF-α and IL-1β, which play an important role in chronic infection, have also been found to be significantly elevated in the BAL fluid of DPB patients. The levels of these two cytokines significantly decrease following treatment with macrolide antibiotics [11, 12]. Yanagihara et al. [13] measured concentrations of inflammatory cytokines, including TNF-α, IL-1β, IFN-γ, IL-2, IL-4 and IL-5, in mice with chronic *P. aeruginosa* respiratory tract infections, a model of DPB. They found that concentration of each cytokine increased significantly 7 days after inoculation with bacteria, and high levels of the cytokines were present even 60 days post-infection. This demonstrates Th1 predominance during the late stage of the disease even though there are no reports showing predominance of the Th1 or Th2 immune

Table 2 - Effects of clarithromycin on the levels of cytokines in the aqueous extract obtained from the mice chronically infected with P. aeruginosa

Cytokines	Cytokine concentration, pg/ml (mean ± SD) Treatment with clarithromycin (n=5)	Control (treatment with saline, n=5)
TNF-α	276.2 ± 119.6 *	1289.9 ± 276.9
IL-1β	554.5 ± 157.7 *	1956.0 ± 356.9
IFN-γ	22.4 ± 7.4	33.9 ± 12.4
IL-2	28.1 ± 12.4	32.7 ± 15.8
IL-4	41.2 ± 8.4	53.8 ± 24.6
IL-5	20.0 ± 7.5	22.0 ± 7.1

** $P < 0.01$ compared with control. TNF, tumor necrosis factor; IL, interleukin; IFN, interferon (From [13] with permission of the Blackwell Publishing Ltd.)*

response in DPB. Treatment of mice with oral clarithromycin for 10 days significantly reduced the concentrations of TNF-α and IL-1β but not other Th1 and Th2 cytokines (Tab. 2) in parallel with a reduction in lymphocyte accumulation in the lungs [13]. This agrees with the clinical efficacy of macrolides in patients with DPB [11, 12].

This mouse model also suggests that TNF-α is essential for the accumulation of lymphocytes in the lung because treatment with an anti-TNF-α antibody significantly reduces both lymphocyte numbers and the level of IL-1β in the lung irrespective of the number of viable bacteria recovered from the lung [13]. Thus, macrolide antibiotics seem to preferentially downregulate lung production of TNF-α and IL-1β relative to IFN-γ, IL-2, IL-4 and IL-5 and to ultimately reduce the accumulation of lymphocytes at the site of the inflammation. Likewise, in asthma patients, roxithromycin reduces peripheral blood leukocyte secretion of IL-3, IL-4, IL-5 and TNF-α into BAL fluid and causes an overall decrease in bronchial responsiveness [14]. *In vitro* studies also demonstrate that 14-membered ring macrolides such as erythromycin and clarithromycin inhibit TNF-α production by lipopolysaccharide-stimulated human monocytes [15, 16].

TNF-α is a potent inducer of the C–C chemokine regulated on activation normal T expressed and secreted (RANTES), and macrophage inflammatory peptides 1α (MIP-1α) and -1β, all of which act through the C–C chemokine receptor 5 [17, 18]. This chemokine/chemokine receptor system is important for mononuclear cell recruitment [19]. Recent studies demonstrate the presence of high levels of RANTES and MIP-1α in the BAL fluid of patients with DPB, and there is a significant correlation between MIP-1α and the number of $CD8^+HLA\text{-}DR^+$ cells in BAL fluid [20].

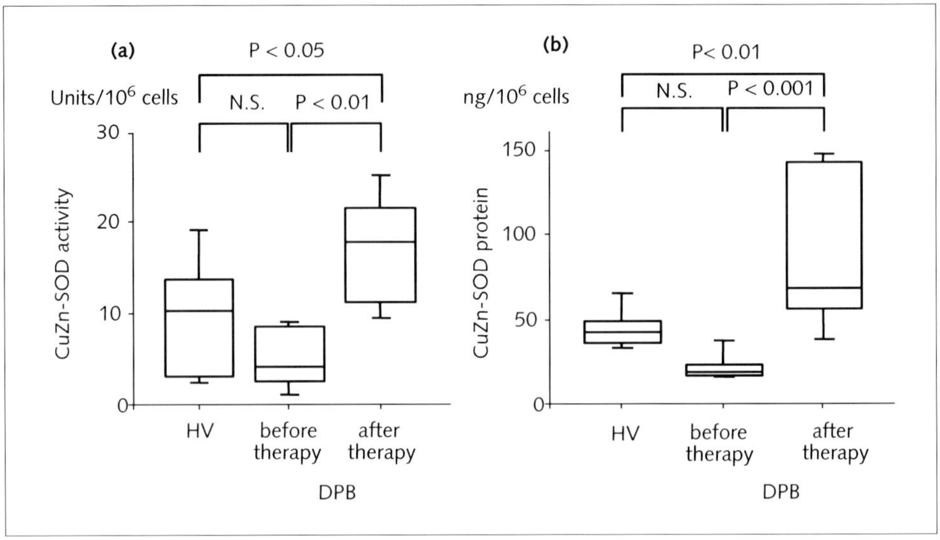

Figure 3
The antioxidant (Cu, Zn-superoxide dismutase) activity (a) and protein level (b) in alveolar macrophages obtained from healthy volunteers (HV) and patients with diffuse panbronchiolitis (DPB) before and after macrolide therapy.
N.S., not significant. From [22] with permission of S. Karger AG, Basel.

The levels of MIP-1α diminished after treatment with macrolide antibiotics [21]. Therefore, macrolide therapy may inhibit mononuclear cell recruitment into the lung by lowering TNF-α levels, which, in turn, reduces the production of C–C chemokines.

Effects of macrolides on the monocyte-macrophage system

The effects of macrolides on the alveolar monocyte–macrophages system have been investigated in patients with DPB [22, 23]. Morikawa and co-workers [22] found enhanced antioxidant activity in alveolar macrophages after long-term erythromycin therapy in patients with DPB (Fig. 3). Katoh et al. [23] also demonstrated alveolar macrophage dysfunction in DPB patients as result of abnormalities in CD44 expression and hyaluronic acid (HA)-binding ability. These abnormalities were normalized after 6–17 months of treatment with erythromycin, roxithromycin or clarithromycin (Fig. 4). These findings demonstrate the enhancing or normalizing effect of macrolides on alveolar macrophage function in human disease. On the

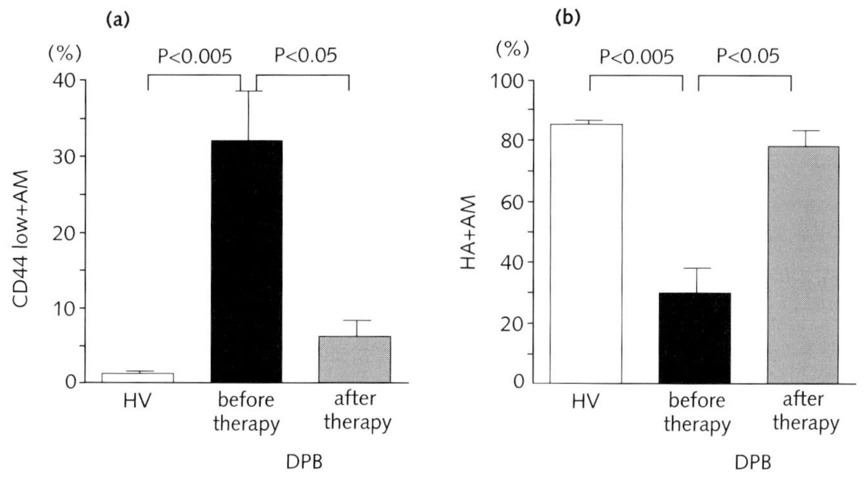

Figure 4
CD44 expression on alveolar macrophages (AM) (a) and hyaluronic acid (HA)-binding activity of AM (b) in healthy volunteers (HV) and patients with diffuse panbronchiolitis (DPB) before and after macrolide therapy.
From [23] with permission of Blackwell Publishing Ltd.

other hand, macrolides have been shown to partially suppress monokine production mostly in *in vitro* systems [15, 16], and when macrolides are administered to rodents, macrophage function, including TNF-α production, is enhanced or inhibited depending on the experimental conditions [13, 24, 25]. Also, roxithromycin has been reported to promote the differentiation of human peripheral monocyte-derived macrophages, while erythromycin has induced the proliferation of mouse peritoneal macrophages [26, 27]. This latter finding appears to be specific to erythromycin because other antibiotics, such as tetracycline, streptomycin, gentamicin, penicillin G, and josamycin, do not induce macrophage proliferation. Thus, macrolides may enhance or normalize the monocyte–macrophage system.

Effects of macrolides on mononuclear cell proliferation and apoptosis

As mentioned previously, high doses (40–200 μg/ml) of 14- or 16-membered ring macrolides markedly suppress the proliferation of polyclonal T-cell mitogens-activated human peripheral T-lymphocytes *in vitro* [10]. In contrast to the inhibitory effect of macrolides, fluoroquinolone antibiotics at therapeutic concentrations (1.56–6.25 μg/ml) are able to enhance peripheral blood lymphocyte (PBL) cell growth stimulated by phytohemagglutinin (PHA) [28, 29] and ciprofloxacin (range 5–80 μg/ml) and other fluoroquinolones potentiate IL-2 synthesis by PHA-stimulated PBLs [30, 31]. These results are supported by several *in vivo* reports on ciprofloxacin-dependent immunomodulation in sublethally irradiated mice [29, 32, 33]. It has also been reported that imipenem, cefodizime and clindamycin are markedly immuno-enhancing antibiotics and cefotaxime, tetracycline, rifampicin, gentamicin, teicoplanin and ampicillin are markedly immuno-depressing antibiotics [34].

Apoptosis is critical for the normal development and tissue homeostasis, including that of the immune system [35], and molecules belonging to the B-cell lymphoma leukemia-2 (Bcl-2)/Bax system and to the Fas/Fas-ligand system play a crucial role in the regulation of the apoptotic process. In particular, Bcl-2 is an intracellular protein that inhibits apoptosis while Bax counteracts the anti-apoptotic function of Bcl-2 by binding to this molecule [36]. Fas is a membrane protein that, when activated by its ligand, induces apoptosis [37, 38]. There is increasing evidence that dysregulations of apoptotic pathways are associated with airway disease, including bronchial asthma. Most of the T-lymphocytes infiltrating the airway of asthmatics are not apoptotic, suggesting that the persistence of airway inflammation may depend upon their continuing proliferation and their increased survival in the bronchial mucosa [39]. In addition, our recent histopathologic study demonstrated low proportion of Fas/Fas-ligand and high proportion of Bcl-2 expressing lymphocytes around respiratory bronchioles of DPB that promotes resistance to apoptosis, leading to the persistence of chronic airway inflammation (unpublished data). Thus, the effect of macrolides has been recently focused on lymphocyte apoptosis (Tab. 3). Roxithromycin augmented the early phase of apoptosis in *Dermatophagoides farinae*-stimulated PBLs at low concentration (1–500 ng/ml), while high concentration of roxithromycin (1 μg/ml) augmented both the early and late phases of apoptosis. However, in both unstimulated and PHA-stimulated PBLs, or in cells from normal subjects, roxithromycin did not affect the induction of apoptosis. Fas ligand but not Fas receptor expression on *D. farinae*-stimulated cells was upregulated after stimulation with 1 μg/ml roxithromycin, while Bcl-2 expression on both unstimulated and *D. farinae*-stimulated PBLs showed a decrease [40]. Other antibiotics, including cefazolin and ampicillin, did not cause significant induction of apoptosis [40]. We also demonstrated that macrolides including clarithromycin, azithromycin and josamycin at higher concentration of 200 μg/ml induced apoptosis and Fas/Fas ligand expression on unstimulated PBLs from normal subjects, while other antibiotics such as beta-lactams, carbapenem and new quinolone did not [41], and that clarithromycin and azithromycin but not josamycin augmented apoptosis of anti-CD3/CD28-activated PBLs from normal subjects at a higher concentration of 100 μg/ml with no significant Fas/Fas-ligand

Table 3 - In vitro *direct effects of macrolides on apoptosis of peripheral blood lymphocyte*

Subjects	Drug/dose	Stimulant	Result	Refs.
Asthma patients	RXM; 1–500 ng/ml	*D. farinae*	augment	[40]
	1 μg/ml	*D. farinae*	augment	
		unstimulated	no change	
		PHA	no change	
Normal subjects	RXM; 1 μg/ml	*D. farinae*, PHA	no change	[40]
	CAM; 200 μg/ml	unstimulated	augment	[41]
	AZM; 200 μg/ml	unstimulated	augment	
	JM; 200 μg/ml	unstimulated	augment	
	CAM; 100 μg/ml	anti-CD3/CD28	augment	[42]
	AZM; 100 μg/ml	anti-CD3/CD28	augment	
	JM; 100 μg/ml	anti-CD3/CD28	no change	
Jurkat T cells	RXM; 10 μg/ml	anti-CD3	augment	[43]
	50 μg/ml	anti-CD3	augment	

RXM, roxithromycin; CAM, clarithromycin; AZM, azithromycin; JM, josamycin; D. farinae, Dermatophagoides farinae*; PHA, phytohemagglutinin*

expression, while clarithromycin and azithromycin inhibited the expression of Bcl-xL protein, which is a notable member of the Bcl-2 family [42]. Jun et al. also found that ciprofloxacin (2.5 and 10 μg/ml) or roxithromycin (10 and 50 μg/ml) induced apoptosis of anti-CD3-activated Jurkat T lymphocytes, and enhanced the expression of Fas ligand and activities of caspase-3 and -8 or the expression of Fas ligand and caspase-3 but not caspase-8 respectively, suggesting that some differences of mechanisms inducing apoptosis of activated Jurkat T cells between quinolone and macrolide could exist [43]. In contrast, moxifloxacin (1 and 10 μg/ml) inhibited both pathways of apoptosis and downregulated the staphylococcal superantigen induced mRNA expression of Fas, Fas ligand, and TNF-RI [44]. The potential therapeutic relevance of these findings should be analyzed cautiously as it may be of relatively low importance compared with the intrinsic antibacterial activities of the fluoroquinolones. Collectively, these results suggest that 14-membered macrolide antibiotics obviously enhance apoptosis of activated lymphocytes, and that induction of Fas/Fas ligand and reduced Bcl-2 or Bcl-xL expression are involved in the increase of apoptosis. This may indicate the therapeutic value for airway diseases such as asthma and DPB.

Conclusion

In this review, we have discussed the anti-lymphocytic and macrophage modulatory activity of macrolide antibiotics. It is clear that 14-membered ring macrolides downregulate the excessive accumulation of lymphocytes at inflammatory sites of chronic airway diseases, including DPB and acute pneumonia caused by influenza virus. In addition, the direct effects of macrolides on mononuclear cell proliferation and apoptosis are evident, while those on cytokine production are still controversial. Discovery of the primary target of the macrolides and its role in mononuclear cell function should provide insight into the immunomodulatory mechanism of macrolides.

References

1 Milik AM, Buechner-Maxwell VA, Sonstein J, Kim S, Seitzman G.D, Beals TF, Curtis JL (1997) Lung lymphocyte elimination by apoptosis in the murine response to intratracheal particulate antigen. *J Clin Invest* 99: 1082–91

2 Mosmann TR, Sad S (1996) The expanding universe of T cell subsets: Th1, Th2 and more. *Immunol Today* 17: 138–46

3 Scott P, Kaufmann SHE (1991) The role of T-cell subsets and cytokines in the regulation of infections. *Immunol Today* 12: 346–8

4 Hsieh CS, Macatonia SE, Tripp CS, O'Garra A, Murphy KM (1993) Development of Th1 $CD4^+$ T cells through IL-12 produced by *Listeria*-induced macrophages. *Science* 260: 547–9

5 Kudoh S, Azuma A, Yamamoto M, Izumi T, Ando M (1998) Improvement of survival in patients with diffuse panbronchiolitis treated with low-dose erythromycin. *Am J Respir Crit Care Med* 157: 1829–32

6 Mukae H, Kadota J, Kohno S, Kusano S, Morikawa T, Matsukura S, Hara K (1995) Increase in activated $CD8^+$ cells in bronchoalveolar lavage fluid in patients with diffuse panbronchiolitis. *Am J Respir Crit Care Med* 152: 613–8

7 Kawakami K, Kadota J, Iida K, Fujii T, Shirai R, Matsubara Y, Kohno S (1997) Phenotypic characterization of T cells in bronchoalveolar lavage fluid (BALF) and peripheral blood of patients with diffuse panbronchiolitis; the importance of cytotoxic T cells. *Clin Exp Immunol* 107: 410–6

8 Yanagihara K, Tomono K, Sawai T, Hirakata Y, Kadota J, Koga H, Tashiro T, Kohno S (1997) Effect of clarithromycin on lymphocytes in chronic respiratory *Pseudomonas aeruginosa* infection. *Am J Respir Crit Care Med* 155: 337–42

9 Sato K, Suga M, Akaike T, Fujii S, Muranaka H, Doi T, Maeda H, Ando M (1998) Therapeutic effect of erythromycin on influenza virus-induced lung injury in mice. *Am J Respir Crit Care Med* 157: 853–7

10 Morikawa K, Oseko F, Morikawa S, Iwamoto K (1994) Immunomodulatory effects of

three macrolides, midecamycin acetate, josamycin, and clarithromycin, on human T-lymphocyte function *in vitro*. *Antimicrob Agents Chemother* 38: 2643–7

11 Sakito O, Kadota J, Kohno S, Abe K, Shirai R, Hara K (1996) Interleukin 1β, tumor necrosis factor alpha, and interleukin 8 in bronchoalveolar lavage fluid of patients with diffuse panbronchiolitis: A potential mechanism of macrolide therapy. *Respiration* 63: 42–8

12 Kadota J, Matsubara Y, Ishimatsu Y, Ashida M, Abe K, Shirai R, Iida K, Kawakami K, Taniguchi H, Fujii T et al (1996) Significance of IL-1beta and IL-1 receptor antagonist (IL-1Ra) in bronchoalveolar lavage fluid (BALF) in patients with diffuse panbronchiolitis (DPB). *Clin Exp Immunol* 103: 461–6

13 Yanagihara K, Tomono K, Kuroki M, Kaneko Y, Sawai T, Ohno H, Miyazaki Y, Higashiyama Y, Maesaki S, Kadota J et al (2000) Intrapulmonary concentrations of inflammatory cytokines in a mouse model of chronic respiratory infection caused by *Pseudomonas aeruginosa*. *Clin Exp Immunol* 122: 67–71

14 Konno S, Asano K, Kurokawa M, Ikeda K, Okamoto K, Adachi M (1994) Antiasthmatic activity of a macrolide antibiotic, roxithromycin: analysis of possible mechanisms *in vitro* and *in vivo*. *Int Arch Allergy Immunol* 105: 308–16

15 Morikawa K, Watabe H, Araake M, Morikawa S (1996) Modulatory effect of antibiotics on cytokine production by human monocytes *in vitro*. *Antimicrob Agents Chemother* 40: 1366–70

16 Iino Y, Toriyama M, Kudo K, Natori Y, Yuo A (1992) Erythromycin inhibition of lipopolysaccharide-stimulated tumor necrosis factor alpha production by human monocytes *in vitro*. *Ann Otol Rhinol Laryngol* (Suppl) 157: 16–20

17 Hornung F, Scala G, Lenardo MJ (2000) TNF-alpha-induced secretion of C-C chemokines modulates C-C chemokine receptor 5 expression on peripheral blood lymphocytes. *J Immunol* 15; 164: 6180–7

18 Lane BR, Markovitz DM, Woodford NL, Rochford R, Strieter RM, Coffey MJ (1999) TNF-alpha inhibits HIV-1 replication in peripheral blood monocytes and alveolar macrophages by inducing the production of RANTES and decreasing C-C chemokine receptor 5 (CCR5) expression. *J Immunol* 163: 3653–61

19 Gerhardt SG, McDyer JF, Girgis RE, Conte JV, Yang SC, Orens JB (2003) Maintenance azithromycin therapy for bronchiolitis obliterans syndrome: results of a pilot study. *Am J Respir Crit Care Med* 168: 121–5

20 Kadota J, Mukae H, Tomono K, Kohno S (2001) High concentrations of beta-chemokines in BAL fluid of patients with diffuse panbronchiolitis. *Chest* 120: 602–7

21 Kadota J, Iida K, Kawakami K, Matsubara Y, Shirai R, Abe K, Tanigichi H, Kaseda M, Kawamoto S, Kohno S (1997) Analysis of inflammatory cell infiltration and its related factors in the lung of patients with diffuse panbronchiolitis. *Jpn J Inflammation* 17: 261–7

22 Morikawa T, Kadota JI, Kohno S, Kondo T (2000) Superoxide dismutase in alveolar macrophages from patients with diffuse panbronchiolitis. *Respiration* 67: 546–51

23 Katoh S, Matsubara Y, Taniguchi H, Fukushima K, Mukae H, Kadota J, Matsukura S,

Kohno S (2001) Characterization of CD44 expressed on alveolar macrophages in patients with diffuse panbronchiolitis. *Clin Exp Immunol* 126: 545–50
24 Kita E, Sawaki M, Mikasa K, Hamada K, Takeuchi S, Maeda K, Narita N (1993) Alterations of host response by a long-term treatment of roxithromycin. *J Antimicrob Chemother* 32: 285–94
25 Sugiyama Y, Yanagisawa K, Tominaga SI, Kitamura S (1999) Effects of long-term administration of erythromycin on cytokine production in rat alveolar macrophages. *Eur Respir J* 14: 1113–6
26 Yoshimura T, Kurita C, Yamazaki F, Shindo J, Morishima I, Machida K, Sumita T, Horiba M, Nagai H (1995) Effects of roxithromycin on proliferation of peripheral blood mononuclear cells and production of lipopolysaccharide-induced cytokines. *Biol Pharm Bull* 18: 876–81
27 Kita E, Sawaki M, Mikasa K, Oku D, Hamada K, Maeda K, Narita N, Kashiba S (1993) Proliferation of erythromycin-stimulated mouse peritoneal macrophages in the absence of exogenous growth factors. *Nat Immun* 12: 326–38
28 Forsgren A, Schlossman SF, Tedder TF (1987) 4-Quinolone drugs affect cell cycle progression and function of human lymphocytes *in vitro*. *Antimicrob Agents Chemother* 31: 768–73
29 Stunkel KG, Hewlett G, Zeiler HJ (1991) Ciprofloxacin enhances T cell function by modulating interleukin activities. *Clin Exp Immunol* 86: 525–31
30 Riesbeck K, Andersson J, Gullberg M, Forsgren A (1989) Fluorinated 4-quinolones induce hyperproduction of interleukin 2. *Proc Natl Acad Sci USA* 86: 2809–13
31 Zehavi-Willner T, Shalit I (1989) Enhancement of interleukin-2 production in human lymphocytes by two new quinolone derivatives. *Lymphokine Res* 8: 35–46
32 Kletter Y, Riklis I, Shalit I, Fabian I (1991) Enhanced repopulation of murine hematopoietic organs in sublethally irradiated mice after treatment with ciprofloxacin. *Blood* 78: 1685–91
33 Shalit I, Kletter Y, Weiss K, Gruss T, Fabian I (1997) Enhanced hematopoiesis in sublethally irradiated mice treated with various quinolones. *Eur J Haematol* 58: 92–8
34 Van Vlem B, Vanholder R, De Paepe P, Vogelaers D, Ringoir S (1996) Immunomodulating effects of antibiotics: literature review. *Infection* 24: 275–91
35 Krammer PH (2000) CD95's deadly mission in the immune system. Nature 407: 789–95
36 Oltvai ZN, Milliman CL, Korsmeyer SJ (1993) Bcl-2 heterodimerizes *in vivo* with a conserved homolog, Bax, that accelerates programmed cell death. *Cell* 74: 609–19
37 Alderson MR, Tough TW, Davis-Smith T, Braddy S, Falk B, Schooley KA, Goodwin RG, Smith CA, Ramsdell F, Lynch DH (1995) Fas ligand mediates activation-induced cell death in human T lymphocytes. *J Exp Med* 181: 71–7
38 Van Parijs L, Biuckians A, Abbas AK (1998) Functional roles of Fas and Bcl-2-regulated apoptosis of T lymphocytes. *J Immunol* 160: 2065–71
39 Vignola AM, Chanez P, Chiappara G, Siena L, Merendino A, Reina C, Gagliardo R, Profita M, Bousquet J, Bonsignore G (1999) Evaluation of apoptosis of eosinophils,

macrophages, and T lymphocytes in mucosal biopsy specimens of patients with asthma and chronic bronchitis. *J Allergy Clin Immunol* 103: 563–73

40 Ogawa N, Sugawara Y, Fujiwara Y, Noma T (2003) Roxithromycin promotes lymphocyte apoptosis in *Dermatophagoides*-sensitive asthma patients. *Eur J Pharmacol* 474: 273–81

41 Ishimatsu Y, Kadota J, Iwashita T, Nagata T, Ishii H, Shikuwa C, Kaida H, Mukae H, Kohno S (2004) Macrolide antibiotics induces apoptosis in human peripheral lymphocytes. *Int J Antimicrob Agents* 24: 49–55

42 Mizunoe S, Kadota J, Tokimatsu I, Kishi K, Nagai H, Nasu M (2004) Clarithromycin and azithromycin induce apoptosis of activated lymphocytes *via* down-regulation of Bcl-xL. *Int Immunopharmacol* 4: 1201–7

43 Jun YT, Kim HJ, Song MJ, Lim JH, Lee DG, Han KJ, Choi SM, Yoo JH, Shin WS, Choi JH (2003) *In vitro* effects of ciprofloxacin and roxithromycin on apoptosis of Jurkat T lymphocytes. *Antimicrob Agents Chemother* 47: 1161–4

44 Konig B, Konig W (2002) Moxifloxacin inhibits staphylococcal superantigen induced apoptosis in T lymphocytes. 12th European Congress of Clinical and Microbiological Infectious Diseases (ECCMID). *Clin Microbiol Inf Dis* 8: 166

Mucoregulatory effects

Macrolides and mucus production

Kiyoshi Takeyama

First Department of Medicine, Tokyo Women's Medical University School of Medicine,
8-1 Kawada-cho, Shinjuku-ku, Tokyo 162-8666, Japan

Introduction

Low-dose, long-term erythromycin therapy is widely used for treating patients with diffuse panbronchiolitis, a chronic inflammatory disease characterized by productive cough and shortness of breath [1]. Because of the reduction of sputum volume in patients with diffuse panbronchiolitis [2], the efficacy of this drug is thought to be based on the reduction of airway secretion. Thus, macrolide antibiotics are now used to treat airway hypersecretory diseases, such as chronic bronchitis, chronic sinusitis, cystic fibrosis and asthma. Recently, the promising evidence for the clinical efficacy of macrolides as a mucoregulatory drug has been accumulating [3–5], and the mechanism of anti-secretory function has been vigorously investigated.

Effect of macrolides on mucus secretion

In 1990, Goswami and co-workers [6] initially demonstrated that erythromycin, but not other antibiotics such as penicillin, ampicillin, tetracycline, and cephalosporin, inhibited both spontaneous and methacholine- and histamine-induced mucus glycoconjugate secretion in cultured explant of human bronchial epithelium and in cultured endometrial adenocarcinoma cells *in vitro*. They concluded that the efficacy of erythromycin was not related to its antibacterial properties. Similarly, a recent study showed that erythromycin and clarithromycin inhibit spontaneous and TNF-α-induced mucus secretion in a dose- and a time-dependent fashion in NCI-H292 cells and in nasal epithelial cells [7]. However, the molecular mechanism of the inhibitory effect of macrolides on mucus secretion is unknown.

Besides the direct effect of macrolides, there is increasing evidence that the immunomodulatory activity of macrolides may contribute to the reduction of mucus hypersecretion. Oral administration of either erythromycin or clarithromycin inhibits lipopolysaccharide (LPS)-induced mucus discharge from goblet cells in guinea pigs, which was associated with the attenuation of LPS-induced recruitment

Antibiotics as Anti-Inflammatory and Immunomodulatory Agents, edited by Bruce K. Rubin and Jun Tamaoki

of neutrophils [8]. In addition, exposure of guinea pigs to inhaled interleukin-8 (IL-8) induces neutrophil recruitment into airway epithelium and upregulates mucus secretion, and these effects are inhibited by pretreatment with 14-member macrolides, erythromycin and roxithromycin [9]. From these results, the efficacy of macrolides on mucus hypersecretion may be associated, at least in a part, with their anti-inflammatory property and anti-neutrophil activity. Further studies are necessary to determine whether macrolides exerted their anti-secretory effect by acting on mucus-producing cells, although it has been suggested that the anti-secretory effect is related to modulating ion channels [10].

Effect of macrolides on mucin synthesis

Mucins are heavily glycosylated, high molecular weight glycoproteins and are known to affect the viscoelasticity of airway mucus. Airway mucins are synthesized by epithelial goblet cells and by mucous cells of the submucosal glands. Each mucus-producing cell expresses specific gel-forming mucin genes; MUC5AC expression is restricted to goblet cells [11, 12] and MUC5B is expressed in mucous cells of submucosal glands [13]. These are the predominant secreted mucins in airway secretion both in healthy and in disease condition [14, 15].

The increase in airway mucus can be explained by mucus-producing cell hyperplasia, which is established and maintained by the upregulation of mucin gene expression. The effect of the 14-membered macrolide antibiotics on mucin gene and protein expression has been evaluated *in vitro* and *in vivo*. A histopathological analysis using Alcian blue/PAS staining, or MUC5AC immunohistochemistry, revealed that oral administration of clarithromycin (5–10 mg/kg) inhibited epithelial mucus production induced by allergic inflammation or by instillation of LPS in rats [7] and by inoculation of *Pseudomonas aeruginosa* in mice [16]. Extracellular signal regulated kinase (ERK)1/2 phosphorylation, which is involved in LPS-induced signal transduction pathway causing mucin expression, was also attenuated by clarithromycin in the lungs of *Pseudomonas aeruginosa*-infected mice (Fig. 1). However, whether clarithromycin exerted the inhibitory effect on mucin synthesis by its direct effect on ERK1/2 phosphorylation or by inhibition of airway inflammation remains unknown.

To assess the direct effect of macrolides on mucin synthesis, the *in vitro* studies have been carried out using cultured airway epithelial cells. Shimizu and colleagues have shown [7] that treatment with either erythromycin or clarithromycin at 10^{-6} M significantly inhibited the constitutive expression of MUC5AC mRNA in human mucoepidermoid cell line, NCI-H292 cells (Fig. 2) and in human nasal epithelial cells. By contrast, the 16-member macrolides josamycin and ampicillin showed no effect on mucin synthesis; these results indicate that the direct inhibitory effects on mucus secretion in 14-member macrolide.

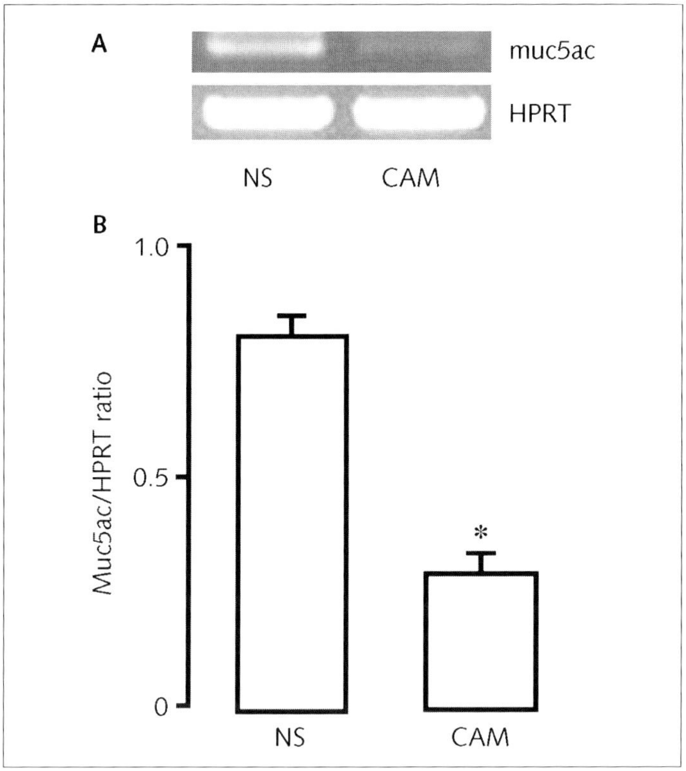

Figure 1
Effect of clarithromycin on muc5ac gene expression
*The levels of muc5ac and hypoxanthine phosphoribosyltransferase (HPRT) mRNA were analyzed competitive RT-PCRs (A), and these levels were determined by densitometry. Data are expressed as ratios of muc5ac to HPRT and as means ± SE of three independent experiments. The result suggests that clarithromycin also reduced muc5ac at the mRNA level (B). *$p < 0.05$ compared with saline-treated mice. (From [16] with permission from the American Physiological Society.)*

Recently, signal transduction of mucin gene expression has been intensively investigated and several candidate pathways have been determined; LPS-induced mucin synthesis is mediated through a c-Src-Ras-MAPK kinase (MEK)1/2-mitogen-activated protein kinase (MAPK)-pp90rsk that leads to activation of nuclear factor (NF)-κB (p65/p50) [17], and a epidermal growth factor receptor (EGFR)–Ras–Raf–ERK signaling pathway is recognized as one of the major pathway to induce mucin gene expression in asthma and chronic obstructive pulmonary disease (COPD) [18, 19]. Takeyama and colleagues reported that pretreatment with both erythromycin

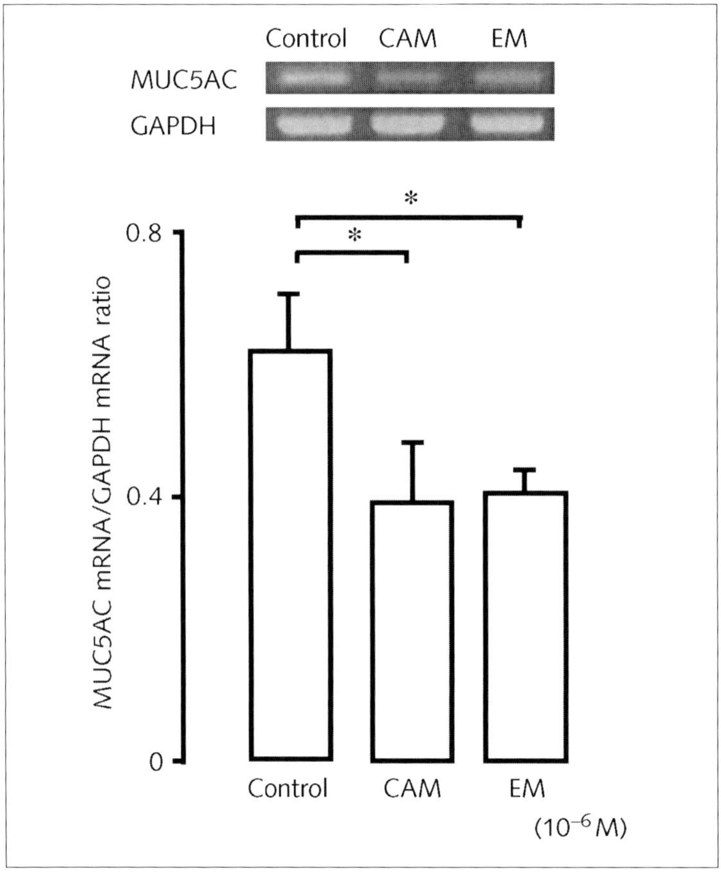

Figure 2
Effects of CAM and EM on MUC5AC messenger RNA (mRNA) expression in cultured NCI-H292 cells
Total RNA was isolated and analyzed for MUC5AC and glyceraldehyde-3-phosphate dehydrogenase (GAPDH) mRNA expression by reverse transcription polymerase chain reaction ($n = 4$). CAM and EM significantly inhibited MUC5AC mRNA expression at 10^{-4} M as demonstrated by the MUC5AC/GAPDH ratio. (From [7] with permission from the American Thoracic Society.)

and clarithromycin attenuated transforming growth factor-α (TGF-α)- and LPS-induced MUC5AC expression in NCI-H292 cells [20]. Both macrolides also attenuated the NF-κB activation without affecting the MEK phosphorylation, indicating that the transcription factor NF-κB is the target molecule for the inhibition of mucin synthesis (Fig. 3). This target is similar to the regulation of IL-8 synthesis [21, 22].

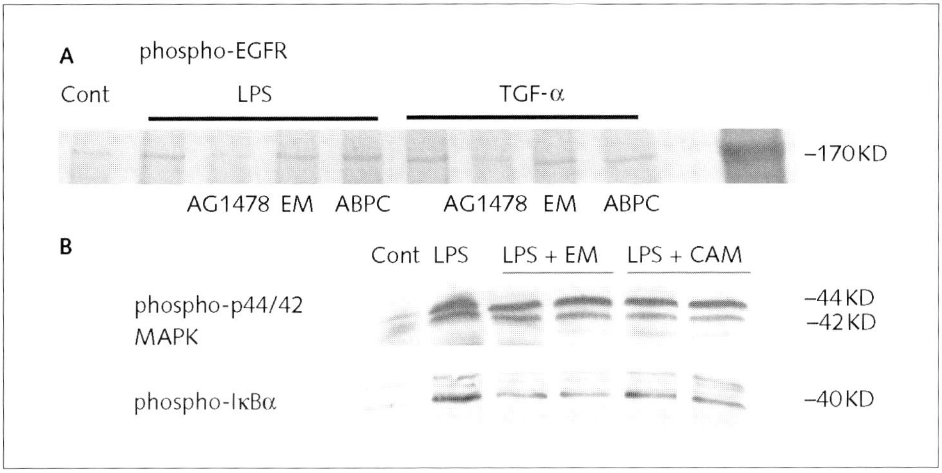

Figure 3
(A) Effect of a selective EGFR tyrosine kinase inhibitor AG1478, erythromycin, and ampicillin on tyrosine phosphorylation of EGFR induced by LPS and TGF-α. Pretreatment with AG1478 inhibited both LPS- and TGF-α-induced EGFR phosphorylation, whereas EM and AMPC was without effect. (B) Effect of macrolide antibiotics on tyrosine phosphorylation of p44/42mapk and IκBα induced by LPS. CAM and EM attenuated the LPS-induced phosphorylation of IκBα, whereas the phosphorylation of p44/42mapk was unchanged (From [13] with permission from the Japan Antibiotics Research Association.)

In addition, 14-membered macrolides inhibit the neutrophil activities such as leukoattractant-activated superoxide generation [23] and neutrophil elastase [24], which can cause mucin synthesis (Fig. 4). Further studies are required to elucidate whether macrolide antibiotics affect other molecules which can regulate mucin production, such as ADAM families and calcium activated chlolide channels.

Clinical effect of macrolides on mucus secretion

Clinical effects of macrolides on mucus secretion have been reported in different airway hypersecretory diseases, such as diffuse panbronchiolitis, purulent rhinitis, chronic bronchitis, cystic fibrosis, and asthma. Tamaoki and colleagues [25] conducted a parallel, double-blind, placebo-controlled study to determine the effects of long-term administration of clarithromycin on sputum production in patients with clinical conditions associated with excessive airway secretions. Treatment with clarithromycin (200 mg/day) for 8 weeks decreased sputum production but did not alter the bacterial density and sputum flora. Similarly, Tagaya et al. [26] demonstrated

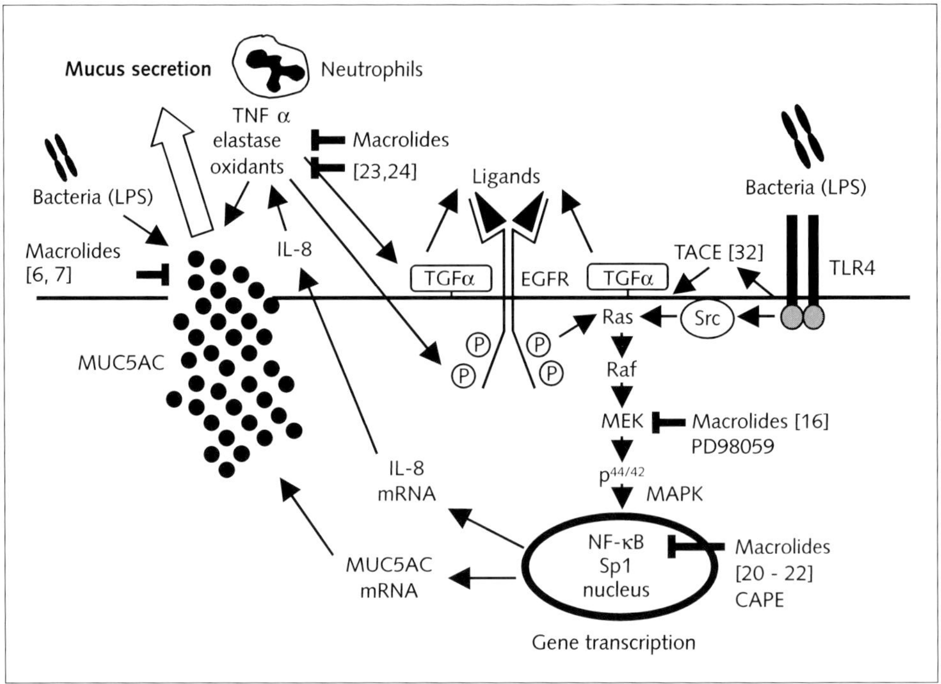

Figure 4
Schematic representation of signal transduction pathway causing mucin gene expression and possible inhibitory sites by macrolides
Mucin gene expression is upregulated by both ligand-dependent and ligand-independent activation of EGFR. LPS caused both tumor necrosis factor α-converting enzyme (TACE)-induced cleavage of TGF-α [32] and activation of Src, which lead to mucin gene expression. Neutrophil elastase is also capable to induce cleavage of TGF-α. The phosphorylation of EGFR initiate the activation of MAPK kinase (MEK) – p44/42^{mapk} signal transduction pathway, leading to activation of transcription factors, NF-κB and Sp1 [19]. Macrolide antibiotics could inhibit the neutrophil activities including oxidative stress [23] and neutrophil elastase [24], and the activation of MEK and NF-κB, resulting in the reduction of airway mucus.

that even short-term (7 days) therapy with clarithromycin (400 mg/day) reduced sputum production in patients with chronic bronchitis or bronchiectasis without apparent respiratory infection. Both studies reported that ampicillin and cefaclor were without effect on mucus secretion, indicating the inhibitory effect on sputum production is not related to their antimicrobial activity and is specific to the 14-membered macrolides. Rubin and co-workers [27] showed that clarithromycin (1,000 mg/day for 2 weeks) resulted in a reduction in the volume of airway secre-

Table 1 - Role of macrolides in airway secretion

1.	Inhibition of mucin gene and protein expression
	inhibition of signaling pathway (MEK)
	inhibition of transcription factor (NF-κB)
	reduction of stimuli for mucin synthesis (elastase, oxidative stress)
	reduction of origin for mucin stimuli (PMN, eosinophil)
2.	Modulation of mucus reology
3.	Activation of mucociliary transport
4.	Modulation of ion channel
5.	Eradication of persistent airway infection

tion in normal subjects and in patients with purulent rhinitis. They also showed that the viscoelasticity of mucus in patients with rhinitis became normal after the treatment. Thus, clinically, macrolides may reduce airway mucus not only by inhibiting the mucus output, but also by facilitating mucus clearance by changing rheology of mucus and by activating ciliary beat [28]. Recently, it has been reported that the volume of gel-forming mucins, MUC5AC and MUC5B, in airway secretion is varied among the hypersecretory diseases. The sputum collected from patients with asthma or with chronic bronchitis contains a large amount of MUC5AC and MUC5B mucins [29]. By contrast, the sputum collected from cystic fibrosis patients contains a small amount of mucins, which was a 93% decrease in MUC5AC and a 70% decrease in MUC5B compared to normal sputum (Fig. 5) [30]. As macrolide therapy is effective in both diseases, the mechanism by which macrolides reduce mucus secretion might be varied among the diseases.

The anti-secretory role of macrolides in asthma is complicated because of the involvement of *Chlamydia pneumoniae* and *Mycoplasma pneumoniae* in the pathogenesis of asthma. Chu and colleagues [31] have reported that the increased expression of substance P was correlated with the epithelial mucus content in the asthmatic patients in which *Mycoplasma pneumoniae* was found. Treatment with macrolides decreased both substance P and mucus expressions, suggesting a possible involvement of antimicrobial as well as anti-inflammatory activity.

Conclusions

Airway mucins act as a physical barrier to many harmful materials. However, in airway inflammatory diseases, once airway mucins are overproduced, these may contribute to the morbidity and mortality associated with these diseases. Macrolide antibiotics, especially in 14-member macrolides exert an anti-secretory effect

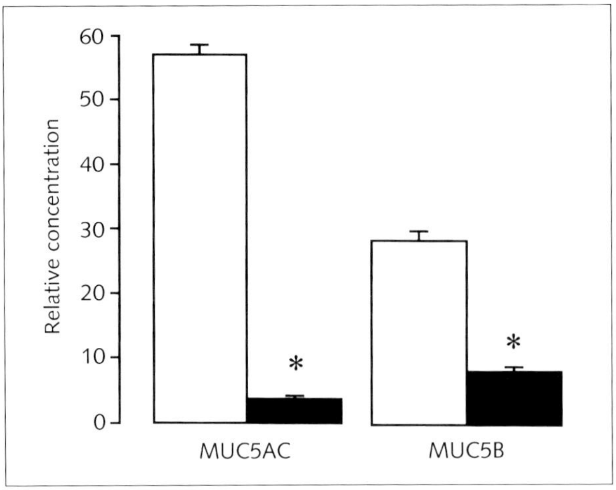

Figure 5
*Serial diluted dot-blot analysis of mucus collected from the end of non-cuffed endotracheal tubes (ETT) in normal control subjects (open bars) and sputum from patients with cystic fibrosis (closed bars). The samples were loaded as volume equivalents from the sputum. The blot was probed with affinity-purified antibodies for MUC5AC and MUC5B. The results for MUC5AC and MUC5B cannot be directly compared with one another because antibody affinity is probably different. *Students* t *test* $P < 0.005$. *(From [30] with permission from the American Toracic Society.)*

through a variety of mechanisms (Tab. 1). The emerging evidence for clinical effectiveness in mucus hypersecretion should encourage further research into understanding the mechanism by which macrolides inhibit signal transduction pathways causing airway mucin gene expression. These efforts may lead to the development of new therapeutic agents for airway diseases.

References

1 Kudoh S, Azuma A, Yamamoto M, Izumi T, Ando M (1998) Improvement of survival in patients with diffuse panbronchiolitis treated with low-dose erythromycin. *Am J Respir Crit Care Med* 157: 1829–32
2 Yamamoto M, Kudoh S, Ina Y, Tamura A (1990) Clinical efficacy of erythromycin for patients with diffuse panbronchiolitis – a double blind study. *Saishin Igaku* 45: 103–8
3 Jaffe A, Bush A (2001) Anti-inflammatory effects of macrolides in lung disease. *Pediatr Pulmonol* 31: 464–73
4 Rubin BK (2002) The pharmacologic approach to airway clearance: mucoactive agents. *Respir Care* 47: 818–22

5 Majima Y (2002) Mucoactive medications and airway disease. *Paediatr Respir Rev* 3: 104–9

6 Goswami SK, Kivity S, Marom Z (1990) Erythromycin inhibits respiratory glycoconjugate secretion from human airways *in vitro*. *Am Rev Respir Dis* 141: 72–8

7 Shimizu T, Shimizu S, Hattori R, Gabazza EC, Majima Y (2003) *In vivo* and *in vitro* effects of macrolide antibiotics on mucus secretion in airway epithelial cells. *Am J Respir Crit Care Med* 168: 581–7

8 Tamaoki J, Takeyama K, Yamawaki I, Kondo M, Konno K (1997) Lipopolysaccharide-induced goblet cell hypersecretion in the guinea pig trachea: inhibition by macrolides. *Am J Physiol* 272: L15–L19

9 Tamaoki J, Nakata J, Tagaya E, Konno K (1996) Effects of roxithromycin and erythromycin on interleukin 8-induced neutrophil recruitment and goblet cell secretion in guinea pig tracheas. *Antimicrob Agents Chemother* 40: 1726–8

10 Irokawa T, Sasaki T, Shimura S, Sasamori K, Oshiro T, Nara M, Tamada T, Shirato K (1999) Cholinomimetic action of macrolide antibiotics on airway gland electrolyte secretion. *Am J Physiol* 276: L951–L957

11 Zuhdi Alimam M, Piazza FM, Selby DM, Letwin N, Huang L, Rose MC (2000) Muc-5/5ac mucin messenger RNA and protein expression is a marker of goblet cell metaplasia in murine airways. *Am J Respir Cell Mol Biol* 22: 253–60

12 Takeyama K, Fahy JV, Nadel JA (2001) Relationship of epidermal growth factor receptors to goblet cell production in human bronchi. *Am J Respir Crit Care Med* 163: 511–6

13 Wickstrom C, Davies JR, Eriksen GV, Veerman EC, Carlstedt I (1998) MUC5B is a major gel-forming, oligomeric mucin from human salivary gland, respiratory tract and endocervix: identification of glycoforms and C-terminal cleavage. *Biochem J* 334: 685–93

14 Hovenberg HW, Davies JR, Herrmann A, Linden CJ, Carlstedt I (1996) MUC5AC, but not MUC2, is a prominent mucin in respiratory secretions. *Glycoconj J* 13: 839–47

15 Davies JR, Svitacheva N, Lannefors L, Kornfalt R, Carlstedt I (1999) Identification of MUC5B, MUC5AC and small amounts of MUC2 mucins in cystic fibrosis airway secretions. *Biochem J* 344: 321–30

16 Kaneko Y, Yanagihara K, Seki M, Kuroki M, Miyazaki Y, Hirakata Y, Mukae H, Tomono K, Kadota J, Kohno S (2003) Clarithromycin inhibits overproduction of muc5ac core protein in murine model of diffuse panbronchiolitis. *Am J Physiol Lung Cell Mol Physiol* 285: L847–L853

17 Li JD, Feng W, Gallup M, Kim JH, Gum J, Kim Y, Basbaum C (1998) Activation of NF-kappaB via a Src-dependent Ras-MAPK-pp90rsk pathway is required for *Pseudomonas aeruginosa*-induced mucin overproduction in epithelial cells. *Proc Natl Acad Sci USA* 95: 5718–23

18 Takeyama K, Dabbagh K, Lee HM, Agusti C, Lausier JA, Ueki IF, Grattan KM, Nadel JA (1999) Epidermal growth factor system regulates mucin production in airways. *Proc Natl Acad Sci USA* 96: 3081–6

19 Perrais M, Pigny P, Copin MC, Aubert JP, Van Seuningen I (2002) Induction of MUC2 and MUC5AC mucins by factors of the epidermal growth factor (EGF) family is medi-

ated by EGF receptor/Ras/Raf/extracellular signal-regulated kinase cascade and Sp1. *J Biol Chem* 277: 32258–67
20 Takeyama K, Tamaoki J, Kondo M, Aoshiba K, Nakata J, Isono K, Nagai A (2001) Effect of macrolide antibiotics on MUC5AC production in human bronchial epithelial cells. *Jpn J Antibiot* 54: 52–4
21 Aoki Y, Kao PN (1999) Erythromycin inhibits transcriptional activation of NF-kappaB, but not NFAT, through calcineurin-independent signaling in T cells. *Antimicrob Agents Chemother* 43: 2678–84
22 Desaki M, Takizawa H, Ohtoshi T, Kasama T, Kobayashi K, Sunazuka T, Omura S, Yamamoto K, Ito K (2000) Erythromycin suppresses nuclear factor-kappaB and activator protein-1 activation in human bronchial epithelial cells. *Biochem Biophys Res Commun* 267: 124–8
23 Anderson R (1989) Erythromycin and roxithromycin potentiate human neutrophil locomotion *in vitro* by inhibition of leukoattractant-activated superoxide generation and autooxidation. *J Infect Dis* 159: 966–73
24 Gorrini M, Lupi A, Viglio S, Pamparana F, Cetta G, Iadarola P, Powers JC, Luisetti M (2001) Inhibition of human neutrophil elastase by erythromycin and flurythromycin, two macrolide antibiotics. *Am J Respir Cell Mol Biol* 25: 492–9
25 Tamaoki J, Takeyama K, Tagaya E, Konno K (1995) Effect of clarithromycin on sputum production and its rheological properties in chronic respiratory tract infections. *Antimicrob Agents Chemother* 39: 1688–90
26 Tagaya E, Tamaoki J, Kondo M, Nagai A (2002) Effect of a short course of clarithromycin therapy on sputum production in patients with chronic airway hypersecretion. *Chest* 122: 213–18
27 Rubin BK, Druce H, Ramirez OE, Palmer R (1997) Effect of clarithromycin on nasal mucus properties in healthy subjects and in patients with purulent rhinitis. *Am J Respir Crit Care Med* 155: 2018–23
28 Takeyama K, Tamaoki J, Chiyotani A, Tagaya E, Konno K (1993) Effect of macrolide antibiotics on ciliary motility in rabbit airway epithelium *in vitro*. *J Pharm Pharmacol* 45:756–8
29 Ordonez CL, Khashayar R, Wong HH, Ferrando R, Wu R, Hyde DM, Hotchkiss JA, Zhang Y, Novikov A, Dolganov G et al (2001) Mild and moderate asthma is associated with airway goblet cell hyperplasia and abnormalities in mucin gene expression. *Am J Respir Crit Care Med* 163: 517–23
30 Henke MO, Renner A, Huber RM, Seeds MC, Rubin BK (2004) MUC5AC and MUC5B mucins are decreased in cystic fibrosis airway secretions. *Am J Respir Cell Mol Biol* 31: 86–91
31 Chu HW, Kraft M, Krause JE, Rex MD, Martin RJ (2000) Substance P and its receptor neurokinin 1 expression in asthmatic airways. *J Allergy Clin Immunol* 106: 713–22
32 Shao MX, Ueki IF, Nadel JA (2003) Tumor necrosis factor alpha-converting enzyme mediates MUC5AC mucin expression in cultured human airway epithelial cells. *Proc Natl Acad Sci USA* 100: 11618–23

Ion channel regulation

Jun Tamaoki

First Department of Medicine, Tokyo Women's Medical University School of Medicine,
8-1 Kawada-cho, Shinjuku-ku, Tokyo 162-8666, Japan

Introduction

Airway hypersecretion is one of the characteristic features of chronic airway inflammation including chronic bronchitis, asthma, bronchiectasis and diffuse panbronchiolitis, and a large amount of secretions stagnated in the respiratory lumen may cause airflow limitation, impairment of mucociliary transport, and recurrent respiratory infection. Airway secretions consist of the mucus synthesized and released by submucosal glands and goblet cells, and the water transported across airway mucosa [1].

Previous studies have shown that long-term administration of macrolide antibiotics provides a marked reduction in the volume of airway secretions without changing sputum flora in some patients with asthma [2], bronchorrhea [3], chronic bronchitis, and diffuse panbronchiolitis [4, 5]. One explanation for the mechanism of efficacy would be the anti-inflammatory effects of macrolides, such as the inhibition of cytokine production [6] and neutrophil migration [7]; another possibility is the direct action on airway secretory cells. Indeed, Goswami et al. [8] first studied nasal mucus glycoconjugate secretion from healthy nonsmoking adults before and after treatment with erythromycin base, penicillin, ampicillin, tetracycline, or cephalosporins. They found that erythromycin, at a concentration of 10 μM, reduced nasal secretion by 35% in both the resting state and when the nose was stimulated with methacholine or histamine, but other antibiotics had no effect on glycoconjugate secretion. Conversely, there is evidence that macrolides affect certain ion channels on airway epithelial cells, and the subsequent alterations in electrolyte transport might also contribute to the anti-secretory effects of macrolides.

Role of ion channels in airway secretion

It is known that secretion of water from the submucosa towards the lumen, and absorption of water to the opposite direction, is generally correlated with secretion of Cl and absorption of Na, respectively, by airway epithelial cells [9]. In fact, at least four types of Cl channels are located on the apical membrane of airway epithe-

Antibiotics as Anti-Inflammatory and Immunomodulatory Agents, edited by Bruce K. Rubin and Jun Tamaoki

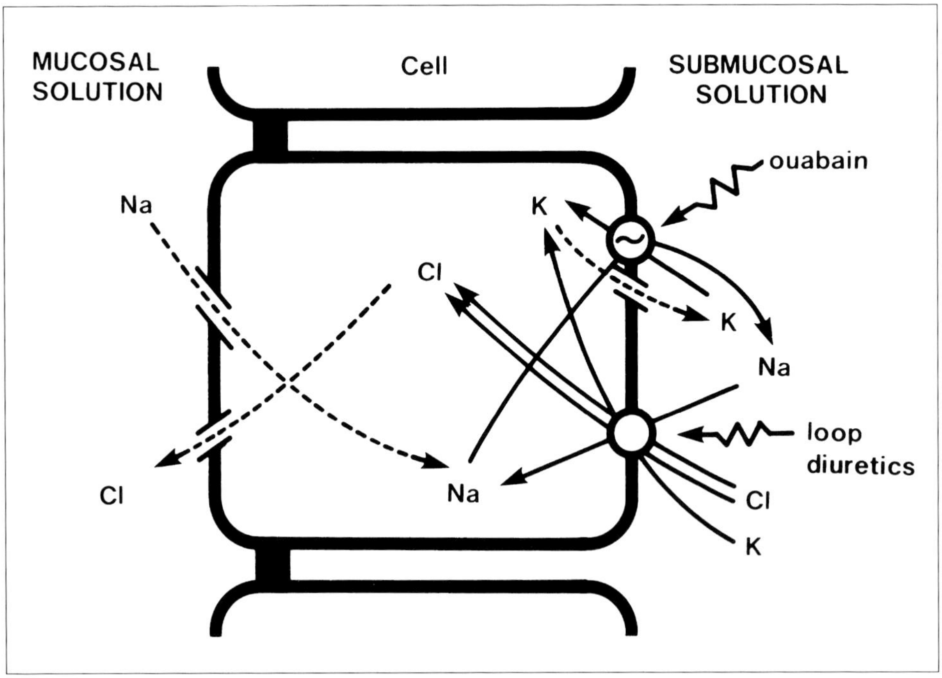

Figure 1
Cellular model of mechanism of electrolyte transport by ciliary epithelium of central airway
At the apical membrane, Na enters and Cl exits the cell through Na channel and Cl channel, respectively. Ouabain and loop diuretics inhibit Na–K–ATPase and Na–K–Cl co-transporter, respectively, located on the submucosal membrane.

lium, i.e., cystic fibrosis transmembrane conductance regulator (CFTR), outwardly rectifying Cl channel (ORCC), Ca^{2+}-activated Cl channel, and volume-sensitive Cl channel [10], and certainly the epithelium can actively secrete Cl through these channels, thereby making the lumen electrically negative. This transepithelial potential difference will drive net diffusion of Na towards the lumen across the tight junctions and other leak pathways within the epithelium (Fig. 1). The resulting transfer of salt creates an osmotic pressure difference, which promotes fluid movement toward the airway lumen. Similar reasoning explains the stimulation of water absorption with the increase in active Na absorption. Thus, the direction and amount of water moved across the epithelium depend, in part, on the balance between these two opposing transport processes [11], and it is possible that airway epithelial Cl secretion is upregulated in patients with airway hypersecretion and, hence, the drugs capable of inhibiting the Cl channel function could be of value in the treatment of such patients.

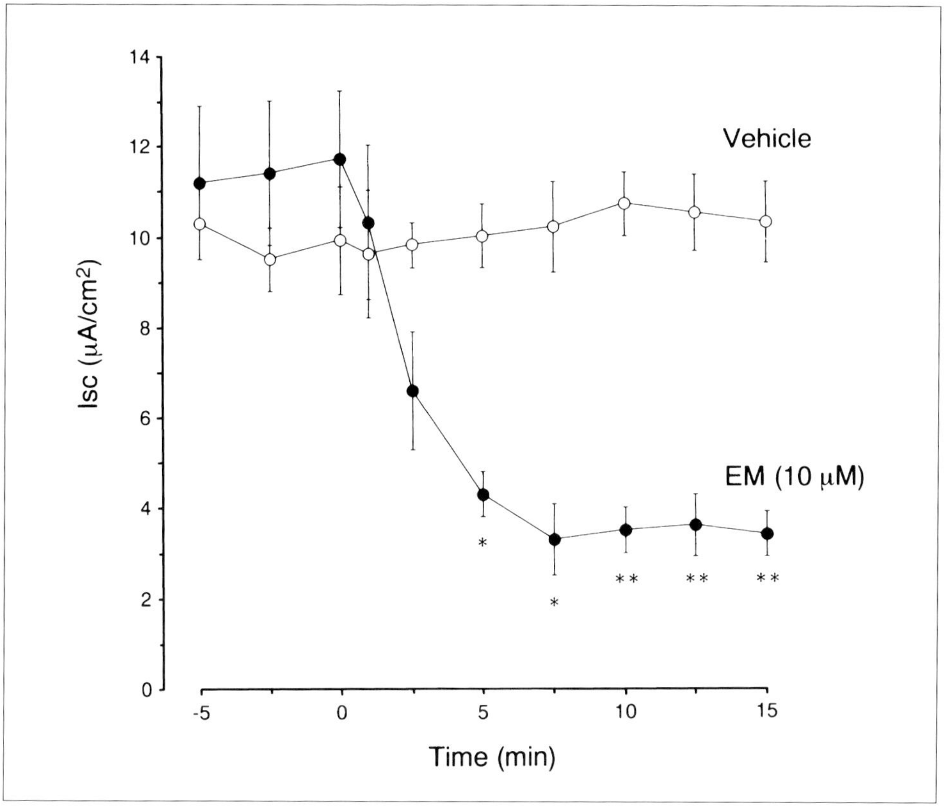

Figure 2
Time course of the effect of erythromycin (EM) on short-circuit current (Isc) of cultured canine tracheal epithelium
EM (10 μM) or its vehicle alone was added to the submucosal solutions in Ussing chamber. Data are means ± SE; n = 7 for each point. $^{*}P < 0.05$, $^{**}P < 0.01$*, significantly different from corresponding values for vehicle.*

Effects on Cl channel *in vitro*

Regarding the effects of macrolides on the airway epithelial Cl channel, electrical properties of cultured canine tracheal epithelium have been measured by Ussing's technique *in vitro* [12]. As shown in Figure 2, erythromycin applied to the submucosal side at concentrations of 10 μM decreased short-circuit current – an electrical parameter that reflects net value of actively transported ions across airway epithelium. This effect was dose-dependent, with the threshold concentration of 3 μM. Sub-

Table 1 - Effect of clarithromycin on bioelectric properties of canine tracheal epithelium in culture

	Isc (μA/cm²)	**PD (mV)**	**G (mS/cm²)**
Baseline	7.6 ± 0.5	2.2 ± 0.4	3.5 ± 0.4
Clarithromycin (M)	7.4 ± 0.7	2.0 ± 0.3	3.7 ± 0.5
Clarithromycin (S)	2.3 ± 0.4 ***	1.0 ± 0.3 **	2.3 ± 0.3 *

*Clarithromycin (100 μM) was added to the mucosal (M) or submucosal solution (S) in Ussing chamber. Values are means ± SE; n = 10. * $P < 0.05$, ** $P < 0.01$, *** $P < 0.001$, significantly different from baseline values. Isc, short-circuit current; PD, potential difference; G, conductance*

sequently, clarithromycin, another 14-membered macrolide, was also found to reduce short-circuit current, transepithelial potential difference, and cell conductance (Tab. 1). This effect was not altered by the Na channel blocker amiloride, but abolished by the Cl channel blocker diphenylamine-2-carboxylate or substitution of Cl in the bathing medium with gluconate, an anion that cannot be transported by airway epithelium. In contrast, the electrical properties were not altered by aminobenzyl penicillin, cefaclor, tetracycline, amikacin, or the 16-membered macrolide josamycin, and slightly reduced by the 15-membered macrolide azithromycin. These *in vitro* findings suggest that 14-membered macrolides may reduce water secretion through a selective inhibition of the airway epithelial Cl channel. Similar findings have also been reported [13, 14], but a discrepancy seems to exist in the concentrations of erythromycin required to produce its *in vitro* and *in vivo* effects. The mean serum concentration following the ingestion of 500 mg erythromycin by the adult volunteers was reported to be 1.6 μM [15], whereas *in vitro* experiments showed that at least 3 μM macrolide is required to decrease Cl secretion. However, because of the species difference, the findings may not necessarily negate its clinical significance. In addition, the serum concentration of macrolide does not accurately reflect the local concentration, since this drug can concentrate intracellularly more than ten-fold [16].

Ikeda and colleagues [17] have shown the effects of antibiotics on Cl secretion by acinar cells isolated from guinea pig nasal gland using a microfluorimetric imaging method and a patch-clamp whole-cell recording. In this experiment, Cl current evoked by acetylcholine was inhibited by roxithromycin and erythromycin but not by josamycin (Figs. 3, 4), indicating again that among macrolides the drugs having 14-membered lactone ring can directly inhibit Cl channel functions. Furthermore, because the acetylcholine-induced Cl current is probably derived from Ca^{2+}-activated Cl channel, this ion channel subtype may be one of the target molecules of macrolides.

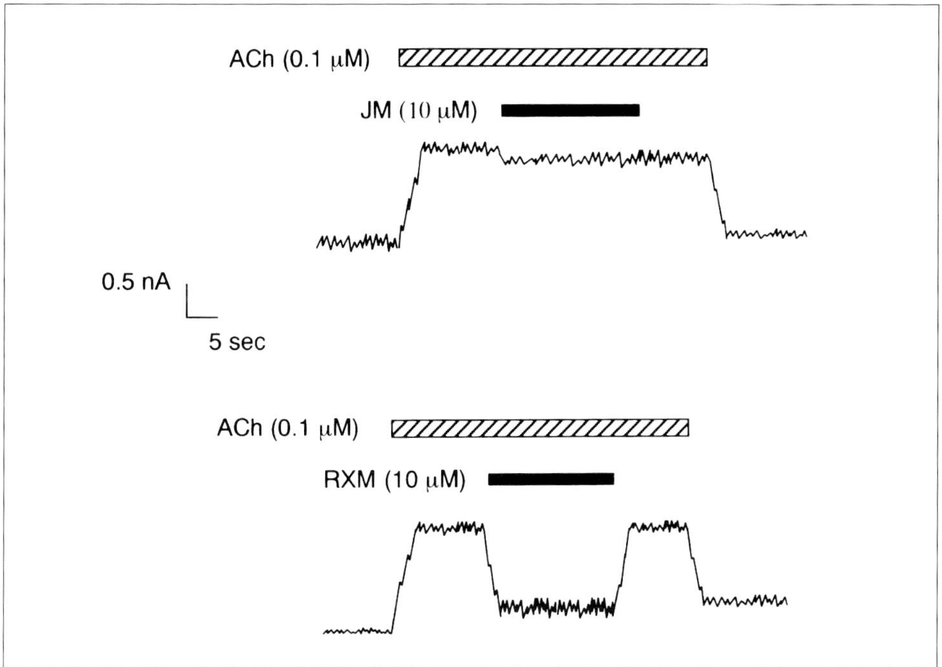

Figure 3
Effects of josamycin (JM) and roxithromycin (RXM) at concentrations of 10 μM on isolated Cl currents in acinar cells of guinea pig nasal gland
After eliciting a control response to acetylcholine (ACh, 0.1 μM), the cells were exposed to each macrolide. The pipette solution was Na-gluconate and the external solution was NaCl without K. A distinct Cl current was isolated when the membrane potential was clamped at 0 mV.

Effects on Cl channel *in vivo*

The effect of macrolides on airway epithelial Cl channel have been studied using excised tissues and cultured cells, but they may not accurately reflect *in vivo* ion transport because of the lack of innervation and blood supply. Therefore, the *in vivo* effects of macrolides on Cl channel were investigated by measuring Cl diffusion potential difference (amiloride-insensitive potential difference, an index of epithelial cellular and paracellular paths available for Cl diffusion) across rabbit tracheal mucosa using a high-impedance voltmeter under open-circuit condition [18] (Fig. 5). As a result, intravenous administration of clarithromycin reduced Cl diffusion potential difference in a dose-dependent manner, whereas aminobenzyl peni-

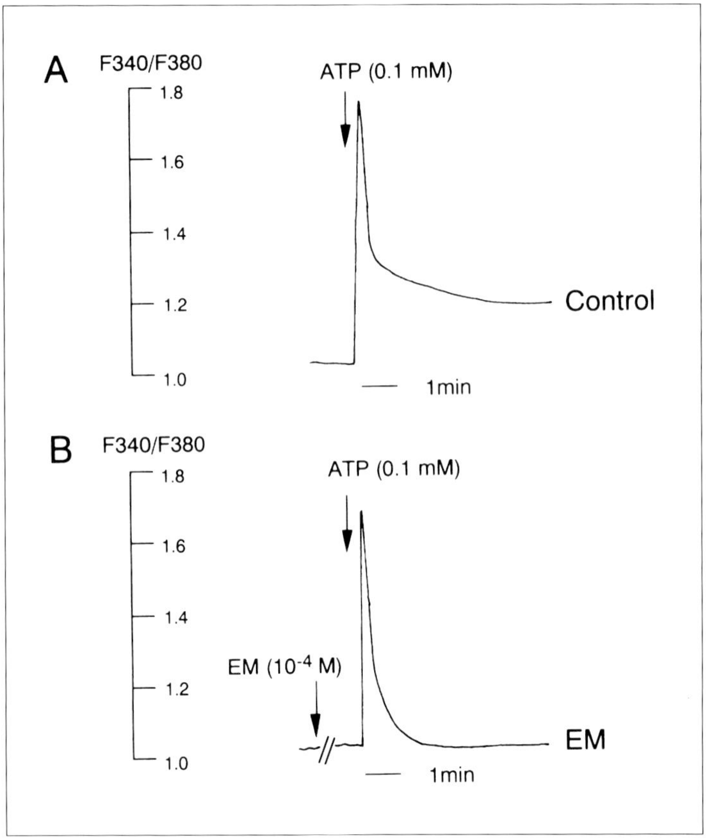

Figure 6
Representative tracings of fura-2 fluorescence ratio in bovine tracheal epithelium exposed to ATP
A, control; B, erythromycin (EM) was added 10 min before ATP. Pretreatment of cells with EM inhibited ATP-induced sustained response but not transient response.

refilling and Ca^{2+} release-activated Ca^{2+} (CRAC) channel that refers to capacitative Ca^{2+} entry or by inactivation of P2X purinoreceptors.

Clinical implication

In clinical studies, the effect of clarithromycin on sputum production and its rheological properties have been examined in patients with chronic lower respiratory tract infections [5]. In this double-blind, placebo-controlled trial, administration of

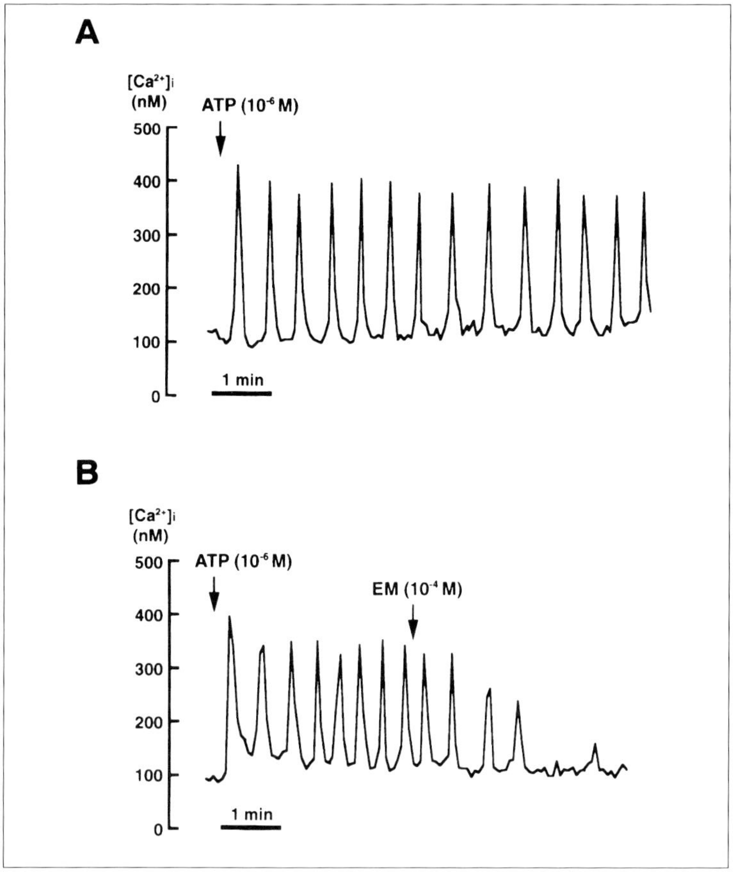

Figure 7
Representative tracings of ATP-induced Ca^{2+} responses in single bovine tracheal epithelium A, low concentration of ATP caused Ca^{2+} oscillations; B, addition of erythromycin (EM) strongly inhibited ATP-induced Ca^{2+} oscillations.

clarithromycin (100 mg twice daily) for 8 weeks almost halved sputum volume, and caused an increase in the percent solids of the sputum, indicating a less hydration. Elastic modulus (G') of the sputum was significantly increased (at 10 Hz), whereas dynamic viscosity (η') remained unchanged. The reduction of sputum production and the corresponding increase in solid composition of the secretions may be associated with the inhibition of airway epithelial Cl secretion. Rubin et al. [23] also showed that in patients with purulent rhinitis clarithromycin (500 mg twice daily) for 2 weeks did not significantly alter sputum viscoelasticity but substantially decreased secretion volume and increased mucociliary transportability. More recent-

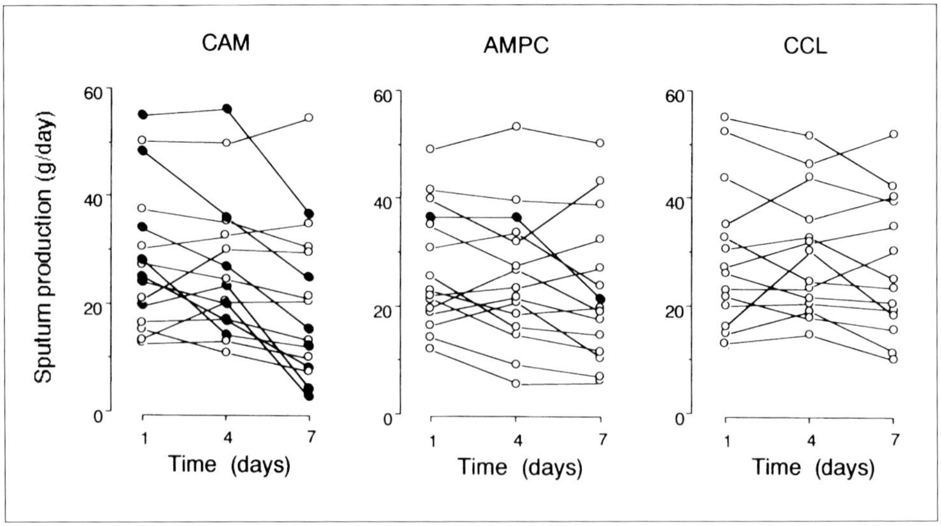

Figure 8
Changes in daily sputum production in patients with chronic bronchitis and bronchiectasis receiving clarithromycin (CAM, n = 16), amoxicillin (AMPC, n = 15) or cefaclor (CCL, n = 14)
Closed circles indicate individuals whose sputum volume decreased by more than 30% of the baseline value (responders).

ly, a double-blind, parallel-group study showed that treatment with clarithromycin (200 mg twice daily) but not by amoxicillin (500 mg three times daily) or cefaclor (250 mg three times daily) for 1 week decreased sputum volume in patients with chronic bronchitis or bronchiectasis without apparent respiratory infection [24] (Fig. 8). Furthermore, this effect was more prominent in the subjects whose sputum Cl concentration was high at the baseline level, and clarithromycin significantly decreased the Cl content at the end of the trial. These results suggest that even short-term administration of macrolide reduces chronic airway hypersecretion, presumably by inhibiting upregulated Cl secretion and the resultant water secretion.

Conclusion

In conclusion, 14-membered macrolides inhibit Cl secretion by airway epithelial Cl channel and Ca^{2+} influx from the extracellular solution. Although subcellular mechanism of these actions warrants further studies, the inhibition of Cl secretion may lead to the reduction of liquid secretion across the airway mucosa toward the lumen, and the inhibition of Ca^{2+} entry may lead to the suppression of inflammatory mediator-induced activation of airway epithelium. It is thus likely that the favorable effects of 14-membered macrolides on chronic airway inflammation might be related, at least in part, to the action on airway epithelial ion channels.

References

1 Nadel JA, Widdicombe JH, Peatfield AC (1985) Regulation of airway secretions, ion transport, and water movement. In: AP Fishman (eds): *The respiratory system*. American Physiological Society, Bethesda, 419–45

2 Suez D, Szefler SJ (1986) Excessive accumulation of mucus in children with asthma: a potential role for erythromycin? A case discussion. *J Allergy Clin Immunol* 77: 330–4

3 Marom ZM, Goswami SK (1991) Respiratory mucus hypersecretion (bronchorrhea): a case discussion. Possible mechanism(s) and treatment. *J Allergy Clin Immunol* 87: 1050–5

4 Kadota J, Sakito O, Kohno S, Sawa H, Mukae H, Oda H, Kawakami K, Fukushima K, Hiratani K, Hara K (1993) A mechanism of erythromycin treatment in patients with diffuse panbronchiolitis. *Am Rev Respir Dis* 147: 153–9

5 Tamaoki J, Takeyama K, Tagaya E, Konno K (1995) Effect of clarithromycin on sputum production and its rheological properties in chronic respiratory tract infections. *Antimicrob Agents Chemother* 39: 1688–90

6 Takizawa H, Ohtoshi T, Kawasaki S, Kohyama T, Sato M, Tanaka M, Kasama T, Kobayashi K, Nakajima J, Ito K (1997) Erythromycin modulates IL-8 expression in human bronchial epithelial cells: studies with normal and inflamed airway epithelium. *Am J Respir Crit Care Med* 156: 266–71

7 Tamaoki J, Takeyama K, Yamawaki I, Kondo M, Konno K (1997) Lipopolysaccharide-induced goblet cell hypersecretion in the guinea-pig trachea: inhibition by macrolides. *Am J Physiol* 272: L15–L19

8 Goswami SK, Kivity S, Marom Z (1990) Erythromycin inhibits respiratory glycoconjugate secretion from human airways *in vitro*. *Am Rev Respir Dis* 141: 72–8

9 Welsh MJ (1987) Electrolyte transport by airway epithelia. *Physiol Rev* 67: 1143–84

10 Boucher RC (1994) State of the art: human airway ion transport. *Am J Respir Crit Care Med* 150: 581–93

11 Widdicombe JH, Kondo M, Mochizuki SJ (1986) Regulation of airway mucosal ion transport. *Int Arch Allergy Appl Immunol* 94: 56–61

12 Tamaoki J, Isono K, Sakai N, Kanemura T, Konno K (1992) Erythromycin inhibits Cl secretion across canine tracheal epithelial cells. *Eur Respir J* 5: 234–8

13 Hirano M, Miwa M, Saito S, Baba R, Takasu A, Iwata S, Hazama A, Okada Y (1998) Effects of macrolides on electrolyte secretion by airway ciliary epithelial cells. *Jpn J Antibiot* 51 (Suppl): 152–4

14 Shinkawa K, Sone S, Takahashi A, Maeda K, Tanoue N, Nakaya Y (2001) Effects of ery-

thromycin and clarithromycin on chloride channels in bronchial epithelial cells. *Jpn J Antibiot* 54 (Suppl): 59–62

15 Anderson R, Fernandes AC, Eftychis HE (1984) Studies on the effects of ingestion of a single 500 mg oral dose of erythromycin stearate on leucocyte motility and transformation and on release *in vitro* of prostaglandin E_2 by stimulated leucocytes. *J Antimicrob Chemother* 14: 41–50

16 Johnson JD, Hand WL, Francis JB, King-Thompson N, Corwin RW (1980) Antibiotic uptake by alveolar macrophages. *J Lab Clin Med* 95: 429–39

17 Ikeda K, Wu D, Takasaka T (1995) Inhibition of acetylcholine-evoked Cl^- currents by 14-membered macrolide antibiotics in isolated acinar cells of the guinea pig nasal gland. *Am J Respir Cell Mol Biol* 13: 449–54

18 Tamaoki J, Takemura H, Tagaya E, Konno K (1995) Effect of clarithromycin on transepithelial potential difference in rabbit tracheal mucosa. *J Infect Chemother* 1: 112–15

19 MacCall E, Li L, Sato H, Shannon TR, Blatter LA, Bers DM (1996) Effects of FK-506 on contraction and Ca^{2+} transients in rat cardiac myocytes. *Circ Res* 79: 1110–21

20 Kanoh S, Kondo M, Tamaoki J, Shirakawa H, Kobayashi H, Nagata N, Konno K (1997) FK506 inhibits ATP-induced intracellular calcium rise in trachaeal epithelium. *Am J Respir Crit Care Med* 155: A609 (Abstr)

21 Kondo M, Konoh S, Tamaoki J, Shirakawa H, Miyazaki S, Nagai A (1998) Erythromycin inhibits ATP-induced intracellular calcium responses in bovine tracheal epithelial cells. *Am J Respir Cell Mol Biol* 19: 799–804

22 Zhao DM, Xue HH, Chida K, Suda T, Oki Y, Kanai M, Uchida C, Ichiyama A, Nakamura H (2000) Effect of erythromycin on ATP-induced intracellular calcium response in A549 cells. *Am J Physiol Lung Cell Mol Physiol* 278: L726–L736

23 Rubin BK, Druce H, Ramirez OE, Palmer R (1997) Effect of clarithromycin on nasal mucus properties in healthy subjects and in patients with purulent rhinitis. *Am J Respir Crit Care Med* 155: 2018–23

24 Tagaya E, Tamaoki J, Kondo M, Nagai A (2002) Effect of a short course clarithromycin therapy on sputum production in patients with chronic airway hypersecretion. *Chest* 122: 213–18

II. Clinical results

The use of macrolides for treatment of diffuse panbronchiolitis

Arata Azuma and Shoji Kudoh

Fourth Department of Internal Medicine, Nippon Medical School, Japan

Introduction

Diffuse panbronchiolitis (DPB) is a chronic airway disease predominantly affecting East Asians. It is pathologically characterized by chronic inflammation diffusely located in the region of respiratory bronchioles and is clinically diagnosed as a special type of sinobronchial syndrome with severe lower airway infection [1]. An unidentified gene in the human leukocyte antigen class I region may predispose Asians to this disease [2]. The prognosis of DPB has improved dramatically over the past 20 years as a result of long-term, low-dose treatment with macrolide antibiotics. The beneficial effects of erythromycin and other 14-membered-ring and 15-membered ring macrolides in the treatment of this disease are considered due to anti-inflammatory rather than antimicrobial mechanisms. Investigations over the past 15 years have revealed many novel effects of macrolides on epithelial cells and inflammatory cells, i.e., neutrophils, lymphocytes, macrophages and dendritic cells. Furthermore, macrolide treatment of DPB has provided new understanding of the pathophysiology and new concepts in the treatment of chronic infectious airway diseases.

Epidemiology, etiology and clinical features of DPB

A nationwide survey in 1980 reported more than 1,000 probable cases of DPB had been collected in Japan [3, 4]. Subsequent clinicopathological conferences extracted 319 clinically definite cases and 82 histologically proven cases of DPB. The male-to-female ratio was 1.4:1, with no remarkable sex predominance noted. Two-thirds of the patients were non-smokers. There was no notable history of inhalation of toxic fumes. According to another previous population-based survey in 1980, the prevalence of physician-diagnosed DPB was 0.00011 among 70,000 employees in the Japanese national railway corporation [5]. Recently, however, the incidence of DBP appears to have decreased.

Antibiotics as Anti-Inflammatory and Immunomodulatory Agents, edited by Bruce K. Rubin and Jun Tamaoki

DPB was also described in other East Asian populations such as the Chinese and Koreans in the 1990s, and there are currently a number of case reports in the literature [6–14], although large surveys have not yet been conducted in China and Korea. Outside Asia, only a limited number of cases have been reported [15–24]; therefore it is currently prudent to conclude that DPB is a chronic airways disease predominantly affecting East Asians.

Neither environmental factors nor infectious agents specific to DPB have been demonstrated thus far [25]. Although the etiology of DPB remains unknown, recent progress in molecular genetics has shed some light on its genetic background. DPB is not a simple genetic disorder, but is considered a multifactorial disease of adulthood. Development of DPB in East Asians, including Asian emigrants, suggests that disease susceptibility may be determined by a genetic predisposition unique to Asians. In fact, human leukocyte antigen (HLA)-B54, an ethnic antigen unique to East Asians, was found to be strongly associated with the disease in Japan [26]. This association was subsequently confirmed at the nucleotide sequence level in a larger case-control study [27]; the odds ratio was 3.4 (95% CI 1.7–7.0). In contrast, Korean patients with DPB exhibited a positive association with another HLA class I antigen, HLA-A11 [28]. Since there is a close relationship between the Japanese and Korean HLA profiles and their genetic background, these observations have raised the possibility that a major disease susceptibility gene is located between HLA-A and HLA-B loci.

Diagnosis

Roentgenographic manifestations

Plain chest X-ray films reveal bilateral, diffuse, small nodular shadows with hyperinflation of the lung. In advanced cases, ring-shaped or tram-line shadows suggesting bronchiectasis appear [29]. High resolution computed tomography (HRCT) is extremely useful for the detection of characteristic pulmonary lesions of DPB (Fig. 1, right) [29–31]. Centrilobular distribution of the lesions is observed, and the disease stage can be evaluated by the number and characteristics of peripheral nodules. In the early stage, only nodular opacities are seen; in later stages, nodules with linear opacities appear that correspond to thickened walls of the second- or third-order bronchial branchings within the secondary pulmonary lobules. In the advanced stage, nodular opacities connected to ring-shaped or ductal opacities develop. In yet more advanced stages, large cystic opacities are accompanied by dilated proximal bronchi exhibiting the appearance of extensive bronchiectasis. These findings suggest that inflammatory lesions may extend from the respiratory bronchioles to the proximal airways.

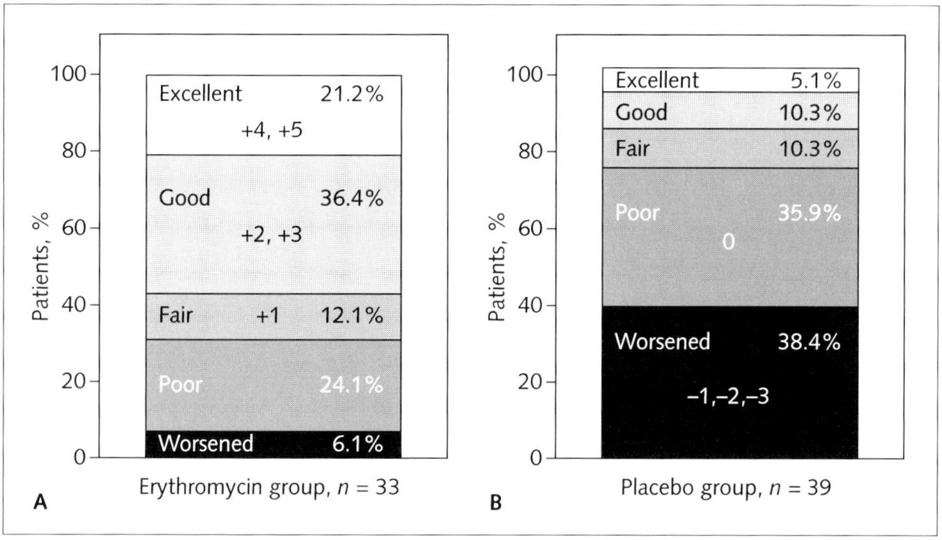

Figure 1
Effects of erythromycin (EM) treatment on diffuse panbronchiolitis: a double-blind study of EM, 600mg/d (A) and placebo (B) for 3 months. Numerical values indicate total scores for six items (degree of dyspnea on exertion, chest radiological findings, PaO2, forced expiratory volume in 1 second, C-reactive protein and sputum volume). 1 point: improved, 0 points: unchanged, –1 point: worsened.

Clinical manifestations

More than 80% of patients have a history of, or are suffering from, chronic paranasal sinusitis [3, 4]. In the second to fifth decade, they usually present with chronic cough and copious purulent sputum production. Exertional dyspnea then develops. Physical examination reveals crackles, wheezes, or both. In half of patients without intervention, sputum volume is greater than 50 ml per day. In a review of 81 histologically proven cases in 1980, 44% had *Hemophilus influenzae* in their sputum at presentation and 22% had *P. aeruginosa* [3, 4]. Less frequently, *Streptococcus pneumoniae* and *Moraxella catarrhalis* are involved in the early stage. The rate of detection of *P. aeruginosa* increases to 60%, on average after four years of monitoring.

Laboratory findings suggest immunological abnormalities and reflect chronic bacterial infection [32]. Cold hemagglutinin titer is continuously increased in most patients without evidence of *Mycoplasma* infection [33]. Serum IgA level is increased, and positive rheumatoid factor is often observed. Other laboratory abnormalities suggesting nonspecific inflammation include mild neutrophilia, increased erythrocyte sedimentation rate, and positive findings for C-reactive protein [3, 4].

Pulmonary function measurements reveal significant airflow limitation relatively resistant to bronchodilators [34]. Decreased forced expiratory volume in 1 second (FEV_1)/forced vital capacity (FVC) less than 70%, decreased vital capacity (VC) less than 80% of predicted value, and residual volume (RV) greater than 50% of predicted value have been used as simple cut-points [3, 4]. Arterial blood gas analysis demonstrates hypoxemia (partial pressure of arterial oxygen [PaO_2] less than 80 mm Hg).

Based on these manifestations of DPB, we usually use the following diagnostic criteria proposed in 1998 [35] by a working group of the Ministry of Health and Welfare of Japan:

1. persistent cough, sputum, and exertional dyspnea
2. history of or current chronic sinusitis
3. bilateral diffuse small nodular shadows on a plain chest X-ray or centrilobular micronodules on chest CT images
4. coarse crackles
5. FEV_1/FVC less than 70% and PaO_2 less than 80 mm Hg
6. cold hemagglutinin titer equal to or higher than 64

Definite cases should meet criteria 1, 2, 3, and at least two of criteria 4, 5, and 6.

Studies of macrolide therapy for DPB in Japan

Over the past 20 years, DPB has changed from a fatal to a curable disease. Before the use of macrolide therapy, the prognosis of patients with DPB was extremely poor. According to a study by a research group of the Ministry of Health and Welfare of Japan in 1981 [36], the 5-year survival rate was approximately 40% from the time of first medical examination. Major bacterial species infecting the airway often changed from *Haemophilus influenzae* to *Pseudomonas aeruginosa* with progression of the disease. After infection by *Pseudomonas*, the prognosis became poor, e.g., the 5-year survival rate in patients with *Pseudomonas aeruginosa* infections was only 8% before entry of macrolide therapy. In 1982, erythromycin was used for the first time to treat DPB. The usefulness of erythromycin for DPB was first suggested by an encounter with one patient with this disease who exhibited marked improvement after treatment with erythromycin. Since at that time it was unusual to cure patients with DPB, we suspected that the medication he had received (600 mg daily erythromycin for > 6 mths) was responsible for his cure. A randomized clinical trial using low-dose erythromycin was therefore begun, and the clinical efficacy of erythromycin in subjects with DPB was first reported in 1984 [37]. In 1987, a paper on 4-years follow-up in treatment of DPB with erythromycin was published [38].

Preliminary Japanese reports from individual institutions revealed at least three important findings that are still accepted today. First, the clinical effects, i.e., decreased amount of sputum and decreases in cough and dyspnea on exertion of macrolide therapy were excellent. Second, there were no significant changes in bacterial species before and after therapy. Third, even in cases involving *Pseudomonas* infection, clinical efficacy was exhibited. In a retrospective study of 52 patients treated with low-dose erythromycin, it was confirmed that this treatment was significantly more effective than that with various conventional agents, i.e., bactericidal beta-lactams, aminoglycosides, new quinolones, or steroids, for 37 patients followed for an average of 43 months [39]. In addition, clinical improvement was excellent after erythromycin therapy as observed over an average follow-up period of 20 months [39]. The Ministry of Health and Welfare research group then performed another retrospective analysis comparing erythromycin therapy with long-term administration of a new quinolone, ofloxacin. In that study as well, erythromycin exhibited better efficacy than treatment with new quinolones [40].

The members of this research group then performed a prospective, double-blind, placebo-controlled study involving daily use of 600 mg of erythromycin for 3 months. In this study, clinical efficacy was evaluated using a scoring system with six items: dyspnea on exertion, amount of sputum, chest radiological findings, arterial oxygen tension, forced expiratory volume in 1 second, and serum C-reactive protein level. The percentage of patients who were "moderately improved" was 57% in the EM group but only 15% in the placebo group. Conversely, the percentages of patients exhibiting aggravation were 6% and 38% in the EM and placebo groups, respectively (Fig. 1) [41]. This double-blind controlled study established the clinical efficacy of erythromycin treatment for DPB in Japan.

Survival curves for DPB patients in three groups based on time at initial diagnosis were significantly improved after the introduction of low-dose erythromycin treatment (Fig. 2) [42]. In the 1970s, the 5-year survival rate was 63%, while between 1980 and 1984 it was 72%. However, following the introduction of erythromycin treatment in 1984, the survival curve significantly improved, with a 5-year survival rate above 90%. Determination of survival rates by patient age showed that erythromycin therapy was more beneficial for older than for younger patients. Furthermore, exploratory analysis of erythromycin therapy revealed no significant difference between survival of patients in the 1970s before erythromycin therapy was established and that of patients who did not receive erythromycin after 1984. Erythromycin treatment has thus clearly contributed to the recent improvement in the prognosis of DPB. Several clinical studies were conducted to confirm the efficacy of macrolides in treatment for DPB. Before long, this favorable effect was confirmed by others [43–46], in all of whose reports clinical efficacy was satisfactory.

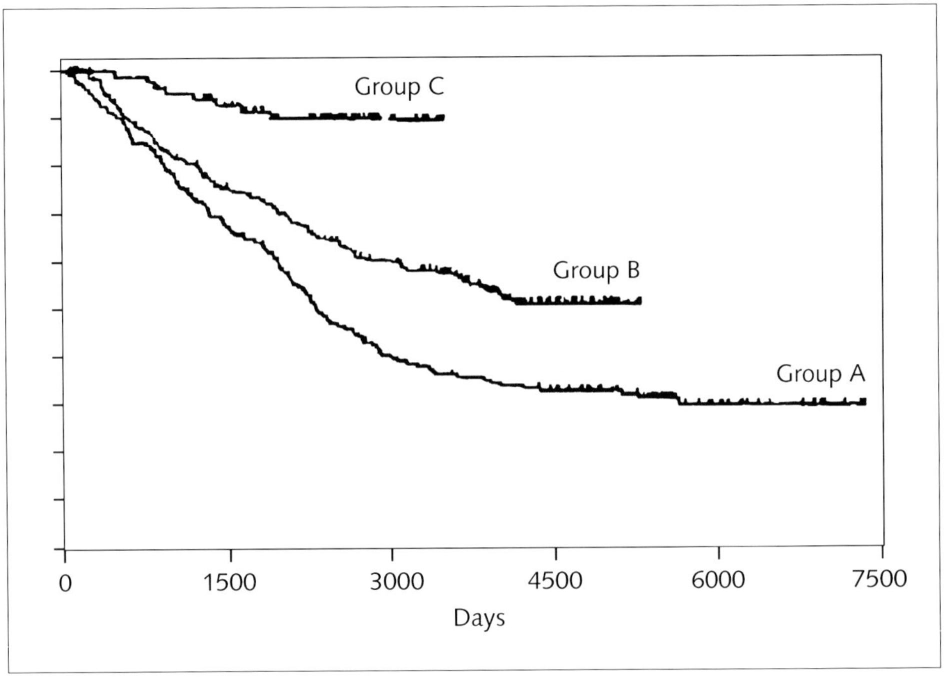

Figure 2
Survival curves according to year of first medical examination for patients with diffuse panbronchiolitis (a:1970-1979, b:1980-1984, c:1985-1990), adapted from Kudoh et al. [8]; with permission

Other macrolides

Recently, 14-membered ring macrolides other than erythromycin have also been used for treatment. Clarithromycin and roxithromycin are semisynthetic 14-membered ring macrolides with modifications in their structures to achieve better gastrointestinal tolerability and tissue penetration than occurs with erythromycin. Clinicians administered these new arrivals for the treatment of DPB in the 1990s and obtained similar clinical benefits [47–50]. These new macrolides were sometimes effective even when erythromycin was not [48]. Azithromycin, a 15-membered ring macrolides had been in limited use in Japan until it was properly available in 2001. It appears to have similar effects on DPB [51], although we have not yet had much experience with its use. Josamycin, a 16-membered ring macrolide, has been ineffective empirically in treating DPB [52].

Recommended treatment protocol

A working group of the Diffuse Lung Disease Committee of the Ministry of Health and Welfare of Japan has proposed clinical guidelines for macrolide therapy for DPB in 2000, based mainly on evidence from the mentioned nonrandomized trials, observational studies, and expert opinion [53]: Macrolides should be started soon after the diagnosis of DPB is made, because clinical response is better in the earlier stage of this disease.

Choice of drug and dose per day

First choice: erythromycin 400 or 600 mg orally. When this is ineffective, it should be stopped because of its adverse effects or drug interactions, and the second choice – clarithromycin 200 or 400 mg orally or roxithromycin 150 or 300 mg orally – should be administered.

Note: 16-membered ring macrolides appear to be ineffective in treating DPB.

Assessment of response and duration of treatment

1. Although clinical response is usually obvious within 2 or 3 months, treatment should be continued for at least 6 months and then overall response should be evaluated.
2. Treatment should be completed after 2 years when clinical manifestations, radiological findings, and pulmonary function measurements are improved and stable without significant impairment of daily activities.
3. Treatment should be restarted if symptoms appear again after cessation of it.
4. When treatment is effective in advanced cases with extensive bronchiectasis or respiratory failure, it should be continued for more than 2 years.

We sometimes use a new quinolone or beta-lactam for a short period, added to a macrolide, when a patient's symptoms progress rapidly with acute exacerbation of DPB.

The role of anti-inflammatory effects of erythromycin in the treatment of chronic airway infection

The change in bacterial species in the sputum of patients with DPB from before to after treatment was investigated (Fig. 3). With conventional treatment with beta-lactams, aminoglycosides, new quinolones, or steroids, counts of *H. influenzae* were

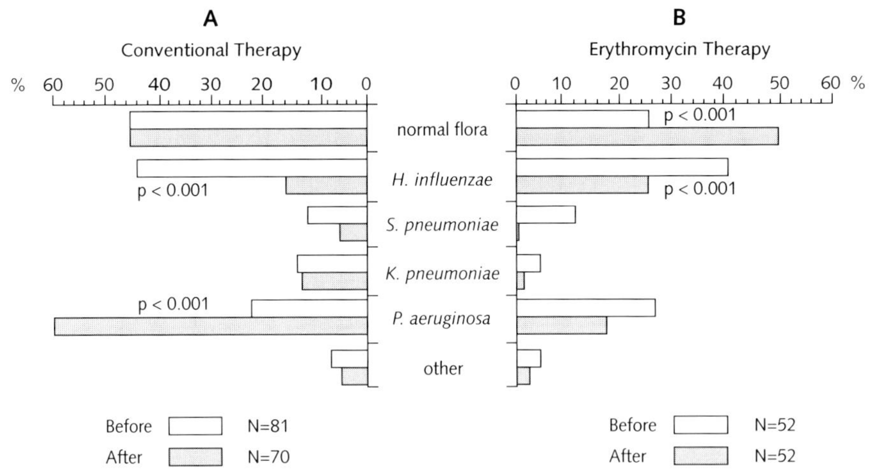

Figure 3
Comparison of bacterial flora in sputum before and after treatment: conventional therapy (A) versus erythromycin treatment (B). (Panel A adapted from Kino et al. [40]; with permission. Panel B adapted from Kudoh et al. [4]; with permission.

decreased and those of *P. aeruginaosa* increased after therapy, but erythromycin therapy reduced counts of both *Haemophilus* and *Pseudomonas* organisms and induced reversion to normal flora. It is believed that the efficacy of bactericidal treatment with antibiotics other than macrolides for chronic airway infection is limited, since chronic airway infection differs from acute airway infection in the type of disease process involved. First, chronic airway infection exists for reasons such as a defect in airway defense mechanisms. Second, chronic airway infection is accompanied by an inflammatory process, which results in the "vicious circle" of effects proposed by Cole [54]. Probably most important is that chronic infection is caused by biofilm organisms. Macrolides may improve prognosis of chronic airway infection by decreasing biofilm formation by *P. aeruginosa*. However, even in the early stage of DPB without *P. aeruginosa* infection, macrolides have been effective for treatment. These results suggest that macrolides primarily exhibit anti-inflammatory effects in addition to decreasing biofilm formation in bacteria.

It is currently believed that erythromycin cuts this vicious circle in chronic airway infection by inhibiting the inflammatory process, as shown in Figure 4. Fur-

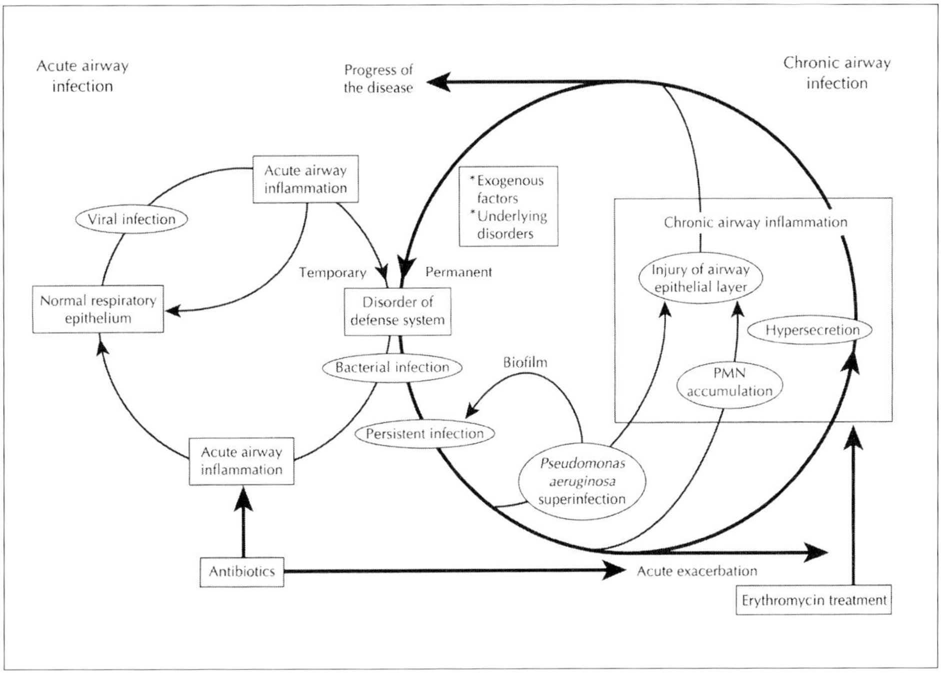

Figure 4
Vicious circle of acute and chronic infection and inflammation of respiratory tract. Erythromycin cuts this circle of chronic airway inflammation.
PMN, polymorphonuclear cells

thermore, it has recently been found that even sub-minimal inhibitory concentrations of 14-membered-ring macrolides exhibit inhibitory effects on biofilm formation and expression of virulence factors (piocyanin, elastase, proteases) of *P. aeruginosa* (see below). Thus, the effects of anti-inflammatory agents like erythromycin are bi-directional, toward host defense mechanisms and bacterial activities.

Macrolides, including erythromycin, are originally bacteristatic antibiotics. In low-dose, long-term erythromycin treatment, however, the mechanism of action of erythromycin involves only bacteristatic effects. First, DPB can improve without elimination of bacteria. Second, improvement can be found even in patients with *P. aeruginosa* infection. Third, the maximal concentration of erythromycin in serum or sputum, which is approximately 1 μg/ml, is lower than the minimum inhibitory concentration for major species of bacteria [55], though the concentration in neutrophils is much higher than this.

Inhibition of hypersecretion

A large amount of sputum is a characteristic manifestation of DPB. Sputum volume markedly decreases after erythromycin therapy. Reduction of sputum volume was the most sensitive parameter of DPB disease activity in the double-blind study mentioned above [41]. Goswami et al. [56] first reported that erythromycin dose-dependently inhibited mucus secretion from airway mucosa *in vitro*, with use of a glycoconjugate marker; however this finding was not reproducible. Tamaoki et al. [57] first observed *in vitro* ion transport in epithelial cells and reported that erythromycin inhibited this transport in dose-dependent fashion when it attached to the serosa. Furthermore, Tamaoki et al. found that this inhibition was due to blockade of chloride channels. Inhibition of mucus and water secretion from epithelial cells may be an important mechanism of action in improving hypersecretion in patients with DPB.

Inhibition of neutrophil activity

Large numbers of neutrophils are found in the bronchoalveolar lavage fluid (BALF) of patients with DPB, frequently reaching 70–80% of BAL cells. After erythromycin treatment, neutrophil elastase levels decrease in both sputum [58] and BALF [59]. Kadota et al. [60] reported that the percentage of neutrophils in BALF markedly decreased after erythromycin therapy, in association with a decrease in neutrophil chemotactic activity. Oishi et al. [61] reported that level of interleukin-8 in BALF markedly decreases along with the number of neutrophils and concentration of neutrophil elastase. Takizawa et al. [62, 63] found that erythromycin dose-dependently inhibited interleukin-8, interleukin-6 and granulocyte-macrophage colony-stimulating factor secretion from epithelial cells *in vitro* using a human airway epithelial cell line. Similar inhibition was found in epithelial cells stimulated by *Haemophilus influenzae* endotoxin [64]. Recently, Desaki et al. [65] reported that erythromycin suppressed activation of nuclear factor (NF)-κB and activator protein-1 in human bronchial epithelial cells and subsequently inhibited interleukin-8 mRNA expression on epithelial cells.

Effects on lymphocytes and macrophages

The characteristic pathological feature of DPB is chronic inflammation with lymphocytes, plasma cells and foamy macrophages in the region of respiratory bronchioles. These foci disappear after erythromycin treatment. A study of bronchoalveolar lavage fluid from a subject with DPB found that the number of memory T cells and activation of $CD8^+$ cells, mainly consisting of cytotoxic T cells, were significantly increased in DPB but were decreased after erythromycin treatment [66]. Keicho et al. [67] reported that proliferation of lymphocytes was inhibited by erythromycin in a dose-dependent fashion, but that erythromycin did not inhibit the expression of interleukin-2 and CD25, a mechanism differing from that of tacrolimus, an immunosuppressant. They concluded that the inhibitory effect of erythromycin on T cells existed in the later activation process, based on the inhibition of T cell responses to interleukin-2.

Furthermore, Keicho et al. [68] and other Japanese investigators have agreed that erythromycin accelerates both the differentiation and proliferation of monocyte–macrophage system cells. Erythromycin has been found to inhibit lipopolysaccharide-induced production of tumor necrosis factor-alpha in human monocytes *in vitro* [69]. However, further clarification of the roles of these effects on lymphocytes and macrophages in eliminating inflammation of the respiratory bronchioles in DPB is needed.

Modulation of bacterial function

Inhibitory effects of macrolides on biofilm formation by *P. aeruginosa*

P. aeruginosa forms a bacterial biofilm by producing alginate when it adheres to mucosa or various medical devices. Macrolide are known to modulate bacterial activity by affecting bioactivity marker of *P. aeruginosa*. Ichimiya et al. reported the effects of certain macrolides on biofilm formation by *P. aeruginosa* [70, 71]. Using scanning electron microscopy (SEM), biofilms on a Teflon sheet with CAM were found to decrease markedly in dose-dependent fashion compared with those on a control Teflon sheet without CAM. Interestingly, this effect was observed even though CAM has no direct bactericidal activity against *P. aeruginosa*. CAM probably inhibited biofilm synthesis *via* inhibition of polysaccharide synthesis. A 1/256-1/64 MIC dose of macrolide inhibited production of alginate, a major component of biofilm, whereas none of RKM, piperacillin, ceftazidime, and ofloxacin inhibited alginate production.

Inhibitory effect of macrolides on the expression of virulence factors of *P. aeruginosa*

The ability of macrolides to inhibit the expression of virulence factors of *P. aeruginosa* at subinhibitory concentrations has been studied.

P. aeruginosa infection is preceded by selective adhesion of bacteria to host target cells via adhesins, including lectins. The production of lectins and of many virulence factors is positively controlled by transcription activators including signaling

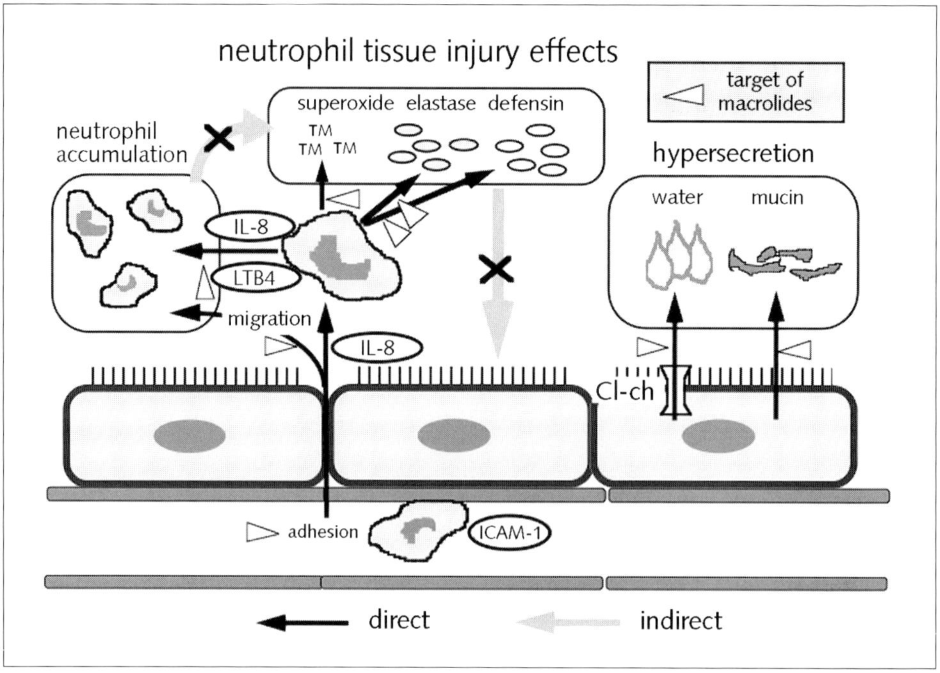

Figure 5
Schematic diagram of airway inflammation and estimated points of action of erythromycin ▷. ICAM-1, intercellular adhesion molecule-1; IL-8, interleukin-8; LTB-4, leukotriene B4.

autoinducers (N-acyl-L-homoserine lactones). Sofer et al. reported that EM at sub-MIC concentrations suppressed the production of *P. aeruginosa* hemagglutinins (including lectins) [72]. In addition, sub-MICs of AZM strongly suppressed the synthesis of elastase, proteases, lecithinase and DNase [73]. CAM and EM were not so effective in doing so. Of these virulence factors, pyocianine, a pigment, is known to suppress superoxide anion production by neutrophils, differentiation and proliferation of lymphocytes, cilliary beating of bronchial epithelial cells, and nitrogen intermediate and cytokine production by alveolar macrophages. EM suppressed the production of pyocianin dose dependently *in vitro* [74]. *In vivo*, the concentration of pyocyanin in sputum from patients with chronic lower respiratory tract infection is reduced after administration of EM.

EM also modulates the effect of piocyanin indirectly. Denning et al. reported that EM suppressed interleukin-8 production by airway epithelial cells induced by piocyanin stimulation [75]. Thus, it is likely that EM prevents the lesion infected with *P. aeruginosa* from tissue injury caused by piocyanin in both direct and indirect fashions.

Effects of erythromycin in the treatment of airway inflammation

The schematic diagram in Figure 5 shows a summary of points of action of erythromycin in the treatment of airway inflammation, prepared based on recent papers. First, erythromycin inhibits hypersecretion due to inhibition of mucus and water secretion from epithelial cells. Second, erythromycin inhibits neutrophil accumulation at sites of inflammation due to inhibition of adhesion of neutrophils to capillary vessels, secretion of interleukin-8 and leukotriene B4 from epithelial cells and neutrophils [76]. These effects reduce the levels of substances injuring tissue, such as elastase and superoxide anion [77], and clearly play important roles in improvement of airway inflammation, although controversies exist concerning the effects of erythromycin on neutrophil activity itself [78–81] and on lymphocytes and macrophages. Furthermore, macrolide inhibited biofilm formation in bacteria and a quorum sensing system (see below) as noted above.

Administration of erythromycin has been established as standard treatment for DPB. Furthermore, erythromycin has been widely used in treating chronic airway inflammation, not only for lower airway diseases (DPB, chronic bronchitis, bronchiectasis, cystic fibrosis and bronchial asthma [82–85]) but also upper airway diseases (chronic sinusitis or exudative otitis media) [86].

Conclusion

In 1994, to begin clarification of the mechanism of action of erythromycin in the treatment of DPB, a study group on the novel effects of macrolides was formally started in Japan. Many clinical and experimental investigations in this field have since been reported; inhibitory effect on bleomycin-induced lung injury in mice, 14-membered-ring macrolides are known to exhibit motilin-like effects on gastrointestinal movement via stimulation of gastrointestinal activity [87, 88]. The anti-inflammatory activities noted here should be considered a third type of effect of 14-membered-ring macrolides. It is expected that the anti-inflammatory effects of erythromycin will be widely beneficial and applied to the treatment of many chronic diseases in the future.

References

1 Homma H, Yamanaka A, Tanimoto S, Tamura M, Chijimatsu Y, Kira S, Izumi T (1983) Diffuse panbronchiolitis: A disease of the transitional zone of the lung. *Chest* 83: 63–9

2 Keicho N, Ohashi J, Tamiya G, Nakata K, Taguchi Y, Azuma A, Ohishi N, Emi M, Park MH, Inoko H et al (2000) Fine localization of a major disease-susceptibility locus for diffuse panbronchiolitis. *Am J Hum Gene* 66: 501–7

3 Izumi T, Doi O, Nobechi A, Homma Y, Kino N, Nakata K, Inatomi K, Homma H (1983) *Annual Report on the study of interstitial lung disease in 1982*. Grant-in Aid from the Ministry of Health and Welfare of Japan, Tokyo, Japan 3–41 [Japanese]
4 Homma H (1986) Diffuse panbronchiolitis. *Jpn J Med* 25(3): 329–34
5 Odaka M, Saito N, Hosoda Y, Chikauchi Y, Morita T, Horikoshi Y, Hanjima T, Chiba Y (1981) *Annual Report on the study of interstitial lung disease in 1980*. Grant-in Aid from the Ministry of Health and Welfare of Japan, Tokyo, Japan 25–8 [Japanese]
6 Kim YW, Han SK, Shim YS, Kim KY, Han YC, Seo JW, Im JG (1992) The first report of diffuse panbronchiolitis in Korea: five case reports. *Intern Med* 31(5): 695–701
7 Chu YC, Yeh SZ, Chen CL, Chen CY, Chang CY, Chiang CD (1992) Diffuse panbronchiolitis: report of a case. *J Formos Med Assoc* 91(9): 912–5
8 Tredaniel J, Cazals-Hatem D, Zalcman G, d'Agay MF, Capron F (1994) Diffuse panbronchiolitis: efficacy of low-dose erythromycin. *Respir Med* 88(6): 479–80
9 Homer RJ, Khoo L, Smith GJ (1995) Diffuse panbronchiolitis in a Hispanic man with travel history to Japan. *Chest* 107(4): 1176–8
10 Hu H, Liu Y, Cai Z, Chen L (1996) A case of diffuse panbronchiolitis. *Chin Med J (Engl)* 109(12): 949–52
11 Zainudin BM, Roslina AM, Fadilah SA, Samad SA, Sufarlan AW, Isa MR (1996) A report of the first three cases of diffuse panbronchiolitis in Malaysia. *Med J Malaysia* 51(1): 136–40
12 Wang H, Sun T, Miao J, Li Y (1998) A definite case of diffuse panbronchiolitis diagnosed by open lung biopsy. *Chin Med J (Engl)* 111(9): 864
13 Tsang KW, Ooi CG, Ip MS, Lam WK, Ngan H, Chan EY, Hawkins B, Ho CS, Amitani R, Tanaka E et al (1998) Clinical profiles of Chinese patients with diffuse panbronchiolitis. *Thorax* 53(4): 274–80
14 Poh SC, Wang YT, Wang WY (2001) Diffuse panbronchiolitis – a case report. *Singapore Med J* 42(6): 271–4
15 Randhawa P, Hoagland MH, Yousem SA (1991) Diffuse panbronchiolitis in North America. Report of three cases and of the literature. *Am J Surg Pathol* 15(1): 43–7
16 Poletti V, Patelli M, Poletti G, Bertanti T, Spiga L (1992) Diffuse panbronchiolitis observed in an Italian male. *Sarcoidosis* 9(1): 67–9
17 Fitzgerald JE, King TE Jr, Lynch DA, Tuder RM, Schwarz MI (1996) Diffuse panbronchiolitis in the United States. *Am J Respir Crit Care Med* 154(2 Pt 1): 497–503
18 Schulte W, Szrepka A, Bauer PC, Guzman J, Costabel U (1999) Diffuse panbronchiolitis. A rare differential diagnosis of chronic obstructive lung disease. *Dtsch Med Wochenschr* 14; 124(19): 584–8 [German]
19 Chantarotorn S, Palwatwichai A, Vattanathum A, Tantamacharik D (1999) Diffuse panbronchiolitis, the first case reports in Thailand. *J Med Assoc Thai* 82(8): 833–8
20 Gulhan M, Erturk A, Kurt B, Gulhan E, Ergul G, Unal P, Capan N (2000) Diffuse panbronchiolitis observed in a white man in Turkey. *Sarcoidosis Vasc Diffuse Lung Dis* 17(3): 292–6

21 Claxton S, Markos J (2000) A case of diffuse panbronchiolitis. *Aust NZ J Med* 30(1): 99–100
22 Martinez JA, Guimaraes SM, Ferreira RG, Pereira CA (2000) Diffuse panbronchiolitis in Latin America. *Am J Med Sci* 319(3): 183–5
23 Fisher MS Jr, Rush WL, Rosado-de-Christenson ML, Goldstein ER, Tomski SM, Wempe JM, Travis WD (1998) Diffuse panbronchiolitis: histologic diagnosis in unsuspected cases involving North American residents of Asian descent. *Arch Pathol Lab Med* 122(2): 156–60
24 Brugiere O, Milleron B, Antoine M, Carette MF, Philippe C, Mayaud C (1996) Diffuse panbronchiolitis in an Asian immigrant. *Thorax* 51(10): 1065–7
25 Corne J (1996) Diffuse panbronchiolitis – a new Japanese export? *Lancet* 30; 348 (9040): 1465–6
26 Sugiyama Y, Kudoh S, Maeda H, Suzaki H, Takaku F (1990) Analysis of HLA antigens in patients with diffuse panbronchiolitis. *Am Rev Respir Dis* 141(6): 1459–62
27 Keicho N, Tokunaga K, Nakata K, Taguchi Y, Azuma A, Bannai M, Emi M, Ohishi N, Yazaki Y, Kudoh S (1998) Contribution of HLA genes to genetic predisposition in diffuse panbronchiolitis. *Am J Respir Crit Care Med* 158(3): 846–50
28 Park MH, Kim YW, Yoon HI, Yoo CG, Han SK, Shim YS, Kim WD (1999) Association of HLA class I antigens with diffuse panbronchiolitis in Korean patients. *Am J Respir Crit Care Med* 159(2): 526–9
29 Akira M, Kitatani F, Lee YS, Kita N, Yamamoto S, Higashihara T, Morimoto S, Ikezoe J, Kozuka T (1988) Diffuse panbronchiolitis: evaluation with high-resolution CT. *Radiology* 168(2): 433–8
30 Nishimura K, Kitaichi M, Izumi T, Itoh H (1992) Diffuse panbronchiolitis: correlation of high-resolution CT and pathologic findings. *Radiology* 184(3): 779–85
31 Hansell DM (2001) Small airways diseases: detection and insights with computed tomography. *Eur Respir J* 17(6): 1294–313
32 Sugiyama Y (1993) Diffuse panbronchiolitis. *Clin Chest Med* 14(4): 765–72
33 Takizawa H, Tadokoro K, Miyoshi Y, Horiuchi T, Ohta K, Shoji S, Miyamoto T, Miyachi S (1986) Serological characterization of cold agglutinin in patients with diffuse panbronchiolitis. *Nihon Kyobu Shikkan Gakkai Zasshi* 24(3): 257–63 [Japanese]
34 Koyama H, Nishimura K, Mio T, Ikeda A, Sugiura N, Izumi T (1994) Bronchial responsiveness and acute bronchodilator response in chronic obstructive pulmonary disease and diffuse panbronchiolitis. *Thorax* 49(6): 540–4
35 Nakata K (1999) *Annual Report on the study of diffuse lung disease in 1998*. Grant-in Aid from the Ministry of Health and Welfare of Japan, Tokyo, Japan 109–111 [Japanese]
36 Kino M (1981) A study on patients with diffuse panbronchiolitis diagnosed by pathological examination. Annual Report of the Research Committee on Interstitial Lung Disease 1981. Ministry of Health and Welfare of Japan, Tokyo, Japan
37 Kudoh S, Kimura H (1984) Clinical effect of low-dose long-term administration of

macrolides on diffuse panbronchiolitis. *Nippon Kyoubu Shikkan Gakkai Zasshi (Jpn J Thorac Dis)* 254 (Suppl): 254

38 Kudoh S, Uetake T, Hagiwara K, Hirayama M, Hus LH, Kimura H, Sugiyama Y (1987) Clinical effects of low-dose long-term erythromycin chemotherapy on diffuse panbronchiolitis. *Nippon Kyobu Shikkan Gakkai Zasshi (Jpn J Thorac Dis)* 25: 632–42

39 Kudoh S, Yamaguchi T, Kurashima A (1988) Long-term therapeutic effects of erythromycin on diffuse panbronchiolitis – A retrospective study. Annual Report of the Research Committee on Interstitial Lung Disease. Ministry of Health and Welfare of Japan, Tokyo, Japan 25–9

40 Yamamoto M, Kondo A, Tamura M, Izumi T, Ina Y, Noda M (1990) Long-term therapeutic effects of erythromycin and new quinolone antibacterial agents on diffuse panbronchiolitis. *Nippon Kyobu Shikkan Gakkai Zasshi (Jpn J Thorac Dis)* 28: 1305–13

41 Yamamoto M, Kudoh S, Ina Y, Tamura A (1990) Clinical efficacy of erythromycin for patients with diffuse panbronchiolitis – A double blind study. *Saishin Igaku* 45: 103–8

42 Kudoh S, Azuma A, Yamamoto M, Izumi T, Ando M (1998) Improvement of survival in patients with diffuse panbronchiolitis treated with low-dose erythromycin. *Am J Respir Critic Care Med* 157: 1829–32

43 Sawaki M, Mikami R, Mikasa K, Kunimatsu M, Ito S, Narita N (1986) The long term chemotherapy with erythromycin in chronic lower respiratory tract infections – first report: comparison with amoxicillin. *Kansenshogaku Zasshi* 60(1): 37–44 [Japanese]

44 Yamamoto M, Kondo A, Tamura M, Izumi T, Ina Y, Noda M (1990) Long-term therapeutic effects of erythromycin and new quinolone antibacterial agents on diffuse panbronchiolitis. *Nihon Kyobu Shikkan Gakkai Zasshi* 28(10): 1305–13 [Japanese]

45 Nagai H, Shishido H, Yoneda R, Yamaguchi E, Tamura A, Kurashima A (1991) Long-term low-dose administration of erythromycin to patients with diffuse panbronchiolitis. *Respiration* 58(3–4): 145–9

46 Fujii T, Kadota J, Kawakami K, Iida K, Shirai R, Kaseda M, Kawamoto S, Kohno S (1995) Long term effect of erythromycin therapy in patients with chronic Pseudomonas aeruginosa infection. *Thorax* 50(12): 1246–52

47 Nakamura H, Fujishima S, Inoue T, Ohkubo Y, Soejima K, Waki Y, Mori M, Urano T, Sakamaki F, Tasaka S et al (1999) Clinical and immunoregulatory effects of roxithromycin therapy for chronic respiratory tract infection. *Eur Respir J* 13(6): 1371–9

48 Mikasa K, Sawaki M, Kita E, Konishi M, Maeda K, Hamada K, Takeuchi S, Masutani T, Sano R, Kunimatsu M et al (1992) Long-term chemotherapy using erythromycin (EM) for chronic lower airway infection: effectiveness of clarithromycin in EM ineffective cases – the fourth report. *Kansenshogaku Zasshi* 66(8): 1097–104 [Japanese]

49 Tamaoki J, Takeyama K, Tagaya E, Konno K (1995) Effect of clarithromycin on sputum production and its rheological properties in chronic respiratory tract infections. *Antimicrob Agents Chemother* 39(8): 1688–90

50 Shirai T, Sato A, Chida K (1995) Effect of 14-membered ring macrolide therapy on chronic respiratory tract infections and polymorphonuclear leukocyte activity. *Intern Med* 34(6): 469–74

51 Kobayashi H, Takeda H, Sakayori S, Kawakami Y, Otsuka Y, Tamura M, Konishi K, Tanimoto S, Fukakusa M, Shimada K et al (1995) Study on azithromycin in treatment of diffuse panbronchiolitis. *Kansenshogaku Zasshi* 69(6): 711–22 [Japanese]

52 Oritsu M (1990) Effectiveness of macrolide antibiotics other than erythromycin. *Therapeutic Research* 11: 545–6 [Japanese]

53 Nakata K, Taguchi Y, Kudoh S (2000) Annual Report on the study of diffuse lung disease in 1999. Grant-in Aid from the Ministry of Health and Welfare of Japan, Tokyo, Japan 111 [Japanese]

54 Cole PCA (1986) "Vicious Circle" hypothesis of the pathogenesis of bronchiectasis. *Eur J Respir Dis* 69: (Suppl): 6–15

55 Nagai H, Shishido H, Yoneda R, Yamaguchi E, Tamura A, Kurashima A (1991) Long-term low-dose administration of erythromycin to patients with diffuse panbronchiolitis. *Respiration* 58: 145–9

56 Goswami SK, Kivity S, Marom Z (1990) Erythromycin inhibits respiratory glycoconjugate secretion from human airway *in vitro*. *Am Rev Respir Dis* 141: 72–8

57 Tamaoki J, Isono K, Sakai N, Kanemura T, Konno K (1992) Erythromycin inhibits Cl secretion across canine tracheal epithelial cells. *Eur Respir J* 5: 234–8

58 Mikami M (1992) Clinical and pathophysiological significance of neutrophil elastase in sputum and the effect of erythromycin in chronic respiratory diseases. *Nippon Kyoubu Shikkan Gakkai Zasshi (Jpn J Thorac Dis)* 29: 72–83

59 Ichikawa Y, Ninomiya H, Koga, Tanaka M, Kinoshita M, Tokunaga N, Yano T, Oizumi K (1992) Erythromycin reduces neutrophils and neutrophil-derived elastolytic-like activity in the lower respiratory tract of bronchiolitis patients. *Am Rev Respir Dis* 146: 196–203

60 Kadota J, Sakito O, Kohno S, Sawa H, Mukae H, Oda H, Kawakami K, Fukushima K, Hiratani K, Hara K (1993) A mechanism of erythromycin treatment in patients with diffuse panbronchiolitis. *Am Rev Respir Dis* 147: 153–9

61 Oishi K, Sonoda F, Kobayashi S, Matsumoto K (1994) Role of interleukin-8 (IL-8) and an inhibitory effect of erythromycin on IL-8 release in the airways of patients with chronic airway diseases. *Infect Immunity* 62: 4145–52

62 Takizawa H, Desaki M, Ohtoshi T, Kawasaki S, Kohyama T, Sato M, Tanaka M, Kasama T, Kobayashi K, Nakajima J et al (1995) Erythromycin suppresses interleukin 6 expression by human bronchial epithelial cells: A potential mechanism of its anti-inflammatory action. *Biochem Biophys Res Commun* 210: 781–6

63 Takizawa H, Desaki M, Ohtoshi T, Kawasaki S, Kohyama T, Sato M, Tanaka M, Kasama T, Kobayashi K, Nakajima J et al (1997) Erythromycin modulates IL-8 expression in normal and inflamed human bronchial epithelial cells. *Am J Respir Critic Care Med* 156: 266–271

64 Khair OA, Devalia JL, Abdelaziz MM, Sapsford RJ, Davis RJ (1995) Effect of erythromycin on *Hemophilus influenzae* endotoxin-induced release of IL-6, IL-8 and sICAM-1 by cultured human bronchial epithelial cells. *Eur Respir J* 8: 1451–7

65 Desaki M, Takizawa H, Ohtoshi T, Kasama T, Kobayashi K, Sunazuka T, Omura S,

Yamamoto K, Ito K (2000) Erythromycin suppresses nuclear factor-kappaB and activator protein-1 activation in human bronchial epithelial cells. *Biochem Biophys Res Commun* 267: 124–8
66 Kawakami K, Kadota J, Iida K, Fujii T, Shirai R, Matsubara Y, Kohno S (1997) Phenotypic characterization of T cells in bronchoalveolar lavage fluid (BALF) and peripheral blood of patients with diffuse panbronchiolitis; the importance of cytotoxic T cells. *Clin Exp Immunol* 107: 410–6
67 Keicho N, Kudoh S, Yotsumoto H, Akagawa KS (1993) Antilymphocytic activity of erythromycin distinct from that of FK506 or cyclosporin A. *J Antibiotics* 46: 1406–13
68 Keicho N, Kudoh S, Yotsumoto H, Akagawa KS (1994) Erythromycin promotes monocyte to macrophage differentiation. *J Antibiotics* 47: 80–9
69 Iino Y, Toriyama M, Kudo K, Natori Y, You A (1992) Erythromycin inhibition of lipopolysaccharide-stimulated tumour necrosis factor-alpha production by human monocytes *in vitro*. *Ann Otol Rhinol Laryngol* 101: 16–20
70 Kobayashi H (1995) Airway biofilm disease: clinical manifestations and therapeutic possibilities using macrolides. *J Infect Chemother* 1: 1–15
71 Ichimiya T, Yamasaki T, Nasu T (1994) *In vitro* effects of antimicrobial agents on Pseudomonas aeruginosa biofilm formation. *J Antimicrob Chemother* 34: 331–41
72 Sofer D, Gilboa-Garber N, Belz A, Garber NC (1999) Subinhibitory erythromycin represses production of Pseudomonas aeruginosa lectins, autoinducer and virulence factors. *Chemotherapy* 45: 335–41
73 Molinari G, Guzman CA, Pesce A, Schito GC (1993) Inhibition of *Pseudomonas aeruginosa* virulence factors by subinhibitory concentrations of azithromycin and other macrolide antibiotics. *J Antimicrob Chemother* 31: 681–8
74 Sato K, Suga M, Nishimura J, Kushima Y, Muranaka H, Ando M (1997) Pyocyanine synthesis by Pseudomonas aeruginosa in chronic airway infection and the effect of erythromycin on its biological activity. *Jpn J Antibiot* 50 (Suppl): 89–91
75 Denning GM, Wollenweber LA, Railsback MA, Cox CD, Stoll LL, Britigan BE (1998) Pseudomonas pyocyanin increases interleukin-8 expression by human airway epithelial cells. *Infect Immun* 66: 5777–84
76 Sakito O, Kadota J, Kohno S, Abe K, Shirai R, Hara K (1996) Interleukin 1 beta, tumor necrosis factor alpha, and interleukin 8 in bronchoalveolar lavage fluid of patients with diffuse panbronchiolitis: A potential mechanism of macrolide therapy. *Respiration* 63: 42–8
77 Umeki S (1993) Anti-inflammatory action of erythromycin. Its inhibitory effect on neutrophil NADPH oxidase activity. *Chest* 104: 1191–3
78 Nelson S, Summer WR, Terry PB, Warr GA, Jakab GJ (1987) Erythromycin induced suppression of pulmonary antibacterial defenses. *Am Rev Respir Dis* 136:1207–12
79 Hirata T, Matsunobe S, Matsui Y, Kado M, Mikiya K, Oshima S (1990) Effect of erythromycin on the generation of neutrophil chemiluminescence *in vitro*. *Nippon Kyoubu Shikkan Gakkai Zasshi (Jpn J Thorac Dis)* 28: 1066–71

80 Oda H, Kadota JI, Kohno S, Hara K (1994) Erythromycin inhibits neutrophil chemotaxis in bronchoalveoli of diffuse panbronchiolitis. *Chest* 106: 1116–23

81 Hojo M, Fujita I, Hamasaki Y, Miyazaki M, Miyazaki S (1994) Erythromycin does not directly affect neutrophil functions. *Chest* 105: 520–3

82 Suez D, Szefler SJ (1986) Excessive accumulation of mucus in children with asthma: a potential role for erythromycin? A case discussion. *J Allergy Clin Immunol* 77: 330–4

83 Miyatake H, Taki F, Taniguchi H, Suzuki R, Takagi K, Satake T (1991) Erythromycin reduces the severity of bronchial hyperresponsiveness in asthma. *Chest* 99: 670–3

84 Konno S, Asano K, Kurokawa M, Ikeda K, Okamoto K, Adachi K (1994) Anti-asthmatic activity of a macrolide antibiotic, roxithromycin: analysis of possible mechanisms *in vitro* and *in vivo*. *Int Arch Allergy Immunol* 105: 308–16

85 Saiman L, Marshall BC, Mayer-Hamblett N, Burns JL, Quittner AL, Cibene DA, Coquillette S, Fieberg AY, Accurso FJ, Campbell PW 3rd; Macrolide Study Group (2003) Azithromycin in patients with cystic fibrosis chronically infected with *Pseudomonas aeruginosa*: a randomized controlled trial. *JAMA* 290(13): 1749–56

86 Enomoto F, Andou I, Nagaoka I, Ichikawa G (2002) Effect of new macrolides on the expression of adhesion molecules on neutrophils in chronic sinusitis. *Auris Nasus Larynx* 29: 267–9

87 Kondo Y, Torii K, Omura S, Itoh Z (1988) Erythromycin and its derivatives with motilin-like biological activities inhibit the specific binding of 125I-motilin to duodenal muscle. *Biochem Biophys Res Commun* 150: 877–82

88 Peeters TL (1993) Erythromycin and other macrolides as prokinetic agents. *Gastroenterology* 105: 1886–99

Macrolides in cystic fibrosis

Adam Jaffé[1] *and Andrew Bush*[2]

[1]Portex Respiratory Medicine Unit, Great Ormond Street Hospital for Children NHS Trust & Institute of Child Health, Great Ormond Street, London WC1N 3JH, UK
[2]Department of Paediatric Respiratory Medicine, Royal Brompton and Harefield NHS Trust, London, UK

Introduction

Cystic fibrosis (CF) is a multisystem disease, affecting primarily the respiratory, gastrointestinal and genitourinary systems. The average age of survival is approximately 30 years with death usually due to respiratory failure. The remarkable similarities between CF and diffuse pan-bronchiolitis (DPB), as discussed in the chapter by Azuma and Kudoh, have led to interest in the use of macrolides in this condition. This chapter will begin by discussing the pathophysiology of CF and then discuss the potential anti-inflammatory and other mechanisms of macrolides which are relevant to CF. Finally the clinical evidence for macrolide use in CF will be reviewed.

Pathophysiology

Approximately 1 in 25 white people carry the abnormal gene, which is localised to the long arm of chromosome 7. It encodes for the cystic fibrosis transmembrane conductance regulator (CFTR), which is a cAMP regulated chloride channel. In addition, CFTR regulates other membrane conductance pathways such as the outwardly rectifying chloride channel, amiloride sensitive sodium channel (ENaC) and basolateral potassium channels. It is also thought to be involved in osmoregulation and in the transport of bicarbonate and may regulate airway surface liquid (ASL) pH.

Although much is known of the molecular pathology of CF, precisely how the defective CFTR results in lung disease is not known. There are a number of mechanisms postulated whereby CFTR abnormalities might predispose to damage to the airways.

Airway surface liquid hydration hypothesis

The tenacious sputum typically seen in patients may be a result of poor hydration of the ASL due to abnormal chloride secretion and hyperabsorption of sodium (the low volume hypothesis). The resultant airway plugging leads to defective ciliary clearance and bacterial infection, which induces inflammatory responses and destruction of the surrounding lung tissue.

Defensin hypothesis

It has been shown *in vitro* that high sodium and chloride concentrations inactivate human β-defensin (HBD)-1, a salt sensitive naturally occurring antimicrobial peptide present in the airway surface [1, 2]. In the high-salt theory, mutant CFTR leads to high sodium and chloride in the ASL and subsequent inactivation of human β-defensin-1, as well as other peptides such as lactoferrin and lysozyme. However, the composition of the airway fluid in CF patients is controversial. If true *in vivo*, then this would partially explain why bacteria are able to infect and multiply in the airways of patients with CF. In addition, HBD-1 and HBD-2 expression is not upregulated in CF epithelium in response to inflammatory stimuli suggesting an intrinsic defect in gene expression [3]. In contrast, β-defensins, particularly HBD-2 levels, are high in BALF of patients with DPB and are thought actively to participate in antimicrobial defence in the respiratory tract [4, 5]. Treatment with macrolides reduce bronchoalveolar lavage fluid (BALF) HBD-2 in DPB. Defensins may therefore be a marker for inflammation and its response to treatment in DPB [5]. The effect of macrolides on HBD in CF is not known.

Primary inflammation hypothesis

Some investigators suggest that infection precedes inflammation in CF [6], but this is controversial. Most studies suggest that the baby born with CF has normal lungs at birth; however, neutrophils and IL-8 have been detected in the lower airway of babies as young as 4 weeks without evidence of infection [7] suggesting that inflammation precedes infection in CF. This hypothesis has been supported by animal studies using CF mice in pathogen-free environments [8] and severe combined immunodeficiency mice who have had subcutaneous implants of CF fetal trachea [9]. It has been suggested that accumulation of protein in the endoplasmic reticulum may lead to calcium release and activation of the transcription factor NF-κB, which in turn stimulates interleukin-8 (IL-8) expression [10]. The relevance to CF is suggested by studies demonstrating endogenous activation of this inflammatory pathway by endoplasmic reticulum overloaded with mutant CFTR [11, 12] which occurs with Class 2 mutations such as ΔF508 [13]. Some studies have demonstrated an increase in basal secretion of IL-8 by CF cells (both epithelial and neutrophil) [14–16], while others have demonstrated an exaggerated secretion following stimulation of the CF cell with stimuli such as TNF-α [17] and *Pseudomonas aeruginosa* (*P. aeruginosa*).

However, these findings are not always consistent and tend to vary between individuals and with the different model systems used to study these effects [18, 19]. At present it is unclear which is the initiating factor in the inflammatory cascade although it is likely, but not certain, that the CF cell exists in a proinflammatory state.

The most characteristic feature of inflammation in the CF lung is the presence of a large number of neutrophils in the airway. This results in an inflammatory cascade with subsequent lung damage [20]. In addition to contributing to the tenacious sputum in CF, neutrophils are also responsible for the production of elastase, oxidants and proteases. The elastase digests elastin in the airway wall, one of many factors resulting in bronchiectasis. It directly causes an increase in mucus secretion, thus worsening airway obstruction and promotes the generation of IL-8 and LTB_4 which are potent chemoattractants, thus recruiting more neutrophils and perpetuating the cycle of inflammation and lung destruction.

Decreased mucin secretion theory

Mucus is a protective coating secreted by the healthy airway. Mucin glycoproteins are the major constituent of the mucous gel, which is responsible for the rheological properties of mucus. The respiratory mucins are under the control of at least eight mucin (MUC) genes. In CF, the sputum is composed of a mixture of mucin polymers, inflammatory cells, inflammatory mediators, DNA from inflammatory cell necrosis and bacteria. It has been previously speculated that the mucus plugging in CF is due to mucus hypersecretion. In order to address this issue, Henke et al. investigated the properties of expectorated sputum in patients with CF and chronic bronchitis [21]. They demonstrated greater mucin-like glycoprotein in those with chronic bronchitis and greater DNA in CF sputum suggesting that DNA probably has a much greater effect on CF sputum properties than mucins. Furthermore, they demonstrated a substantial reduction in the gel forming mucins MUC5AC and MUC5B in the CF airways relative to normal mucus. They suggested that this might be due to a relative increase in other components of CF sputum, or perhaps a primary defect in mucin secretion in CF.

Cell-receptor hypothesis

Adherence of *Pseudomonas* to airway epithelial cells is critical for establishing infection. The mechanism for this is controversial. Davies et al. have demonstrated *Pseudomonas* binding by a specific cell-surface receptor, the asialylated glycoprotein, asialoGM1 [22]. Recently, investigators have shown that binding of *P. aeruginosa* to this receptor induces interleukin (IL)-8 secretion *via* the nuclear factor (NF) κB signalling pathway [23]. CFTR itself has been shown to be a receptor for *P. aeruginosa*, and after binding there is internalisation and destruction of the bacterium [24]. An alternative hypothesis is that the abnormal pH of the ASL [25, 26] results in an increase in asialoGM1 receptors which act as binding sites for *P. aeruginosa* [27, 28]. This would explain the preferential infection of CF airways by *P. aeruginosa*. However, recent evidence suggest that at least at a later stage, these bacteria form hypoxic macrocolonies in the airway, and are not in direct contact with the epithelial cells [29]. It is likely that chronic infection of the airways is a process in which initial adherence to epithelial cells is one important step. Factors initiating infection are probably different from those perpetuating chronic infection.

Clinical features in cystic fibrosis

Most patients with CF die from lung disease, which is compounded by recurrent and persistent infection with *Staphylococcus aureus*, *Haemophilus influenzae* and ultimately chronic infection with *P. aeruginosa* in more than 80% of patients. This is similar to DPB where many different bacterial species may initially infect the airway, but *P. aeruginosa* ultimately sets up a chronic infection of the airway with associated biofilm formation. Another similarity between CF and DPB is that sinus disease is a common feature. In CF, this may be associated with nasal polyps and cause chronic rhinitis and headaches necessitating surgical removal of polyps in some instances. Other clinical pathological features characteristic of CF, such as pancreatic insufficiency, distal ileal obstructive syndrome and infertility, are not seen in patients with DBP.

Because there has only been recent confirmation of clinical effectiveness of macrolides in CF, there is a paucity of work exploring the mechanisms of action in CF. Most of the work has been done in patients with DPB, models of DPB or with *in vitro* systems, which have been extrapolated to CF.

Proposed mechanisms of action in CF

Signalling pathways and chemokine release

An early step in the inflammatory process is the signalling to effector cells *via* proinflammatory molecules of the various cytokine and chemokine families. As discussed above, neutrophils predominate in the airways of patients with CF and activation results in parenchymal lung damage through the production of elastase. Rather than having a direct modulating effect on the neutrophil itself, macrolides may influence neutrophil chemotactic activity indirectly by affecting chemoattractants such as IL-8. Work in animal models and in patients with DPB has demonstrated a reduction in neutrophil influx into the airway following treatment with erythromycin. It has been shown that macrolides modulate the production of specific cytokines such as interleukin (IL)-1β, IL-6, IL-8 [4, 30–35], TNF-α [36, 37] and granulocyte-macrophage colony stimulating factor [32]. It is likely that NF-κB is required for transcription of all these cytokines.

An interesting finding with regard to CF is that blood neutrophils demonstrate different cytokine profiles from those found in the CF airway [15]. Corvol et al. demonstrated that IL-8 spontaneously released from CF airway and blood neutrophils was significantly higher than controls. Furthermore, CF airway neutrophils produced more IL-8 compared to the CF blood neutrophils. LPS did not enhance cytokine release in either type of neutrophils. Interestingly, dexamethasone was able to reduce IL-8 production by CF blood neutrophils but had no effect on airway neutrophil cytokine production. These observations may have some relevance to the effects of macrolides in CF.

Whether the anti-inflammatory mechanisms of macrolides are relevant to CF is unknown, but one important possibility is that IL-8 secretion is modulated by these antibiotics. In an open study, treatment with low dose erythromycin for one month in six CF patients decreased IL-8 in sputum [38]. Reduced IL-8 expression in cells obtained from bronchoscopy was also demonstrated in five adult patients with DPB, asthma and bronchiectasis following macrolide therapy [34]. The same group demonstrated a reduction in mRNA levels and IL-8 release in cultured media containing erythromycin or clarithromycin, suggesting that macrolides exert a direct effect on the airway epithelial cell. Of potential clinical relevance to CF, Wallwork et al. demonstrated that clarithromycin was as effective as prednisolone in reducing the production of IL-5, IL8 and granulocyte-macrophage colony-stimulating factor in nasal tissue cultured from patients with chronic rhinosinusitis [39]. There is *in vitro* evidence to suggest that macrolides may repress IL-8 gene expression by suppression of both activator protein-1 (AP-1) binding sites and the transcription factor NF-κB [40–42]. This may be particularly important in CF if there is an intrinsic increase in NF-κB activation in CF epithelial cells both in a basal state and following stimulation with *P. aeruginosa*. This hypothesis is further supported by a study from Escotte et al. who demonstrated that fluticasone inhibited constitutive and LPS-induced IL-6 and IL-8 production *via* the Iκ-α/β kinase pathway in both CF and non-CF bronchial epithelial cells [43]. The effect of macrolides on the signalling pathway, particularly NF-κB, in CF deserves further attention.

Direct neutrophil effect

In addition to the indirect effects of macrolides on neutrophils discussed above, there is evidence that macrolides may affect neutrophil function in many ways.

Endothelial and airway adhesion

Intercellular adhesion molecule (ICAM)-1, plays an important role in the adhesion of neutrophils to airway epithelium [44] and is overexpressed in CF airway epithelium [18, 45, 46], a process which may be regulated by NF-κB. Thus any treatment which reduces neutrophil adherence to either epithelial or endothelial cells may downregulate the inflammatory cascade. To support this, studies on cultures of human bronchial epithelial cells stimulated with *Haemophilus influenzae* endotoxin have shown that erythromycin causes a reduction in IL-6, IL-8, soluble ICAM-1 and decreased neutrophil migration and adhesion to epithelial cells [31]. Using a similar model, Kawasaki et al. demonstrated that roxithromycin inhibited neutrophil adhesion to epithelial cells [32]. Furthermore, macrolides such as erythromycin reduce the expression of integrins CD11b/CD18 in neutrophils stimulated by lipopolysaccharide (LPS) and inhibit their oxidative burst [47]. Another study demonstrated a significant reduction of CD11b/CD18 on the surface of whole blood cells following treatment with erythromycin [48]. In a model using cultured fibroblasts, clarithromycin decreased expression of several adhesion molecules such as ICAM-1, vascular cell adhesion molecule (VCAM)-1 and lymphocyte function-associated antigen-3 (LFA-3). Importantly, Li et al. recently demonstrated that erythromycin inhibited VCAM-1 mRNA and neutrophil airway infiltration in a mouse model of lung fibrosis and therefore may have a role in the prevention of fibrosis [49].

Migration

As discussed previously, the increased presence of neutrophils in CF airway may increase lung damage. Macrolides may have an effect on migration either directly by affecting secretion of chemoattractants such as IL-8, or they may have a direct effect on neutrophils.

In the only study to date in CF, Brennan et al. demonstrated that CF-derived blood neutrophils had significantly increased migration to IL-8 compared to non-CF neutrophils [50]. They subsequently treated eight CF children with a 4-week course of oral erythromycin and demonstrated no effect on neutrophil migration. However, it is likely that the length of treatment was too short to exert an anti-inflammatory effect; and secondly, it is known that newer macrolides such as clarithromycin or azithromycin exert a greater anti-inflammatory effect.

Elastase production and oxidant burst

Various studies have demonstrated impairment of the oxidative burst of neutrophils both *in vitro* [51–55] and *in vivo* by different classes of macrolides [56]. Macrolides may have a role in preventing superoxide generation by neutrophils and thus limit lung tissue damage in children with lung conditions in which there is a predominantly inflammatory and oxidant component such as CF.

Apoptosis

Aoshiba et al. demonstrated that erythromycin increased cyclic AMP levels in neutrophils *in vitro* which led to acceleration of apoptosis at 24 h in a dose dependent manner [57]. This was confirmed *in vivo* by Culic et al. who administered a 3 day course of azithromycin to 12 healthy subjects and demonstrated a time dependent increase in apoptosis in peripheral blood neutrophils, which was still ongoing 4 weeks following the last dose [56]. While not extensively investigated, it is thought that CF neutrophil function, including apoptosis differ from those of healthy subjects. The effect of macrolides on apoptosis in CF neutrophils remains to be examined.

Effect on *Pseudomonas aeruginosa*

An interesting finding is that the clinical improvement in patients with DPB is independent of whether they are chronically infected with *P. aeruginosa*. Since the effect is below the minimum inhibitory concentration (MIC) for *P. aeruginosa* it has been suggested that the effect of macrolides in DPB on *Pseudomonas* is anti-inflammatory rather that antibacterial [30, 58, 59]. There are many potential mechanisms by which macrolides may affect *Pseudomonas* in CF.

Adherence

It has been suggested that the oropharyngeal barrier is an innate host defence mechanism to prevent *Pseudomonas* colonising and infecting the airway. In a clinical trial, Baumann et al. investigated the effects of azithromycin on adherence of *P. aeruginosa* to buccal epithelial cells [60]. They demonstrated a 70% decrease of adherence following oral azithromycin given twice weekly for 3 months in 11 children with CF, and concluded that azithromycin may prevent early infection with *Pseudomonas*. This clinical study confirms previous *in vitro* suggestions that macrolides decrease *Pseudomonas* adherence to silicon filters [61], acid damaged murine trachea [62] and human type IV basement collagen [63, 64].

Mucoid conversion

Preliminary data suggest that treatment with azithromycin impairs the change from non-mucoid to the mucoid phenotype in BALB/c mice infected with alginate embedded *P. aeruginosa* [65]. This finding needs to be confirmed but may have implications for the treatment of patients with CF.

Biofilm and mucus rheology

As discussed elsewhere in this book, mucoid *P. aeruginosa* produces a biofilm from alginate production, which makes eradication difficult. It has been suggested that in CF, this

monas in the lower airway [93]. Erythromycin stimulates endogenous NO production by a protein kinase A dependant mechanism [94]. However, inducible NO synthase production is suppressed by erythromycin and clarithromycin, probably *via* AP-1 and NF-κB [95]. In addition erythromycin causes release of NO from non-adrenergic, non-cholinergic neurones, a system thought to modulate airway inflammation [96]. It is unknown to what extent macrolides affect NO production in the CF airway.

Bronchoconstriction

As described above, ET-1, a potent vasoconstrictor, may be modulated by macrolides [97]. This may help in expectorating sputum in CF patients. Interestingly, in a recent long-term study discussed later in this Chapter, 17% of CF subjects experienced wheezing following azithromycin [98]. The mechanism is not clear, but the authors have suggested that less viscous mucus in the airway exacerbated wheeze.

Airway remodelling

The airway changes seen with disease progression such as dilatation, fibrosis and neovascularisation may be modulated by macrolides. 14-member macrolides appear to reduce tumour angiogenesis by an unknown mechanism [99] and therefore it is possible that bronchial neovascularisation in CF could be reduced. Roxithromycin inhibits TNF-α induced vascular endothelial growth factor (VEGF) production [100]. Angiogenesis may also be inhibited indirectly *via* effects on IL-8, which seems to be angiogenic as well as proinflammatory. A rapamycin analogue inhibited epidermal growth factor induced proliferation in a murine model of lung inflammation and remodelling [101]. If this effect were the same in the human lower airway, this could have profound implications for prevention of remodelling, a process which occurs in children with CF [102].

Bioactive phospholipids

It is possible that macrolides protect epithelial cells from a number of toxic and proinflammatory lipids. Cell injury causes the release of cell membrane phospholipid derived arachidonic and is converted into platelet activating factor (PAF), leukotrienes, prostaglandins and thromboxane A_2. Many of these membrane-derived phospholipids modulate direct and leukocyte mediated damaging effects on the epithelium of the airway, which are inhibited by macrolides [103]. Ketolides have been shown to be cytoprotective against the effects of bioactive phospholipids, lysophosphatidylcholine, PAF and lyso-PAF on nasal epithelial strips obtained from normal volunteers [104]. It is unknown to what extent macrolides protect airway epithelium in CF.

Antibacterial effects in CF

Typical infections

In addition to *Pseudomonas*, organisms such as *Staphylococcus aureus* and *Haemophilus influenzae* regularly infect the lower airway in patients with CF. Macrolides are potentially beneficial in CF due to their broad-spectrum antibacterial properties. Interestingly, no clinical trial in CF has yet demonstrated a clinically significant effect on microbiology, despite evidence for clinical improvement [98, 105].

Atypical infections

Non-tuberculous mycobacterial (NTM) pulmonary infections are increasingly recognised in patients with CF. As macrolides are part of many standard treatment regimens for these organisms, it has been suggested that this might be one mechanism of a beneficial action. This is compounded by the fact that true NTM infection may be difficult to distinguish from sputum contamination.

Clinical evidence in CF

Efficacy

The index case, which sparked our interest in the potential of macrolides in CF, was a 17 year old male with ΔF508/GF551D genotype [106].

From the age of 15 he began to deteriorate clinically with poor lung function and low oxygen saturation necessitating oxygen therapy. At 16 years, he was put on the waiting list for a heart-lung transplant. Because of reports of the effectiveness of macrolide use in panbronchiolitis in Japan, azithromycin 500 mg daily was commenced. In the following months he improved spectacularly with trebling of lung function, from forced vital capacity (FVC) 840 mls (26%) to 2,420 mls (65%), forced expiratory volume in one second (FEV_1) from 300 mls (11% predicted) to 940 mls (26%), and an increase in oxygen saturation breathing air from 65% to 93%. He subsequently came off the heart–lung transplant waiting list, and remained off for a further 6 years until he subsequently required lung transplantation.

Subsequently, in an open study, children with CF who had end-stage CF lung disease or chronic airflow limitation unresponsive to conventional therapy were treated with long-term azithromycin [107]. In this study, children were treated with at least 3 months of daily azithromycin. Lung function in the 6 months before treatment was compared to their post treatment average, together with oxygen saturation and height and weight z-scores. Seven children (median age 12.1 year [range 5.8 to 16.8]) were studied, all of whom were chronically infected with *P. aeruginosa*. Median treatment length was 6 months (range 0.3–1.2). The FVC% rose by median 11.3% (–5.5 to 24.7) from 62.8 (33.9 to 95.9) to 70.3 (58.3 to 112.6). The median FEV_1% also rose by 11.0% (–3.6 to 13.4, $p < 0.03$) from 47.5 (12.2 to 75.4) to 49.5 (23.2 to 88.8). Clearly, the limitation of this study was that it did not have a control group, however, historical controls and clinical experience (with all the danger of relying on such suboptimal comparisons) suggested deterioration would be more likely if only conventional treatment had been given. No other significant differences were observed. The improvements seen were similar to those reported in patients with DPB treated with long-term macrolides [30, 58]. Similar cases were reported from other centres throughout the world in abstract form [108–110].

Ordõnez et al. reported a single blinded pilot study with clarithromycin [111]. They treated 10 CF patients (aged 12–26 years) with placebo for 3 weeks followed by 6 weeks of clarithromycin. In addition to measuring pulmonary function following the two treatment arms, they measured inflammatory markers in induced sputum. They demonstrated no improvement in lung function following clarithromycin and concluded that it is not effective in improving airway obstruction. In addition they demonstrated no change in sputum neutrophil numbers, IL-8, free neutrophil elastase, TNF-α and myeloperoxidase. They could not identify a correlation between inflammatory markers and lung function. However, it is likely that clarithromycin was not given for sufficiently long, or to enough patients, to exert any detectable effect in this study. It is recognised that in DPB, clinical benefit is often not seen until at least 6 weeks of therapy. They did report that one subject had an improvement of 800 ml (11%) following treatment but did not report other individual responses. This might suggest that macrolides exert their effects in an, as yet undefined, subset of patients, something seen in the initial study [107], and subsequent work [98, 105].

In 2002, Wolter and colleagues published the first randomised double-blind placebo controlled study in adults [112]. Sixty subjects were randomised to receive a placebo or 250 mg azithromycin daily for 3 months. The overall difference in change in FEV_1% predicted was significantly better in the treatment group. Over the treatment period, the placebo group had a significantly greater decline in FEV_1% (–3.62%) and FVC% (–5.73%) compared to the azithromycin group in which lung function was maintained. In addition, the azithromycin group had fewer intravenous antibiotic courses and fewer infective exacerbations. The quality of life and dyspnoea scores significantly improved in the azithromycin group, especially in patients with *S. aureus* in their sputum. Overall, there was no change in bacteriology following treatment. They also demonstrated that C-reactive protein (CRP) levels in the placebo arm remained constant, but fell significantly in the azithromycin group. However, unfortunately by chance, the azithromycin group contained more women with overall worse lung function than the placebo group necessitating adjustments in the statistical modelling. The relevance of this is that the fall in CRP was strongly related to the baseline CRP, which was higher in the treatment group. Because of this, it is difficult to draw firm conclusions of the effect of azithromycin on CRP. In addition, because the two groups were not matched, the azithromycin group having poorer lung function, the implication of this being that azithromycin may have a role in those patients with worse lung disease. Furthermore, the very rapid decline in lung function in the placebo group makes the data difficult to interpret. The authors recognised that treatment of patients with severe and permanent lung disease may limit any potential benefit from anti-inflammatory treatment. They recommended trials in children to allow for stratification of results based on severity of lung disease.

In a study which addressed this very question, Equi et al. conducted the first randomised, placebo-controlled study of azithromycin in children with CF [105]. The study lasted for 15 months. Subjects received daily azithromycin (bodyweight ≤ 40 kg: 250 mg ≥ 40 kg: 500 mg) or placebo for 6 months. Following a 2-month washout period, treatments were then crossed over. The following were exclusion criteria: abnormal clotting; abnormal liver function tests three times laboratory upper limit; history of deafness in patient or first degree relative; previous *Burkholderia cepacia*; organ transplantation; oral steroids in the preceding 2 weeks; commencement of rhDNase in the 2 months prior to recruitment. Previous chronic infection with *P. aeruginosa* was not a specific entry criterion. Subjects were reviewed at nine specific visits. The primary outcome measure was the relative change in FEV_1 between the active and placebo groups. Subjects underwent a 3 min standardised 15 cm step exercise test, quality of life assessment and hearing tests. Sputum and cough swabs were cultured for the common CF pathogens. In addition, sputum was assessed for total IL-8 and neutrophil elastase.

The median relative difference between azithromycin and placebo was 5.4% in percentage predicted FEV_1 (95% CI 0.8–10.5). Of note, there was marked individual variation. Thirteen of 41 patients had an improvement in FEV1 by more than 13%. Five patients had deterioration in FEV_1 of 13%. The azithromycin limb demonstrated an increase in mean percentage predicted FVC at every time point with a median relative difference of 3.9% between the two groups.

Because of *in vitro* evidence of an interaction between dornase alpha and azithromycin [113], the results were analysed for an effect of dornase alpha. It should be noted that this was a retrospective analysis suggested by an anonymous peer reviewer for *The Lancet*, not a predetermined analysis. The median relative difference between azithromycin and placebo for FEV_1 was 11.5% (95% CI 5.3–16.5) for the 26 patients not treated with dornase alpha and –3.6% (–22 to 3.9) for those receiving dornase alpha. However, there was marked individual variation. In addition, it is likely that those patients on dornase alpha represented the group who were chronically infected with *P. aeruginosa* and were likely to have worse lung function. In addition, the study was not powered to detect any subgroup effects. It is difficult, therefore, to conclude that the lack of benefit in lung function in the dornase alpha group was solely due to the possible drug interaction. In addition, this effect was not seen in the USA study published subsequently [98]. Further investigation is needed before any firm recommendation not to combine azithromycin and dornase alpha can be sustained.

In addition to the effect on lung function, the azithromycin group had significantly fewer courses of oral antibiotics. No affect was seen on sputum IL-8 and neutrophil elastase, microbiological profiles and 3 min step test. There was no difference in quality of wellbeing scores, in contrast to Wolter's study. In summary, this study demonstrated a significant improvement in lung function following 6 months of daily azithromycin and the authors concluded that a 4–6 month trial is justified in those children with CF who do not respond to conventional treatment. Despite evidence of clinical improvement, both the optimal dosing regimen and mechanisms of action remain unclear.

To assess the effect of azithromycin taken three times per week, the Cystic Fibrosis Foundation sponsored a study incorporating 23 centres throughout the USA [98]. It was a double-blind randomised placebo-controlled trial in both children (≥ 6 years) and adults chronically infected with *P. aeruginosa* with FEV_1% ≥ 30%. Subjects received placebo or azithromycin (≤ 40 kg: 250 mg ≥ 40 kg: 500 mg) three times per week for 6 months. 185 subjects were recruited and 87 were randomised to the treatment limb. They demonstrated a significant relative change in FEV_1% predicted of 6.2% (0.094 litres), and 5.00% in FVC% predicted between the two groups, similar to that seen in the studies by Wolter and Equi, which disappeared 4 weeks after ceasing the study. The azithromycin group had a less risk of experiencing an exacerbation (defined by the need for intravenous antibiotics or at least 7 days of quinolones), with a reduction in the number of subjects hospitalised. In addition, the treatment group gained 0.7 kg more than the placebo group. While there was no significant improvement in total CF quality of life scores, there was a trend to improvement in the treatment group. No effect on elastase and IL-8 was seen similar to Equi et al. Although not as yet published, subsequent subgroup analysis has suggested that, in contrast to the study by Equi et al., no effect of dornase alpha was seen (Saiman, NACF Conference, Anaheim 2003). Furthermore, those patients homozygous for ΔF508 had the most marked response. As a result of this study, the Cystic Fibrosis Foundation is now recommending the use of macrolides in CF patients over the age of 6 who are chronically infected with *P. aeruginosa* (www.CFF.org).

To date there has been one further published study which reviewed one UK CF centre's experience with azithromycin [114]. In their 21 months experience, 20 adults with chronic *P. aeruginosa* and declining lung function (> 10% fall in FEV_1 over 12 months) were commenced on daily azithromycin 250 mg for more than 3 months. Comparisons were made with 20 patients, all chronically infected with *P. aeruginosa*, who had stable lung function (< 5% change in FEV_1 and FVC) over the previous year. They demonstrated a significant increase in FEV_1% predicted from 50.2% to 59.1% ($p = 0.001$) and FVC% predicted from 64.5% to 76.1% ($p = 0.002$). The control group demonstrated a small but insignificant fall in lung function. In addition, the treated group gained more weight than controls (3.9 kg *versus* 1.3 kg ($p = 0.040$) and had a 48% reduction in the frequency of intravenous antibiotics. As this was not a controlled study, the limitations are similar to the work we initially reported as described above [107].

At present there is little evidence to guide dosage of macrolides in patients with CF. Our own personal practice is to consider azithromycin in patients who continue to deteriorate despite conventional CF treatment. Because of the marked individual variation, patients receive daily azithromycin for 6 months and then reviewed. If no improvement is seen then the drug is stopped. If there is an improvement in the clinical state then the options are: a) stop treatment; b) continue every day; c) reduce to three times per week.

Safety

While the clinical trials have demonstrated another potential therapy for patients with CF, it is important with any new therapeutic development that safety remains paramount. There is now definite clinical evidence of benefit, yet the mechanisms by which this occurs in CF is unknown. In addition, the dosing regimen is unknown. Because of the very long half-life of azithromycin, it continues to accumulate within body tissue and does not plateau. In the study by Ordõnez et al., no subjects had an increase in liver function tests during the 6 weeks of clarithromycin [111]. One subject withdrew during the 3-week placebo treatment due to gastrointestinal complaints. A few subjects suffered minor gastrointestinal symptoms during the treatment period. In Wolter's study of daily azithromycin for 3 months, 16 adverse effects were seen in 15 patients causing three subjects to discontinue treatment [112] – seven of these were in the placebo group. One subject withdrew because of an urticarial reaction thought "likely" to be related to azithromycin. Neutropenia in another subject in the treatment limb and "swelling" in a subject in the placebo limb were considered "possibly" related to the study drug. Two further events (one in each limb) of a drug rash were considered "possibly" to be related to the study drug. In Equi's study in children, there were no subjective reports of side effects [105]. Twelve of 190 subjects failed hearing tests in minor ways but were normal when tested 2 months later. One subject had a rise in aspartate transaminase and alanine transaminase at the end of 6 months of azithromycin. After stopping the drug, the enzyme levels halved. No adverse effects were seen on clotting.

In the American study, there were 85 (96%) adverse effects (21 serious) in the azithromycin group, and 94 (96%) in the placebo group (32 serious) [98]. Most serious side effects were related to CF exacerbations. More subjects in the azithromycin group exhibited nausea, diarrhoea and wheezing (17% *versus* 4%) despite improved lung function. It is unknown why this last symptom occurred; it had not been reported in previous studies. It is possible that mobilisation of less viscous mucus into the airway may have exacerbated wheeze, although mucus rheology was not specifically studied. It is also possible that an increase in airflow caused wheeze, as seen in seriously ill asthmatics that begin to wheeze following improvement with therapy. As discussed above, macrolides reduce ET-1 expression and thus may augment bronchodilatation which makes the reported wheezing even more puzzling. No difference in haematology, liver function or hearing was seen between the two groups.

Thus it is seen that both daily and three times per week azithromycin is both effective and safe, at least in the medium term. It is unknown if treatment beyond 6 months is safe although in our practice, CF patients have been treated with azithromycin for many years without side effects; similarly, safe, long-term treatment is reported in HIV positive children treated for non-tuberculous mycobacterial infection.

Future research directions

From the preceding discussion, it is clear that azithromycin is beneficial in at least some individuals with CF. However many questions remain unanswered. In order to refine treatment strategies, it is important to unravel which are the important underlying anti-inflammatory and potential antibacterial mechanisms in CF. This might allow the testing of "designer macrolides" with enhanced activity in those areas. In addition, the emergence of resistant organisms directly linked to increase macrolide use highlights the need for microbiological surveillance [115].

It is clear that there is great individual variability in the response to macrolides in CF and future research should attempt to identify the factors that predict a good response. Potential factors include age, genotype (particularly ΔF508), chronic infection with *P. aeruginosa*, type of *P. aeruginosa*, presence of other common CF pathogens e.g., *Staphylococcus aureus*, concomitant therapy e.g., dornase alpha. Another important area, which deserves further work, is identification of the optimal dosing regimen in order to minimise potential adverse reactions.

Until these questions are answered, we believe that azithromycin has the potential to improve clinical status in those patients not responding to conventional CF treatment. In these cases, we recommend instigating a 6-month trial of therapy. Our personal practice is to commence with daily treatment for 6 months, and if there is an objective improvement, consider a reduction to three times per week for prolonged periods.

An increased understanding both of why particular patients do and do not respond and how macrolides work at the cellular and molecular level will ultimately assist in the development of specifically targeted, novel macrolides as therapy for individual CF patients in the future.

Conflict of interest

Adam Jaffé has previously been sponsored by Pfizer and has been awarded a Pfizer Academic Travel Scholarship for his work in CF Gene Therapy. Andrew Bush and Adam Jaffé have a research program funded in part by Pfizer. Pfizer manufacture azithromycin.

References

1 Smith JJ, Travis SM, Greenberg EP, Welsh MJ (1996) Cystic fibrosis airway epithelia fail to kill bacteria because of abnormal airway surface fluid. *Cell* 85: 229–36

2 Goldman MJ, Anderson GM, Stolzenberg ED, Kari UP, Zasloff M, Wilson JM (1997) Human β-Defensin-1 is a salt-sensitive antibiotic in lung that is inactivated in cystic fibrosis. *Cell* 88: 553–60

3 Dauletbaev N, Gropp R, Frye M, Loitsch S, Wagner TO, Bargon J (2002) Expression of human beta defensin (HBD-1 and HBD-2) mRNA in nasal epithelia of adult cystic fibrosis patients, healthy individuals, and individuals with acute cold. *Respiration* 69: 46–51

4 Ashitani J, Mukae H, Nakazato M, Ihi T, Mashimoto H, Kadota J, Kohno S, Matsukura S (1998) Elevated concentrations of defensins in bronchoalveolar lavage fluid in diffuse panbronchiolitis. *Eur Respir J* 11: 104–11

5 Hiratsuka T, Mukae H, Iiboshi H, Ashitani J, Nabeshima K, Minematsu T, Chino N, Ihi T, Kohno S, Nakazato M (2003) Increased concentrations of human β-defensins in plasma and bronchoalveolar lavage fluid of patients with diffuse panbronchiolitis. *Thorax* 58: 425–30

6 Armstrong DS, Grimwood K, Carzino R, Carlin JB, Olinsky A, Phelan PD (1995) Lower respiratory infection and inflammation in infants with newly diagnosed cystic fibrosis. *BMJ* 310: 1571–2

7 Khan TZ, Wagener JS, Bost T, Martinez J, Accurso FJ, Riches DW (1995) Early pulmonary inflammation in infants with cystic fibrosis. *Am J Respir Crit Care Med* 151: 1075–82

8 Zahm JM, Gaillard D, Dupuit F, Hinnrasky J, Porteous D, Dorin JR, Puchelle E (1997)

Early alterations in airway mucociliary clearance and inflammation of the lamina propria in CF mice. *Am J Physiol* 272: C853–C859

9 Tirouvanziam R, de Bentzmann S, Hubeau C, Hinnrasky J, Jacquot J, Peault B, Puchelle E (2000) Inflammation and infection in naive human cystic fibrosis airway grafts. *Am J Respir Cell Mol Biol* 23: 121–7

10 Pahl HL, Sester M, Burgert HG, Baeuerle PA (1996) Activation of transcription factor NF-kappaB by the adenovirus E3/19K protein requires its ER retention. *J Cell Biol* 132: 511–22

11 DiMango E, Ratner AJ, Bryan R, Tabibi S, Prince A (1998) Activation of NF-kappaB by adherent *Pseudomonas aeruginosa* in normal and cystic fibrosis respiratory epithelial cells. *J Clin Invest* 101: 2598–605

12 Weber AJ, Soong G, Bryan R, Saba S, Prince A (2001) Activation of NF-kappaB in airway epithelial cells is dependent on CFTR trafficking and Cl- channel function. *Am J Physiol Lung Cell Mol Physiol* 281: L71–L78

13 Miele L, Cordella Miele E, Xing M, Frizzell R, Mukherjee AB (1997) Cystic fibrosis gene mutation (deltaF508) is associated with an intrinsic abnormality in Ca^{2+}-induced arachidonic acid release by epithelial cells. *DNA Cell Biol* 16: 749–59

14 Tabary O, Escotte S, Couetil JP, Hubert D, Dusser D, Puchelle E, Jacquot J (2001) Relationship between IkappaBalpha deficiency, NFkappaB activity and interleukin-8 production in CF human airway epithelial cells. *Pflugers Arch* 443 (Suppl 1): S40–S44

15 Corvol H, Fitting C, Chadelat K, Jacquot J, Tabary O, Boule M, Cavaillon JM, Clement A (2003) Distinct cytokine production by lung and blood neutrophils from children with cystic fibrosis. *Am J Physiol Lung Cell Mol Physiol* 284: L997–1003

16 Eidelman O, Srivastava M, Zhang J, Leighton X, Murtie J, Jozwik C, Jacobson K, Weinstein DL, Metcalf EL, Pollard HB (2001) Control of the proinflammatory state in cystic fibrosis lung epithelial cells by genes from the TNF-alphaR/NFkappaB pathway. *Mol Med* 7: 523–34

17 Venkatakrishnan A, Stecenko AA, King G, Blackwell TR, Brigham KL, Christman JW, Blackwell TS (2000) Exaggerated activation of nuclear factor-kappaB and altered IkappaB-beta processing in cystic fibrosis bronchial epithelial cells. *Am J Respir Cell Mol Biol* 23: 396–403

18 Aldallal N, McNaughton EE, Manzel LJ, Richards AM, Zabner J, Ferkol TW, Look DC (2002) Inflammatory response in airway epithelial cells isolated from patients with cystic fibrosis. *Am J Respir Crit Care Med* 166: 1248–56

19 Pizurki L, Morris MA, Chanson M, Solomon M, Pavirani A, Bouchardy I, Suter S (2000) Cystic fibrosis transmembrane conductance regulator does not affect neutrophil migration across cystic fibrosis airway epithelial monolayers. *Am J Pathol* 156: 1407–16

20 Davis PB, Drumm M, Konstan MW (1996) Cystic fibrosis. *Am J Respir Crit Care Med* 154: 1229–56

21 Henke MO, Renner A, Huber RM, Seeds MC, Rubin BK (2004) MUC5AC and MUC5B mucins are decreased in cystic fibrosis airway secretions. *Am J Respir Cell Mol Biol* 31: 86–91

22 Davies J, Dewar A, Bush A, Pitt T, Gruenert D, Geddes DM, Alton EW (1999) Reduction in the adherence of *Pseudomonas aeruginosa* to native cystic fibrosis epithelium with anti-asialoGM1 antibody and neuraminidase inhibition. *Eur Respir J* 13: 565–70

23 Li J, Johnson XD, Iazvovskaia S, Tan A, Lin A, Hershenson MB (2003) Signaling intermediates required for NF-kappa B activation and IL-8 expression in CF bronchial epithelial cells. *Am J Physiol Lung Cell Mol Physiol* 284: L307–L315

24 Pier GB, Grout M, Zaidi TS, Olsen JC, Johnson LG, Yankaskas JR, Goldberg JB (1996) Role of mutant CFTR in hyper susceptibility of cystic fibrosis patients to lung infections. Science 271: 64–7

25 Poschet JF, Boucher JC, Tatterson L, Skidmore J, Van Dyke RW, Deretic V (2001) Molecular basis for defective glycosylation and *Pseudomonas* pathogenesis in cystic fibrosis lung. *Proc Natl Acad Sci USA* 98: 13972–7

26 Imundo L, Barasch J, Prince A, Al Awqati Q (1995) Cystic fibrosis epithelial cells have a receptor for pathogenic bacteria on their apical surface. *Proc Natl Acad Sci USA* 92: 3019–23

27 Saiman L, Prince A (1993) *Pseudomonas aeruginosa* pili bind to asialoGM1 which is increased on the surface of cystic fibrosis epithelial cells. *J Clin Invest* 92: 1875–80

28 Davies JC, Stern M, Dewar A, Caplen NJ, Munkonge FM, Pitt T, Sorgi F, Huang L, Bush A, Geddes DM et al (1997) CFTR gene transfer reduces the binding of *Pseudomonas aeruginosa* to cystic fibrosis respiratory epithelium. *Am J Respir Cell Mol Biol* 16: 657–63

29 Worlitzsch D, Tarran R, Ulrich M, Schwab U, Cekici A, Meyer KC, Birrer P, Bellon G, Berger J, Weiss T et al (2002) Effects of reduced mucus oxygen concentration in airway *Pseudomonas* infections of cystic fibrosis patients. *J Clin Invest* 109: 317–25

30 Sakito O, Kadota J, Kohno S, Abe K, Shirai R, Hara K (1996) Interleukin 1 beta, tumor necrosis factor alpha, and interleukin 8 in bronchoalveolar lavage fluid of patients with diffuse panbronchiolitis: a potential mechanism of macrolide therapy. *Respiration* 63: 42–8

31 Khair OA, Devalia JL, Abdelaziz MM, Sapsford RJ, Davies RJ (1995) Effect of erythromycin on *Haemophilus influenzae* endotoxin-induced release of IL-6, IL-8 and sICAM-1 by cultured human bronchial epithelial cells. *Eur Respir J* 8: 1451–7

32 Kawasaki S, Takizawa H, Ohtoshi T, Takeuchi N, Kohyama T, Nakamura H, Kasama T, Kobayashi K, Nakahara K, Morita Y et al (1998) Roxithromycin inhibits cytokine production by and neutrophil attachment to human bronchial epithelial cells *in vitro*. *Antimicrob Agents Chemother* 42: 1499–502

33 Mukae H, Kadota J, Ashitani J, Taniguchi H, Mashimoto H, Kohno S, Matsukura S (1997) Elevated levels of soluble adhesion molecules in serum of patients with diffuse panbronchiolitis. *Chest* 112: 1615–21

34 Takizawa H, Desaki M, Ohtoshi T, Kawasaki S, Kohyama T, Sato M, Tanaka M, Kasama T, Kobayashi K, Nakajima J et al (1997) Erythromycin modulates IL-8 expression in normal and inflamed human bronchial epithelial cells. *Am J Respir Crit Care Med* 156: 266–71

35 Yamanaka Y, Tamari M, Nakahata T, Nakamura Y (2001) Gene expression profiles of human small airway epithelial cells treated with low doses of 14- and 16-membered macrolides. *Biochem Biophys Res Commun* 287: 198–203
36 Iino Y, Toriyama M, Kudo K, Natori Y, Yuo A (1992) Erythromycin inhibition of lipopolysaccharide-stimulated tumor necrosis factor alpha production by human monocytes *in vitro*. *Ann Otol Rhinol Laryngol* (Suppl) 157: 16–20
37 Schultz MJ, Speelman P, Zaat S, van Deventer SJ, van der Poll T (1998) Erythromycin inhibits tumor necrosis factor alpha and interleukin 6 production induced by heat-killed *Streptococcus pneumoniae* in whole blood. *Antimicrob Agents Chemother* 42: 1605–9
38 Everard ML, Sly P, Brenan S, Ryan G (1997) Macrolide antibiotics in diffuse panbronchiolitis and in cystic fibrosis [letter]. *Eur Respir J* 10: 2926
39 Wallwork B, Coman W, Feron F, Mackay-Sim A, Cervin A (2002) Clarithromycin and prednisolone inhibit cytokine production in chronic rhinosinusitis. *Laryngoscope* 112: 1827–30
40 Abe S, Nakamura H, Inoue S, Takeda H, Saito H, Kato S, Mukaida N, Matsushima K, Tomoike H (2000) Interleukin-8 gene repression by clarithromycin is mediated by the activator protein-1 binding site in human bronchial epithelial cells. *Am J Respir Cell Mol Biol* 22: 51–60
41 Kikuchi T, Hagiwara K, Honda Y, Gomi K, Kobayashi T, Takahashi H, Tokue Y, Watanabe A, Nukiwa T (2002) Clarithromycin suppresses lipopolysaccharide-induced interleukin-8 production by human monocytes through AP-1 and NF-kappa B transcription factors. *J Antimicrob Chemother* 49: 745–55
42 Desaki M, Takizawa H, Ohtoshi T, Kasama T, Kobayashi K, Sunazuka T, Omura S, Yamamoto K, Ito K (2000) Erythromycin suppresses nuclear factor-kappaB and activator protein-1 activation in human bronchial epithelial cells. *Biochem Biophys Res Commun* 267: 124–8
43 Escotte S, Tabary O, Dusser D, Majer-Teboul C, Puchelle E, Jacquot J (2003) Fluticasone reduces IL-6 and IL-8 production of cystic fibrosis bronchial epithelial cells *via* IKK-beta kinase pathway. *Eur Respir J* 21: 574–81
44 Tosi MF, Stark JM, Smith CW, Hamedani A, Gruenert DC, Infeld MD (1992) Induction of ICAM-1 expression on human airway epithelial cells by inflammatory cytokines: effects on neutrophil-epithelial cell adhesion. *Am J Respir Cell Mol Biol* 7: 214–21
45 Hubeau C, Lorenzato M, Couetil JP, Hubert D, Dusser D, Puchelle E, Gaillard D (2001) Quantitative analysis of inflammatory cells infiltrating the cystic fibrosis airway mucosa. *Clin Exp Immunol* 124: 69–76
46 De Rose V, Oliva A, Messore B, Grosso B, Mollar C, Pozzi E (1998) Circulating adhesion molecules in cystic fibrosis. *Am J Respir Crit Care Med* 157: 1234–9
47 Lin HC, Wang CH, Liu CY, Yu CT, Kuo HP (2000) Erythromycin inhibits beta2-integrins (CD11b/CD18) expression, interleukin-8 release and intracellular oxidative metabolism in neutrophils. *Respir Med* 94: 654–60
48 Okubo Y (1997) Macrolides reduce the expression of surface Mac-1 molecule on neutrophil. *Kurume Med J* 44: 115–23

49 Li Y, Azuma A, Takahashi S, Usuki J, Matsuda K, Aoyama A, Kudoh S (2002) Fourteen-membered ring macrolides inhibit vascular cell adhesion molecule 1 messenger RNA induction and leukocyte migration: role in preventing lung injury and fibrosis in bleomycin-challenged mice. *Chest* 122: 2137–45
50 Brennan S, Cooper D, Sly PD (2001) Directed neutrophil migration to IL-8 is increased in cystic fibrosis: a study of the effect of erythromycin. *Thorax* 56: 62–4
51 Labro MT, el Benna J, Babin-Chevaye C (1989) Comparison of the *in vitro* effect of several macrolides on the oxidative burst of human neutrophils. *J Antimicrob Chemother* 24: 561–72
52 Anderson R (1989) Erythromycin and roxithromycin potentiate human neutrophil locomotion *in vitro* by inhibition of leukoattractant-activated superoxide generation and autooxidation. *J Infect Dis* 159: 966–73
53 Villagrasa V, Berto L, Cortijo J, Perpina M, Sanz C, Morcillo EJ (1997) Effects of erythromycin on chemoattractant-activated human polymorphonuclear leukocytes. *Gen Pharmacol* 29: 605–9
54 Hand WL, Hand DL, King-Thompson NL (1990) Antibiotic inhibition of the respiratory burst response in human polymorphonuclear leukocytes. *Antimicrob Agents Chemother* 34: 863–70
55 Wenisch C, Parschalk B, Zedtwitz-Liebenstein K, Weihs A, el Menyawi I, Graninger W (1996) Effect of single oral dose of azithromycin, clarithromycin, and roxithromycin on polymorphonuclear leukocyte function assessed *ex vivo* by flow cytometry. *Antimicrob Agents Chemother* 40: 2039–42
56 Culic O, Erakovic V, Cepelak I, Barisic K, Brajsa K, Ferencic Z, Galovic R, Glojnaric I, Manojlovic Z, Munic V et al (2002) Azithromycin modulates neutrophil function and circulating inflammatory mediators in healthy human subjects. *Eur J Pharmacol* 450: 277–89
57 Aoshiba K, Nagai A, Konno K (1995) Erythromycin shortens neutrophil survival by accelerating apoptosis. *Antimicrob Agents Chemother* 39: 872–7
58 Nagai H, Shishido H, Yoneda R, Yamaguchi E, Tamura A, Kurashima A (1991) Long-term low-dose administration of erythromycin to patients with diffuse panbronchiolitis. *Respiration* 58: 145–9
59 Dupont MJ, Lapointe JR (1995) Effect on *Pseudomonas aeruginosa* alginate expression of direct plating and culture of fresh cystic fibrosis sputum on to *pseudomonas* isolation agar containing subinhibitory concentrations of roxithromycin and rifampicin. *J Antimicrob Chemother* 36: 231–6
60 Baumann U, Fischer JJ, Gudowius P, Lingner M, Herrmann S, Tummler B, von der HH (2001) Buccal adherence of *Pseudomonas aeruginosa* in patients with cystic fibrosis under long-term therapy with azithromycin. *Infection* 29: 7–11
61 Nakashio S, Susa C, Qiu S, Kijima A, Iwasawa H, Shimomura H, Kanemitsu K, Hori S, Mizushima Y, Shimada J (1993) Antimicrobial activity of clarithromycin and its effect on bacterial adherence to medical material. *Jpn J Antibiot* 46: 428–36

62 Yamasaki T (1990) Adherence of *Pseudomonas aeruginosa* to mouse tracheal epithelium – the effect of antimicrobial agents. *J Jpn Assoc Infect Dis* 64: 575–83
63 Tsang KW, Ng P, Ho PL, Chan S, Tipoe G, Leung R, Sun J, Ho JC, Ip MS, Lam WK (2003) Effects of erythromycin on *Pseudomonas aeruginosa* adherence to collagen and morphology *in vitro*. *Eur Respir J* 21: 401–6
64 Tsang KW, Shum DK, Chan S, Ng P, Mak J, Leung R, Shum IH, Ooi GC, Tipoe GL, Lam WK (2003) *Pseudomonas aeruginosa* adherence to human basement membrane collagen *in vitro*. *Eur Respir J* 21: 932–8
65 Kobayashi O, Moser C, Jensen PO, Hoiby N (2000) Azithromycin treatment inhibits induction of mucoid phenotype in susceptible BALB/c mice with chronic *Pseudomonas aeruginosa* lung infection. Proceedings of XIIIth International Cystic Fibrosis Congress, Stockholm, Sweden 164
66 Kobayashi H (1995) Biofilm disease: its clinical manifestation and therapeutic possibilities of macrolides. *Am J Med* 99: 26S–30S
67 Hoiby N, Koch C (1990) *Pseudomonas aeruginosa* infection in cystic fibrosis and its management. *Thorax* 45: 881–4
68 Ichimiya T, Takeoka K, Hiramatsu K, Hirai K, Yamasaki T, Nasu M (1996) The influence of azithromycin on the biofilm formation of *Pseudomonas aeruginosa in vitro*. *Chemotherapy* 42: 186–91
69 Goswami SK, Kivity S, Marom Z (1990) Erythromycin inhibits respiratory glycoconjugate secretion from human airways *in vitro*. *Am Rev Respir Dis* 141: 72–8
70 Tamaoki J, Nakata J, Takeda Y, Takemura H, Tagaya E, Konno K (1996) Effect of macrolide antibiotics on airway goblet hypersecretion in guinea pigs. *Kansenshogaku Zasshi* 70: 591–6
71 Shimizu T, Shimizu S, Hattori R, Gabazza EC, Majima Y (2003) *In vivo* and *in vitro* effects of macrolide antibiotics on mucus secretion in airway epithelial cells. *Am J Respir Crit Care Med* 168: 581–7
72 Dupont MJ, Lapointe JR (1990) Quantitative effect of roxithromycin and rifampicin on mucoid cultures from directly plated sputum of cystic fibrosis patients chronically colonized with *Pseudomonas aeruginosa*. *Drugs Exp Clin Res* 16: 597–605
73 Shibuya Y, Wills PJ, Cole PJ (2001) The effect of erythromycin on mucociliary transportability and rheology of cystic fibrosis and bronchiectasis sputum. *Respiration* 68: 615–19
74 Tai S, Sudo E, Sun F, King M, Sextro W, von der Hardt H, Baumann U (1999) Effect of azithromycin treatment on sputum rheology in cystic fibrosis patients. *Pediatr Pulmonol* (Suppl) 19: 265
75 App EM, Konig K, Duffner U, Baumann U, King M, von der Hardt H (2000) The effects of azithromycin therapy on sputum inflammation in CF lung disease. *Am J Respir Crit Care Med*(Suppl) 161: A758
76 Kaneko Y, Yanagihara K, Seki M, Kuroki M, Miyazaki Y, Hirakata Y, Mukae H, Tomono K, Kadota J, Kohno S (2003) Clarithromycin inhibits overproduction of

muc5ac core protein in murine model of diffuse panbronchiolitis. *Am J Physiol Lung Cell Mol Physiol* 285: L847–L853

77 Tateda K, Ishii Y, Matsumoto T, Furuya N, Nagashima M, Matsunaga T, Ohno A, Miyazaki S, Yamaguchi K (1996) Direct evidence for antipseudomonal activity of macrolides: exposure-dependent bactericidal activity and inhibition of protein synthesis by erythromycin, clarithromycin, and azithromycin. *Antimicrob Agents Chemother* 40: 2271–5

78 Tateda K, Comte R, Pechere JC, Kohler T, Yamaguchi K, Van Delden C (2001) Azithromycin inhibits quorum sensing in *Pseudomonas aeruginosa*. *Antimicrob Agents Chemother* 45: 1930–3

79 Altschuler EL (1998) Azithromycin, the multidrug-resistant protein, and cystic fibrosis [letter]. *Lancet* 351: 1286

80 Lallemand JY, Stoven V, Annereau JP, Boucher J, Blanquet S, Barthe J, Lenoir G (1997) Induction by antitumoral drugs of proteins that functionally complement CFTR: a novel therapy for cystic fibrosis? [letter]. *Lancet* 350: 711–12

81 Sermet-Gaudelus I, Kessler R, Stoven V, Annereau JP, Thuillier L, Bieth J, Bonnefond JP, Dommergues JP, VanDeVenne C, Weizenblum C et al (1998) Dramatic improvement of cystic fibrosis during and after antitumorous chemotherapy: a report of three cases. *Pediatr Pulmonol* (Suppl) 17: 219–20

82 Gant TW, O'Connor CK, Corbitt R, Thorgeirsson U, Thorgeirsson SS (1995) *In vivo* induction of liver P-glycoprotein expression by xenobiotics in monkeys. *Toxicol Appl Pharmacol* 133: 269–76

83 Tamaoki J, Isono K, Sakai N, Kanemura T, Konno K (1992) Erythromycin inhibits Cl secretion across canine tracheal epithelial cells. *Eur Respir J* 5: 234–8

84 Tagaya E, Tamaoki J, Kondo M, Nagai A (2002) Effect of a short course of clarithromycin therapy on sputum production in patients with chronic airway hypersecretion. *Chest* 122: 213–18

85 Advenier C, Sarria B, Naline E, Puybasset L, Lagente V (1990) Contractile activity of three endothelins (ET-1, ET-2 and ET-3) on the human isolated bronchus. *Br J Pharmacol* 100: 168–72

86 Blouquit S, Sari A, Lombet A, D'herbomez M, Naline E, Matran R, Chinet T (2003) Effects of endothelin-1 on epithelial ion transport in human airways. *Am J Respir Cell Mol Biol* 29: 245–51

87 Pradal U, Delmarco A, Cipolli M, Cazzola G (2001) Chloride transport may be restored by long-term azithromycin treatment in patients with cystic fibrosis. *Pediatr Pulmonol* (Suppl) 20: 280–1

88 Gillie DJ, Barker PM (2001) Effect of clarithromycin on *in vivo* ion transport by CFTR –/– mouse nasal epithelium. *Pediatr Pulmonol* (Suppl) 22: 259

89 Equi A, Davies JC, Geddes DM, Bush A, Hyde SC, Alton EWFW (2002) Effect of azithromycin on *in vivo* ion transport in cystic fibrosis patients. *Am J Respir Crit Care Med* (Suppl) 165: B37

90 App EM, Konig A, Lam R, Duszyk M, King M, Duffner K (2001) Macrolides stimulate transepithelial anion secretion in epithelial cells. *Pediatr Pulmonol* (Suppl) 22: 204–5
91 Middleton PG, Geddes DM, Alton EW (1996) Trimethoprim and tetracycline inhibit airway epithelial sodium absorption. *Am J of Respir Crit Care Med* 154: 18–23
92 Meng QH, Springall DR, Bishop AE, Morgan K, Evans TJ, Habib S, Gruenert DC, Gyi KM, Hodson ME, Yacoub MH et al (1998) Lack of inducible nitric oxide synthase in bronchial epithelium: a possible mechanism of susceptibility to infection in cystic fibrosis. *J Pathol* 184: 323–31
93 Gaston B, Ratjen F, Vaughan JW, Malhotra NR, Canady RG, Snyder AH, Hunt JF, Gaertig S, Goldberg JB (2002) Nitrogen redox balance in the cystic fibrosis airway: effects of antipseudomonal therapy. *Am J Respir Crit Care Med* 165: 387–90
94 Mitsuyama T, Hidaka K, Furuno T, Hara N (1997) Neutrophil-induced endothelial cell damage: inhibition by a 14-membered ring macrolide through the action of nitric oxide. *Int Arch Allergy Immunol* 114: 111–15
95 Tamaoki J, Kondo M, Kohri K, Aoshiba K, Tagaya E, Nagai A (1999) Macrolide antibiotics protect against immune complex-induced lung injury in rats: role of nitric oxide from alveolar macrophages. *J Immunol* 163: 2909–15
96 Culic O, Erakovic V, Parnham MJ (2001) Anti-inflammatory effects of macrolide antibiotics. *Eur J Pharmacol* 429: 209–29
97 Takizawa H, Desaki M, Ohtoshi T, Kawasaki S, Kohyama T, Sato M, Nakajima J, Yanagisawa M, Ito K (1998) Erythromycin and clarithromycin attenuate cytokine-induced endothelin-1 expression in human bronchial epithelial cells. *Eur Respir J* 12: 57–63
98 Saiman L, Marshall BC, Mayer-Hamblett N, Burns JL, Quittner AL, Cibene DA, Coquillette S, Fieberg AY, Accurso FJ, Campbell PW III (2003) Azithromycin in patients with cystic fibrosis chronically infected with *Pseudomonas aeruginosa*: a randomized controlled trial. *JAMA* 290: 1749–56
99 Yatsunami J, Fukuno Y, Nagata M, Tominaga M, Aoki S, Tsuruta N, Kawashima M, Taniguchi S, Hayashi S (1999) Antiangiogenic and antitumor effects of 14-membered ring macrolides on mouse B16 melanoma cells. *Clin Exp Metastasis* 17: 361–7
100 Yatsunami J, Tsuruta N, Hara N, Hayashi S (1998) Inhibition of tumor angiogenesis by roxithromycin, a 14-membered ring macrolide antibiotic. *Cancer Lett* 131: 137–43
101 Fujitani Y, Trifilieff A (2003) *In vivo* and *in vitro* effects of SAR 943, a rapamycin analogue, on airway inflammation and remodeling. *Am J Respir Crit Care Med* 167: 193–8
102 Hilliard TN, Madden N, Nicholson AG, Alton EWFW, Davies JC, Bush A (2003) Airway inflammation and remodelling in children with cystic fibrosis. *Thorax* 58 (Suppl III): iii64
103 Feldman C, Anderson R, Theron AJ, Ramafi G, Cole PJ, Wilson R (1997) Roxithromycin, clarithromycin, and azithromycin attenuate the injurious effects of bioactive phospholipids on human respiratory epithelium *in vitro*. *Inflammation* 21: 655–65
104 Feldman C, Anderson R, Theron A, Mokgobu I, Cole PJ, Wilson R (1999) The effects

of ketolides on bioactive phospholipid-induced injury to human respiratory epithelium *in vitro*. *Eur Respir J* 13: 1022–8
105 Equi A, Balfour-Lynn I, Bush A, Rosenthal M (2002). Long term azithromycin in children with cystic fibrosis: a randomised, placebo-controlled crossover trial. *Lancet* 360: 978–84
106 Jaffe A, Bush A (2001) Anti-inflammatory effects of macrolides in lung disease. *Pediatr* 31: 464–73
107 Jaffe A, Francis J, Rosenthal M, Bush A (1998) Long-term azithromycin may improve lung function in children with cystic fibrosis. *Lancet* 351: 420
108 Anstead MI, Kuhn RJ, Hartford LH, Craigmyle L, Halsey S, Kanga JF (1999) Effect of chronic azithromycin on lung function in cystic fibrosis. *Pediatr Pulmonol* (Suppl) 19: 283
109 Hallberg K, Gronowitz E, Strandvik B (2000) Azithromycin improves pulmonary symptoms in patients with CF. Proceedings of XIIIth International Cystic Fibrosis Congress, Stockholm, Sweden 165
110 Hampton E, Lindsay F, Pagan J, Singleton P (2000) An observational report of the use of azithromycin in cystic fibrosis. *Proceedings of XIIIth International Cystic Fibrosis Congress, Stockholm, Sweden* 165
111 Ordonez CL, Stulbarg M, Grundland H, Liu JT, Boushey HA (2001) Effect of clarithromycin on airway obstruction and inflammatory markers in induced sputum in cystic fibrosis: A pilot study. *Pediatr Pulmonol* 32: 29–37
112 Wolter J, Seeney S, Bell S, Bowler S, Masel P, McCormack J (2002) Effect of long term treatment with azithromycin on disease parameters in cystic fibrosis: a randomised trial. *Thorax* 57: 212–16
113 Ripoll L, Reinert P, Pepin LF, Lagrange PH (1996) Interaction of macrolides with alpha dornase during DNA hydrolysis. *J Antimicrob Chemother* 37: 987–91
114 Pirzada OM, McGaw J, Taylor CJ, Everard ML (2003) Improved lung function and body mass index associated with long-term use of Macrolide antibiotics. *J Cystic Fibrosis* 2: 69–71
115 Prunier AL, Malbruny B, Laurans M, Brouard J, Duhamel JF, Leclercq R (2003) High rate of macrolide resistance in *Staphylococcus aureus* strains from patients with cystic fibrosis reveals high proportions of hypermutable strains. *J Infect Dis* 187: 1709–16

Macrolides and upper airway/sinus disease

Kazuhiko Takeuchi[1], *Yuichi Majima*[1] *and Qutayba Hamid*[2]

[1]Department of Otorhinolaryngology, Mie University School of Medicine, 2-174 Edobashi, Tsu, Mie 514-8507, Japan
[2]McGill University, Canada

Introduction

Historically, macrolides were first used for chronic sinusitis as immunomodulatory mediators in Japan. In Japan, 14-membered-ring macrolide antibiotics had been routinely used for the treatment of diffuse panbronchiolitis (DPB) since Kudoh reported that long-term, low-dose oral administration of erythromycin (EM) was effective for the disease in 1987 [1]. DPB is a disease of unclear etiology, characterized by chronic inflammation in the respiratory bronchioles. DBP is not uncommon in Japan, but is rare elsewhere. More than 75% of diffuse panbronchiolitis patients have chronic sinusitis, and chronic sinusitis associated with diffuse panbronchiolitis improves during macrolide treatment.

EM was originally recovered from a soil sample from the Philippine archipelago. It is the metabolic product of a strain of *Streptomyces erythreus*, discovered by McGuire and co-worker in 1952. Clarithromycin (CAM), roxithromycin (RXM) and azithromycin (AZM) are new semi-synthetic derivatives of EM. Not only EM but also other 14- and 15- membered-ring macrolides have been proven to be effective for diffuse panbronchiolitis. Macrolide is now widely used for chronic sinusitis and otitis media in Japan. In this chapter, we will discuss clinical results of macrolide for chronic sinusitis, nasal polyps, and otitis media with effusion.

Chronic sinusitis

Clinical efficacy of erythromycin

The first study on the usefulness of macrolide in treating chronic sinusitis not associated with diffuse panbronchiolitis was published in 1991 by Kikuchi et al. [2]. Twenty-six adult patients with chronic sinusitis whose symptoms persisted in spite of Caldwell-Luc operation and conservative therapy were treated with 400~600 mg of EM per day for 7.9 months on average. Rhinorrhea was reduced in 60%, post-

Antibiotics as Anti-Inflammatory and Immunomodulatory Agents, edited by Bruce K. Rubin and Jun Tamaoki

nasal drip 50%, nasal obstruction in 60%, hyposmia in 11.8% and sense of dullness in the head in 100%. This therapy was effective even when EM-resistant bacteria such as *Haemophilus influenzae* were present. Thus, it was thought that the effectiveness of erythromycin was due to mechanisms other than antibacterial activity.

Efficacy of other new macrolides including clarithromycin (CAM), roxithromycin (RXM) and azithromycin (AZM)

CAM is at least as effective as EM in the treatment of chronic sinusitis. Hashiba et al. [3] investigated the clinical efficacy of long-term administration of CAM in intractable cases of chronic sinusitis. Forty-five adult patients were treated with 400 mg/day for 8–12 weeks. Improvement of symptoms and rhinoscopic findings was noted in 71.1% of the patients. For periods of up to 12 weeks, clinical efficacy depended upon the duration of treatment. Administration for more than 12 weeks might further improve clinical results. No significant side effects were noted during the course of CAM treatment.

RXM 150 mg for 3 months is effective for chronic sinusitis. Kimura et al. [4] studied the clinical efficacy of RXM administered at the daily dosage of one tablet (150 mg) for 3 months in 30 patients with chronic sinusitis. Subjective and objective symptoms disappeared or decreased markedly, especially postnasal drip and nature of discharge in 87% of the patients. All symptoms significantly decreased, except for the sensation of foul odor. Symptoms improved even in those cases in which *Haemophilus influenzae* was detected. Thus RXM also produces clinically benefits through immunological or anti-inflammatory mechanisms.

The effects of different macrolides have been compared in a few studies. Kita et al. [5] compared EM with RXM. In this study, 71 patients with chronic sinusitis were treated with 600 mg EM or 150 mg RXM daily for 3 months. There were no significant differences between the effectiveness of EM and that of RXM. Hashiba et al. [6] compared EM with CAM. In this study, EM and CAM were randomly assigned to patients. Adults were given a total of 600 mg of EM administered as three doses daily, or 400 mg of CAM administered as two doses daily. Children were given 10~15 mg/kg of EM as three doses daily or 200 mg of CAM as two doses daily, when body weight exceeded 30 kg, and 100 mg of CAM as two doses when body weight was less than 30 kg. Clinical efficacy was assessed by symptoms and rhinoscopic findings after 12 weeks administration. 68% of adults treated with CAM, and 38% of adults given EM, demonstrated improvement. In children, 77% of CAM patients and 40% of EM patients demonstrated improvement. In both adults and children, a significant difference between the two groups was shown. They concluded that the clinical efficacy of CAM exceeded that of EM in the treatment of chronic sinusitis.

Table 1 - Recommended doses of macrolides for adults and children ([8])

	erythromycin	clarithromycin	roxithromycin
Adults	400~600 mg	200~400 mg	200~400 mg
children	8~12 mg/kg	4~8 mg/kg	

Felstead et al. [7] compared patients with upper respiratory tract infections treated with 1.5 g azithromycin in five or six doses over 5 days with patients treated with 10 g erythromycin in 40 doses over 10 days. The majority of the patients had sinusitis. Clinical cure was recorded in 83% of azithromycin- and 79% of erythromycin-treated patients.

Thus, new macrolides are as effective for chronic sinusitis as EM. Because of lack of information, superiority of one macrolide over others is not clear at present. This should be investigated in future studies.

Dose and length of administration

Recommended doses of macrolides [8] are summarized in Table 1. These doses were given empirically, as erythromycin had been administered at a daily dose of 400–600 mg to diffuse panbronchiolitis patients [9].

Shinkawa et al. [10] examined when to stop long-term administration of RXM in patients with chronic sinusitis. The clinical effect was observed in approximately 60% of the patients at 6 months and the clinical effect did not increase even if RXM was administered for 13 months on average. From these results, they concluded that the long-term administration of RXM should not be continued more than 6 months. This contrasts with the cases of diffuse panbronchiolitis, in which macrolides are given for years. Since chronic sinusitis is not a life-threatening disease as diffuse panbronchiolitis, it is important to minimize the duration of administration in order to prevent adverse effects.

Symptoms of chronic sinusitis recur after discontinuation of macrolides in some patients. In such a case, macrolides should be restarted, which often gives favorable results.

There is evidence to show that macrolide administration for much shorter periods is effective. For example, MacLeoad et al. reported that CAM 500 mg twice daily for 14 days was effective for adult chronic sinusitis patients [11]. However, we believe that longer administration will give more favorable results, because improvement of symptoms is expected until 6 months after the start of administration [10].

When macrolides are effective, rhinorrhea decreases first among subjective symptoms. Improvement in X-ray or CT findings usually takes longer. Thus it is not necessary to continue macrolide administration until abnormal opacification in paranasal sinuses are completely resolved [7].

Indication

In Japan, macrolide treatment for chronic sinusitis spread rapidly without thorough information on its indication. As a result, it was found that some patients responded to macrolides but others did not. Clinical investigation of usefulness of macrolides revealed the limitation of the therapy. It was found that there was a poor response to macrolides; in sinusitis patients associated with type I allergic reaction in pathophysiology, where ostiomeatal complexes are completely occluded, large nasal polyps, and with acute exacerbation during long-term macrolide therapy [8].

There is evidence to show that macrolide is not effective for sinusitis where allergic reactions play a role. Suzuki et al. suggests that macrolide therapy is indicated for patients without atopy or smear/tissue/peripheral blood eosinophilia [12]. They studied the immunological and histopathological factors that affect the prognosis of chronic rhinosinusitis under long-term, low-dose macrolide therapy. Patients with normal levels of serum IgE showed a significantly higher symptomatic improvement rate than those with high levels of serum IgE. The symptomatic improvement rate was inversely correlated with the eosinophil counts in the peripheral blood, in the nasal smear and in the sinus mucosa. These results suggest that macrolide therapy is indicated for patients without atopy or smear/tissue/peripheral blood eosinophilia. Meanwhile, the CT score failed to correlate with the symptomatic improvement rate. Thus, the severity of the disease is unlikely to be a prognostic factor of chronic rhinosinusitis under long-term low-dose macrolide therapy [12].

Patients with severe ostiomeatal complexes occlusion or with large nasal polyps do not respond to macrolide therapy. Hirano et al. [13] administered two types of newly developed 14-membered macrolides to 31 patients with chronic sinusitis for 2 to 3 months and the clinical efficacy was compared with effects of macrolides on the potency of ostiomeatal unit revealed by CT scan and the size of nasal polyps. Non-responders to macrolide treatment tended to show a complete obstruction of the ostiomeatal complexes. The presence of large nasal polyps in the middle nasal meatus seems to be a critical factor of resistance to the macrolides.

Pediatric patients with chronic sinusitis respond to macrolides as well as adults patients. Since symptoms of pediatric sinusitis patients tend to fluctuate, their symptoms and their infection aggravate when they catch a cold. Macrolides should be temporarily discontinued and cephalosporin antibiotics should be started.

In chronic sinusitis, diseases with different etiologies are included and the pathophysiological processes may differ among different countries and different time periods. In developing countries, for example, they have more infectious sinusitis cases than developed nations. Previously, all sinusitis was thought to be infectious. It is now clear, however, that the majority of patients with chronic sinusitis do not have an infectious disorder, and this has led to the need for more appropriate terminology to describe the myriad of conditions that make up chronic sinusitis. There are four major pathophysiological processes that have been described as causing chronic sinusitis [14]. These include chronic infectious sinusitis, chronic inflammatory sinusitis, hyperplastic eosinophilic sinusitis, and allergic fungal sinusitis [14]. Most of the clinical investigations in Japan were conducted in patients with chronic inflammatory sinusitis. The effectiveness of macrolide needs to be determined in each of the four types of chronic sinusitis under stricter classification.

Postoperative use

The macrolide can be used postoperatively. Moriyama et al. evaluated the effect of EM therapy after endoscopic sinus surgery for chronic sinusitis [15]. The subjects analyzed in their retrospective study are cases who had previously undergone surgery for chronic pan-sinusitis. They are classified into two groups: one group received a postoperative long-term, low-dose EM regimen and the other had not received this treatment. Greater improvement of symptoms is achieved in the EM group than in the non-EM group.

Another report suggests the use of macrolides when sinus surgery was not effective. Cervin et al. [16] tested the efficacy of long-term, low-dose EM therapy in 17 patients with chronic sinusitis persistent after sinus surgery. All patients were treated with EM 250 mg twice daily or CAM 250 mg once daily and were assessed after 3 months. Responders were reassessed after 12 months of treatment. As a result, 12 out of 17 patients responded to treatment. However, placebo-controlled studies are needed to validate the potential of this treatment.

Possible mechanisms

It is generally agreed that macrolides exert their effects for chronic sinusitis not by antibacterial mechanism but by anti-inflammatory mechanism. Suppression of neutrophil recruitment and mucus secretion may be two major mechanisms by which macrolide antibiotics exert their effect on chronic sinusitis. Suzuki et al. assessed neutrophil numbers in the nasal smear and IL-8 level in the nasal discharge before and after long-term low-dose RXM administration in patients with chronic sinusitis. Neutrophils and the IL-8 level in the nasal discharge were decreased after the treatment. These findings suggest that long-term low-dose RXM administration inhibits the positive feedback mechanism of neutrophil recruitment and IL-8 production by the recruited neutrophils, which is considered to be an essential cause of the prolongation of sinusitis [17].

Fujita et al. [18] examined the effects of macrolides on interleukin-8 secretion from human nasal epithelial cells. They examined the *in vivo* effects of EM and CAM. Fifteen patients with chronic sinusitis received macrolide treatment (CAM 400 mg/day) for 1 to 3 months. The number of neutrophils and IL-8 concentrations in the nasal discharges of these patients decreased significantly at 1–2 months after the treatment. *In vitro* effects of EM and CAM on IL-8 secretion were examined in nasal epithelial cells cultured at the air–liquid interface. These results suggest that macrolide treatment inhibits neutrophil infiltration and IL-8 secretion in nasal epithelium *in vivo*.

Shimizu et al. [19] examined the *in vivo* effects of macrolide antibiotics on mucus hypersecretion. They induced hypertrophic and metaplastic changes of goblet cells in rat nasal epithelium by intranasal instillation of ovalbumin (OVA) in OVA-sensitized rats and by intranasal LPS instillation. Oral administration of CAM significantly inhibited OVA- and LPS-induced mucus production and neutrophil infiltration, whereas josamycin and ampicillin showed no effect. Kim et al. [20] examined the *in vitro* effect of roxithromycin on MUC2 gene expression in cultured epithelial cells. Roxithromycin suppressed MUC2 gene transcriptional activity in a dose-dependent manner in HM3-MUC2 cells. Roxithromycin also decreased MUC2 gene transcriptional activity induced by PMA in a dose-dependent manner. NF-κB activation, but not AP-1 activation, was significantly suppressed by roxithromycin in HM3-MUC2 cells. Thus, roxithromycin suppresses MUC2 gene expression in epithelial cells and this suppression is probably *via* inhibition of NF-κB activation.

The effects of macrolides on rheological properties of nasal mucus are examined by Rhee et al. [21]. To determine the effects of oral administration of clarithromycin (CAM) on rheological properties, they measured the spinability, dynamic viscoelasticity, and solid composition of human nasal mucus from patients with chronic sinusitis before and after administration of CAM for 4 weeks. After administration of CAM, the spinability and percentage solid composition of nasal mucus increased respectively, whereas the ratio of the viscosity to the elasticity of nasal mucus after the administration of CAM decreased in all of the mucus samples. These results suggest that treatment with CAM may modulate the rheological properties of nasal mucus in patients with chronic sinusitis.

Biofilm formed by bacteria has recently been shown to be involved in making infectious diseases intractable. There are two reports from one Institute concerning the inhibitory activity of macrolides [22, 23]. Kondoh et al. [22] observed that biofilms on a Teflon sheet with CAM decreased markedly in a dose-dependent manner as compared with a control Teflon sheet without CAM. Ozeki et al. [23] indirectly measured and evaluated the inhibitory effect of RXM on biofilm formation on the inside of a plastic test tube by determining the number of living bacteria in the biofilm. As a result, RXM was found to inhibit biofilm formation even though it does not have antimicrobial activity against *Pseudomonas aeruginosa*. They concluded that there is a possibility that macrolides are effective against infectious otolaryngological diseases when biofilm formation is likely to be pathologically involved, even if the detected bacteria are not sensitive to the drug.

Nasal polyps

Clinical efficacy of macrolides for nasal polyps

Nasal polyposis is a chronic inflammatory disease of the nasal mucosa with inflammatory cell infiltration and structural modifications of the epithelium (secretory hyperplasia and squamous metaplasia) and lamina propria (basement membrane thickening, extracellular matrix accumulation and fibrosis).

As mentioned before, patients with large nasal polyps respond less well to macrolide treatment. However, there is evidence to show that macrolides are effective for reducing the size of nasal polyps. Ichimura et al. reported that roxithromycin was effective in reducing nasal polyp size [24]. In order to assess the efficacy of this treatment for nasal polyps, they administered RXM (1 tablet: 150 mg a day) for at least 8 weeks to 20 patients with nasal polyps associated with chronic sinusitis. It was effective in controlling nasal polyps with the overall incidence of improvement being 52%. The incidence of improvement increased with time after the start of medication in both groups. Smaller polyps were more likely to decrease in size, but some larger polyps also markedly decreased in size. Associated allergic conditions and the extent of eosinophilic infiltration had no relation to the treatment result. They speculate that the mechanism of effectiveness of RXM is through its suppressive potency in cytokine production from inflammatory cells. Yamada et al. also reported that macrolide treatment decreased the size of nasal polyps [25].

Possible mechanisms

According to Yamada T et al. [25], the reduction in IL-8 may be an important aspect of the effect of macrolide treatment on nasal polyps in chronic rhinosinusitis. They administered CAM for 8 to 12 weeks (400 mg/day) to patients with nasal polyps due to chronic rhinosinusitis and measured the IL-8 level in nasal lavage from them. The IL-8 levels in nasal lavage from patients with nasal polyps were reduced during macrolide treatment. There was significant correlation between decreased IL-8 levels in nasal lavage and the clinical effect of macrolides on the size of the nasal polyps. In the group whose polyps were reduced in size, the IL-8 levels dramatically decreased, and were significantly higher before macrolide treatment than those in the group whose polyps showed no change.

Another possible mechanism by which macrolide exerts its effect on nasal polyps may be by fibroblasts in nasal polyps. Nonaka et al. [26] reported that roxithromycin inhibits the growth of nasal polyp fibroblast. Fibroblasts are resident cells thought to play an important role in the development of fibrosis in the nasal polyps. Nasal polyp fibroblast lines were generated from untreated patients, and those who were treated with RXM (300 mg/day) for 1 month before biopsy. Nasal polyp fibroblast lines that were treated with RXM exhibited a lower proliferating rate *in vitro* as compared to those that were not treated with RXM. Treatment of nasal polyp fibroblast lines with RXM suppressed the proliferation of fibroblasts in a dose-dependent manner. They demonstrated that RXM directly suppressed nasal polyp fibroblast proliferation, and that this effect of RXM on fibroblast growth was persistent, indicating that RXM may prevent the progression of nasal polyposis by inhibiting the development of fibrosis.

Otitis media with effusion

Clinical efficacy of macrolides

Incidence of otitis media with effusion is high (54%) in the adult patients with sinobronchial syndrome defined as having both sinusitis and lower respiratory tract diseases [27]. Sixteen patients with both sinobronchial syndrome and otitis media with effusion were given low-dose and long-term EM therapy (erythromycin base, 600 mg/d for more than 4 months); of these, 13 became effusion-free and most subjects showed improvement in the symptoms of sinobronchial syndrome. EM therapy thus seems to be effective for the treatment of sinobronchial syndrome and associated otitis media with effusion.

Iino then administered EM to children with chronic otitis media with effusion [28]. In their study, 25 children with chronic otitis media with effusion received low-dose and long-term EM treatment, and middle ear effusion was resolved in 18 out of the 25 patients. However, there is no control group in the study. Iino also [29] used CAM to 95 pediatric patients with otitis media with effusion and analyzed factors influencing the resolution of the effusions. As a result, associations with sinusitis, absence of adenoid hypertrophy, and age of three and over, were the factors for the effectiveness of CAM. On the other hand, gender, association with allergic diseases, or bilaterality did not influence the efficacy of CAM therapy. Therefore, macrolide antibiotics are recommended for older pediatric patients with otitis media with effusion who are associated with chronic sinusitis.

Mechanisms of action

Enomoto et al. [30] examined the effect of EM on leukocyte accumulation and expression of adhesion molecules L-selectin and Mac-1, using a rat experimental model. Administration of EM inhibited leukocyte (neutrophil) accumulation in the middle ear cavity after LPS stimulation. Moreover, EM downregulated L-selectin expression and inhibited interleukin (IL)-8-induced upregulation of Mac-1 on peripheral blood neutrophils. These findings suggest that EM may improve otitis media with effusion by inhibiting neutrophil accumulation in the middle ear cavity through modulating the expression of adhesion molecules L-selectin and Mac-1 on peripheral blood neutrophils.

Conclusion

We are given the strong impression that patients' symptoms and rhinoscopic findings improve with macrolide therapy. However, these assessments were not performed blinded, so an investigator bias cannot be ruled out. Therefore, well designed control studies are desirable to prove the clinical efficacy of macrolides for these upper airway diseases.

References

1 Kudoh S, Uetake T, Hagiwara K, Hirayama M, Hus LH, Kimura H, Sugiyama Y (1987) Clinical effects of low-dose long-term erythromycin chemotherapy on diffuse panbronchiolitis. *Jpn J Thorac Dis* 25: 632–42

2 Kikuchi S, Suzaki H, Aoki A, Ito O, Nomura Y (1991) Clinical effect of long-term low dose erythromycin therapy for chronic sinusitis. *Pract Otol (Kyoto)* 84: 41–7

3 Hashiba M, Baba S (1996) Efficacy of long-term administration of clarithromycin in the treatment of intractable chronic sinusitis. *Acta Otolaryngol* (Suppl) 525: 73–8

4 Kimura N, Nishioka K, Nishizaki K, Ogawa T, Naitou Y, Masuda Y (1997) Clinical effect of low-dose, long-term roxithromycin chemotherapy in patients with chronic sinusitis. *Acta Med Okayama* 51: 33–7

5 Kita H, Takezawa H, Isobe M, Kataura A (1995) Long-term low dose erythromycin (EM) and roxithromycin (RXM) therapy for chronic sinusitis. *Pract Otol (Kyoto)* (Suppl) 84: 62–9

6 Hashiba M, Baba S, Tohnai A, Okajima H, Matsuda T, Hondo J, Ohya H, Yokota A, Koto A, Ito M et al (1997) Clinical efficacy of long-term macrolides therapy for chronic sinusitis-comparison between erythromycin and clarithromycin. *Pract Otol (Kyoto)* 90: 717–27

7 Felstead SJ, Daniel R (1991) Short-course treatment of sinusitis and other upper respiratory tract infections with azithromycin: a comparison with erythromycin and amoxycillin. European Azithromycin Study Group. *J Int Med Res* 19: 363–72

8 Hashiba M, Suzaki H, Furuta S, Yanagi K, Ohyama M, Baba S (1998) Guideline for

macrolide treatment of chronic paranasal sinusitis (draft). *Jpn J Antibiot* 51 (Suppl A): 86–9
9 Nagai H, Shishido H, Yoneda R, Yamaguchi E, Tamura A, Kurashima A (1991) Long-term low-dose administration of erythromycin to patients with diffuse panbronchiolitis. *Respiration* 58: 145–9
10 Shinkawa A, Sakai M (1996) Study on the duration of long-term administration of roxithromycin (RXM) in patients with chronic sinusitis. *ORL Tokyo* 39 (Suppl 1): 108–11
11 MacLeod CM, Hamid QA, Cameron L, Tremblay C, Brisco W (2001) Anti-inflammatory activity of clarithromycin in adults with chronically inflamed sinus mucosa. *Adv Ther* 18: 75–82
12 Suzuki H, Ikeda K, Honma R, Gotoh S, Oshima T, Furukawa M, Takasaka T (2000) Prognostic factors of chronic rhinosinusitis under long-term low-dose macrolide therapy. *ORL J Otorhinolaryngol Relat Spec* 62: 121–7
13 Hirano K, Ikeda K, Shimomura A, Ohshima T, Kondo Y, Takasaka T (1995) Clinical investigation of effectiveness and ineffectiveness in patients with chronic sinusitis in the treatment of newly developed macrolides. *ORL Tokyo* 38 (Suppl 3): 251–7
14 Steinke JW, Borish L (2003) Clarification of terminology in patients with eosionopilic and noneosinophilic hyperplastic rhinosinusitis. Author reply. *J Allergy Clin Immunol* 112: 222–3
15 Moriyama H, Yanagi K, Ohtori N, Fukami M (1995) Evaluation of endoscopic sinus surgery for chronic sinusitis: post-operative erythromycin therapy. *Rhinology* 33: 166–70
16 Cervin A, Kalm O, Sandkull P, Lindberg S (2002) One-year low-dose erythromycin treatment of persistent chronic sinusitis after sinus surgery: clinical outcome and effects on mucociliary parameters and nasal nitric oxide. *Otolaryngol Head Neck Surg* 126: 481–9
17 Suzuki H, Shimomura A, Ikeda K, Oshima T, Takasaka T (1997) Effects of long-term low-dose macrolide administration on neutrophil recruitment and IL-8 in the nasal discharge of chronic sinusitis patients. Tohoku J Exp Med 182: 115–24
18 Fujita K, Shimizu T, Majima Y, Sakakura Y (2000) Effects of macrolides on interleukin-8 secretion from human nasal epithelial cells. Eur Arch Otorhinolaryngol 257: 199–204
19 Shimizu T, Shimizu S, Hattori R, Gabazza EC, Majima Y (2003) *In vivo* and *in vitro* effects of macrolide antibiotics on mucus secretion in airway epithelial cells. *Am J Respir Crit Care Med* 168: 581–7
20 Kim DY, Takeuchi K, Ishinaga H, Kishioka C, Suzuki S, Basbaum C, Majima Y (2004) Roxithromycin suppresses mucin gene expression in epithelial cells. *Pharmacology* 72: 6–11
21 Rhee CS, Majima Y, Arima S, Jung HW, Jinn TH, Min YG, Sakakura Y (2000) Effects of clarithromycin on rheological properties of nasal mucus in patients with chronic sinusitis. *Ann Otol Rhinol Laryngol* 109: 484–7
22 Kondoh K, Hashiba M, Baba S (1996) Inhibitory activity of clarithromycin on biofilm synthesis with Pseudomonas aeruginosa. *Acta Otolaryngol* (Suppl) 525: 56–60

23 Ozeki M, Miyamoto N, Hashiba M, Baba S (1996) Inhibitory effect of roxithromycin on biofilm formation of *Pseudomonas aeruginosa. Acta Otolaryngol* (Suppl) 525: 61–3
24 Ichimura K, Shimazaki Y, Ishibashi T, Higo R (1996) Effect of new macrolide roxithromycin upon nasal polyps associated with chronic sinusitis. *Auris Nasus Larynx* 23: 48–56
25 Yamada T, Fujieda S, Mori S, Yamamoto H, Saito H (2000) Macrolide treatment decreased the size of nasal polyps and IL-8 levels in nasal lavage. *Am J Rhinol* 14: 143–8
26 Nonaka M, Pawankar R, Tomiyama S, Yagi T (1999) A macrolide antibiotic, roxithromycin, inhibits the growth of nasal polyp fibroblasts. *Am J Rhinol* 13(4): 267–72
27 Iino Y, Sugita K, Toriyama M, Kudo K (1993) Erythromycin therapy for otitis media with effusion in sinobronchial syndrome. *Arch Otolaryngol Head Neck Surg* 119: 648–51
28 Iino Y, Sugita K, Ishitoya J, Nakai A, Ambe K, Masuda T, Shimizu H, Toriyama M (1992) Erythromycin treatment for otitis media with effusion in children. *Pract Otol (Kyoto)* 85: 713–20
29 Iino Y (2001) Efficacy of macrolide therapy for children with serous otitis media. *Jpn J Antibiot* 54 (Suppl) C:23–5
30 Enomoto F, Ichikawa G, Nagaoka I, Yamashita T (1998) Effect of erythromycin on otitis media with effusion in experimental rat model. *Acta Otolaryngol* (Suppl) 539: 57–60

Benefits of macrolides in the treatment of asthma

Rose Jung[1], Mark H. Gotfried[2,3] and Larry H. Danziger[3]

[1]Department of Clinical Pharmacy, University of Colorado Health Science Center, Denver, CO, USA; [2]Department of Medicine, University of Arizona, Phoenix, Arizona, USA; [3]Department of Pharmacy Practice, University of Illinois at Chicago, Chicago, IL, USA

Introduction

Macrolides are a widely used class of antimicrobials that feature one or more deoxy- or amino-sugars bound to a 14-, 15-, or 16-membered macrocyclic lactone ring. Although their anti-inflammatory properties have been recognized since the 1950s, these characteristics did not create a great deal of interest until macrolide therapy was documented to reduce symptoms and improve survival in patients with diffuse panbronchiolitis (DBP) [1]. Reports of clinical success in this disease, characterized by progressive airflow limitation and recurrent respiratory tract infections, suggested potential benefits of their long-term application in a variety of chronic inflammatory pulmonary diseases, such as asthma.

Asthma affects approximately 5% of the population, and is associated with increased risks of long-term morbidity and mortality [2]. Persistent airway inflammation, a central feature of asthma, results in airway hyperresponsiveness and repeated episodes of airway obstruction. It is commonly accepted that inflammatory cell infiltration with secretion of proinflammatory cytokines plays a key role in the pathogenesis of asthma. Given the role of airway inflammation, the prompt initiation of an anti-inflammatory agent is considered the mainstay of therapy [2]. A growing body of experimental and clinical evidence clearly indicates that the 14- and 15-membered ring macrolide antibiotics possess distinct immunomodulatory properties capable of attenuating inflammation of the respiratory tract.

The macrolides have the unique ability to accumulate in high concentrations intracellularly, most likely accounting in some part for many of their immunodulatory effects. Although the unified mechanism by which the macrolides exert their anti-inflammatory or immunomodulatory effect remain elusive, numerous studies have documented many of the cellular processes modulated by this class of agents. Macrolides have been shown to inhibit the production of proinflammatory cytokines, the activation of nuclear transcription factors, neutrophil oxidative burst, endothelin-1 release, intracellular adhesion molecules (ICAM)-1 expression, mucus hypersecretion, neutrophil migration, and eosinophilic inflammation [3]. These diverse immunomodulatory and/or anti-inflammatory properties of the macrolides

Antibiotics as Anti-Inflammatory and Immunomodulatory Agents, edited by Bruce K. Rubin and Jun Tamaoki

hold promise as being beneficial in the treatment of asthma. Over the last decade a growing body of evidence suggests that the macrolides reduce airway hyperresponsiveness and improve pulmonary function in patients with asthma [4]. The purpose of this Chapter is to present the available *in vitro* and *in vivo* evidence of the immunomodulatory/anti-inflammatory properties of macrolides and examine their potential usefulness in the treatment of asthma.

In vitro studies

Downregulation of proinflammatory cytokine production

Cytokines are involved in the coordination of the inflammatory process as either proinflammatory such as tumor necrosis factor (TNF)-α, interleukin (IL)-6, IL-8, IL-12, and interferon (IFN)-γ or anti-inflammatory such as IL-10, IL-12 and IL-18. Macrolides have been shown to impede the production of proinflammatory cytokines. Erythromycin has been reported to reduce the *Haemophilus influenzae* endotoxin-induced IL-6 and IL-8 expression from cultured human bronchial epithelial cells [5]. Similarly, Northern Blot analysis indicated that erythromycin A exhibited a dose-dependent decrease in IL-6 mRNA expression in BEAS-2B human bronchial epithelial cells [6]. Roxithromycin has also been shown to suppress the IL-1β-induced IL-6 and IL-8 and granulocyte-macrophage colony stimulating factor (GM-CSF) production in Beta-1A human bronchial epithelial cells [7].

The ability of macrolides to affect the production of a wide variety of cytokines was also evaluated in whole blood or in peripheral blood leucocytes *ex vivo*. Erythromycin produced a dose-dependent decrease in heat killed *Streptococcus pneumoniae* (HKSP)-induced production of TNF-α and IL-6 in human whole blood *in vitro* and in whole blood obtained from healthy subjects after a 30 min infusion of erythromycin 1 g [8]. At the higher concentrations of erythromycin, the release of IL-1, IL-12, and IFN-γ was also affected. Using leucocytes isolated from patients with asthma, Konno et al. demonstrated that roxithromycin suppressed the mitogen-activated secretion of T cell cytokine IL-2, IL-3, IL-4 and monocyte cytokine TNF-α in a dose-dependent manner [9]. Although the molecular interactions involved in the inflammatory cascade are complex, reduction in the levels of proinflammatory cytokines by macrolides provides some insight into their anti-inflammatory activities.

Inhibition of activation of transcriptional factors

The stimulation of cells with various cytokines initiates the inflammatory process by activating transcription factors. Once bound to the promoter region of genes, these factors act on genes that encode inflammatory cytokines, chemokines, adhesion molecules, and other proteins that induce and amplify inflammation. Although there are numerous transcription factors, only nuclear factor-κB (NF-κB) and activator protein-1 (AP-1) have been evaluated in relation to macrolides [10–14].

NF-κB is essential for the transcription of genes that encode a number of proinflammatory molecules including TNF-α, ICAM-1, inducible nitric oxide synthase (iNOS), IL-6, and IL-8. Erythromycin has been shown to downregulate IL-8 cytokine gene expression in Jurkat T cells by inhibiting NF-κB activation through interference with non-calcineurin-dependent signaling pathways [10]. Similar observations were found when nuclear and cytoplasmic extracts were analyzed from a human monocytic leukemia cell line (U-937), Junkat cells (a T-cell line), a pulmonary epithelial cell line (A549), and peripheral blood mononuclear cells [11]. The pretreatment of U937, A549, and Jurkat cells with clarithromycin was shown to suppress production of proinflammatory cytokines *via* inhibition of TNF-α-induced NF-κB activation in a concentration-dependent manner. Although the exact mechanism of this has not yet been determined, it has been theorized that since the macrolides readily diffuse into intracellular fluids they may inhibit NF-κB activation by interfering with the generation of reactive oxygen intermediates.

In addition to NF-κB, macrolides also inhibit activation of AP-1. Clarithromycin has been shown to repress TNF-α-induced IL-8 gene transcription by inhibiting AP-1 binding to the IL-8 gene promoter in human bronchial epithelial cells [12]. Erythromycin suppressed the phorbol myristate acetate (PMA)-induced activation of both NF-κB and AP-1 in human bronchial epithelial cells [13]. Similar findings were also noted in monocytes and THP-1 cells when clarithromycin modified lipopolysaccharide (LPS)-induced binding of both AP-1 and NF-κB and reduced IL-8 production [14]. These results suggest that both AP-1 and NF-κB are key factors for IL-8 gene transcription and macrolides can displace the binding of these transcriptional factors, reducing the expression of inflammatory cytokines.

Attenuation of neutrophil accumulation

Recent evidence indicates that the neutrophil is a source of various proinflammatory cytokines and chemokines. Several studies have demonstrated that IL-8 plays a pivotal role as a potent neutrophil chemotactic and activating factor [15–17]. Once migrated within the respiratory tract, neutrophils secrete IL-8 which stimulates neutrophilic airway inflammation. The stimulation of human bronchial epithelial cells with LPS, IFN-δ, IL-1β, and TNF-α upregulates neutrophil adhesion to the epithelial cells following interaction with ICAM-1 [3, 7]. The macrolides have consistently been shown to inhibit neutrophil accumulation in pulmonary alveoli by inhibiting the expression of ICAM-1 and the release of neutrophil chemoattractant chemokines such as IL-8.

As previously mentioned, macrolides reduce IL-8 production by bronchial epithelial cells thus inhibiting neutrophil chemotaxis [5, 18]. In addition, roxithromycin has been shown to inhibit the expression of ICAM-1 and neutrophil adhesion to cultured epithelial cells in a concentration-dependent fashion [7]. These findings suggest that roxithromycin indirectly modulated the recruitment of neutrophils to inflamed sites by suppressing the expression of ICAM-1. Clarithromycin also significantly reduced LPS-induced expression of ICAM-1 expression in a concentration-dependent manner when added to cultured rat tracheal epithelial cells [3].

The inhibitory effect of macrolides on LPS-induced neutrophil infiltration was confirmed in guinea pig airways [19]. Pretreatment with oral clarithromycin (10 mg/kg) inhibited the LPS-induced neutrophil recruitment at 3, 6, and 9 h after LPS inhalation. Erythromycin also significantly suppressed acute neutrophil influx into the lung and the expression of ICAM-1 in an established murine model of experimental extrinsic allergic alveolitis [20].

Reduction in nitric oxide-induced lung injury

Inappropriately generated or overproduced nitric oxide (NO) by the inducible form of NO synthase (iNOS) may cause lung inflammation and injury [21]. This may occur as a result of viral or bacterial infections which induce antigen-antibody immune complexes [22]. The deposition of immune complexes stimulates alveolar macrophages or neutrophils to release IL-1β and TNF-α. The release of these cytokines, in turn, leads to an inflammatory response, resulting in lung injury. The effect of macrolides on NO-induced lung injury may be mediated by inhibition of the production of these cytokines that upregulate iNOS activity.

Erythromycin has been noted to significantly inhibit the release of IL-1β and TNF-α [23]. Furthermore, treatment of rat pulmonary alveolar macrophages with erythromycin, clarithromycin, roxithromycin, and josamycin decreased the immune-complex-induced production of NO and iNOS mRNA. However, these macrolides had no effect on IL-1β and TNF-α-induced NO release. These findings indicate that macrolides inhibit immune complex-induced IL-1β and TNF-α release and subsequent iNOS gene expression and NO release. These results were confirmed in a rat model of immune complex-induced lung injury [23]. Erythromycin dosed at 50 mg/kg significantly reduced the immune-complex-induced increase in exhaled NO concentration and reduced neutrophil accumulation within the alveolar spaces.

Impaired neutrophil oxidant burst

Reactive oxidant production by polymorphonuclear neutrophils (PMN) is known to damage tissue and cause bronchial hyperresponsiveness. The reported effect of macrolides on superoxide production in studies is confusing. When compared to erythromycin, spiramycin, oleandomycin, and josamycin, roxithromycin strongly decreased the PMN oxidative burst [24, 25]. However, another study found that both erythromycin and roxithromycin selectively inhibited superoxide generation by activated neutrophils [26]. Clarithromycin has also demonstrated dose-dependent inhibition of superoxide production by activated neutrophils and was also documented to have a membrane-stabilizing effect [27]. These data seem to indicate that the macrolides may stabilize the epithelial cell membrane by inhibiting PMN oxidative burst and may be a factor in reducing bronchial hyperresponsiveness.

Reduced production of endothelin-1

Endothelin-1 is a potent vasoconstrictor and has bronchoconstrictor effects. It has been reported to stimulate mucus secretion and to cause mucosal edema, thereby contributing to airway inflammation as a result of the increased presence of eosinophils and neutrophils in respiratory tract tissues. Bronchial smooth muscle cells have been shown to possess specific binding sites for endothelin-1. Interestingly, asthmatics are thought to release large amounts of biologically active endothelin-1. Similar to corticosteroids, both erythromycin and clarithromycin have been shown to suppress endothelin-1 release and expression by human bronchial epithelial cells [28]. These effects on endothelin-1 are likely to be one of the more important mechanisms by which bronchoconstriction and pulmonary inflammation is reduced with treatment of macrolide in asthmatic patients.

Modification of the rheologic properties of mucus

It has been established that the rheologic properties of mucus greatly impacts mucociliary clearance and consequently airway inflammation. Mucins are macromolecular glycoproteins that impart viscoelastic properties to mucus and are mediated by mucin genes, MUC4, MUC5AC, and MUC5B. In a human bronchial epithelial cell model the addition of erythromycin or clarithromycin has been noted to inhibit LPS-induced MUC5AC gene expression and modulate transforming growth factor (TGF)-α-induced and LPS-induced phosphorylation of inhibitor of NF-κB (I-κBα). These data indicate that the macrolides may not only decrease mucin production but may also alter the properties of mucus, potentially improving its clearance from airways in patients with inflammation of the respiratory tract.

Animal studies

Inhibition of inflammatory response

The macrolides have been documented to have potent anti-inflammatory effects in various animal models. In rat a carrageenan pleurisy model of acute inflammation, roxithromycin, azithromycin, or clarithromycin all were documented to reduce edema formation [29]. In this model of acute inflammation, roxithromycin appeared to be more effective in reducing edema formation than either azithromycin or clarithromycin [29]. In another study roxithromycin was noted to produce a marked effect on edema formation in carrageenan and poly-L-arginine hind paw edema models as well as a croton oil inflamed ear model [30]. These studies seem to indicate that in various animal models roxithromycin has the most potent anti-inflammatory activity of the macrolides, followed by azithromycin and clarithromycin, and with erythromycin being least effective.

Decreased mucus secretion

Treatment with macrolides has been documented to reduce mucus hypersecretion in various animal models. This effect is likely not a result of a direct effect upon mucus-producing goblet cells, but rather may be associated with their anti-inflammatory activities. A daily dose of clarithromycin or erythromycin for 1 week prior to exposure with nebulized LPS in pathogen-free guinea pigs resulted in a dose-dependent decrease in LPS-induced goblet cell mucus secretion [19]. Similar to these results, pretreatment with single daily doses of roxithromycin or erythromycin for 1 week prior to IL-8 inhalation, in a rodent model, inhibited goblet cell mucus secretion [23]. In this study the increase in the numbers of neutrophils in the tracheal mucosa was noted to coincide with increased mucus discharge. These data suggest that the macrolides may be mediating airway goblet cell mucus secretion by the inhibition of cytokines.

Human data/clinical experience

Unrelated to their known antibacterial properties, the 14- and 15-membered macrolide antibiotics possess anti-inflammatory activity that may contribute to the clinical benefits noted in patients with airway inflammation. As early as the 1950s, the macrolide antibiotics (troleandomycin and erythromycin) had been studied and documented to be of value in corticosteroid dependent asthmatics [31]. Since then, reductions in steroid use in steroid-dependent asthmatic patients (of greater than 50%) [32–36], hospital admissions [35, 37, 38], airway hyperresponsiveness [34], less evidence of adrenal suppression [37] as well as improved spirometry test results (e.g., FEV_1 and FVC) [32, 33, 38] and asthma control [37, 38] have been documented in various trials using the older as well as the newer macrolide antibiotics.

Spector et al. studied 74 corticosteroid-dependent patients (treated with methylprednisolone) with severe asthma and chronic bronchitis in a double-blind crossover trial comparing troleandomycin to placebo [39]. Sixty-six percent of those patients studied showed a considerable improvement in pulmonary function measurements, sputum production, in the need for bronchodilators, as well as subjective evaluations. Much of this effect was ascribed to the troleandomycin induced inhibition of theophylline and methylprednisolone metabolism by the hepatic cytochrome P-450 complex in these patients [40]. However, various open label studies with troleandomycin in methylprednisolone-dependent asthmatics (children and adults) have demonstrated a greater reduction in corticosteroid doses than would be predicted by hepatic inhibition of corticosteroid metabolism alone. Rosenberg et al. reported that in methylprednisolone-dependent asthmatics treatment with troleandomycin resulted in a reduction in corticosteroid doses greater than would have been predicted by inhibition of methylprednisolone metabolism by the liver [41]. Spahn et al. have also suggested that the beneficial effects of the macrolides are not only a result of the inhibition of the clearance of methylprednisolone but also a result of their anti-inflammatory properties [42].

Miyatake et al. [43] evaluated 23 asthmatic patients not receiving steroid therapy who were treated with a 10 week course of low dose erythromycin (200 mg three times daily). These investigators documented a significant decrease in bronchial hyperresponsiveness to histamine challenge in these patients (as measured by PC20 – 20% fall in FEV_1). Similar results have been observed with clarithromycin and roxithromycin [44, 45]. In a randomized, double-blind, placebo-controlled, crossover study, Amayasu and co-workers [44] measured bronchoconstriction after a methacholine challenge in 17 adult patients with mild-to-moderate bronchial asthma who received placebo or clarithromycin, 200 mg twice daily for 8 weeks. The investigators reported a statistically significant reduction in the clarithromycin treated patients *versus* placebo in blood and sputum eosinophil counts, and sputum eosinophilic cationic protein, as well as in the suppression of bronchial hyperresponsiveness after 8 weeks of treatment. These investigators also reported that overall the symptom score significantly decreased after clarithromycin treatment in the clarithromycin treated patients. Ekici et al. evaluated 11 patients with mild asthma who received 250 mg azithromycin orally, twice weekly for 8 weeks. They noted a significant increase in the PC20 of methacholine challenged patients whereas the FEV_1 and FVC were unaffected [46].

Gotfried et al. conducted a double-blind, randomized, placebo-controlled, pilot study [47] to evaluate the efficacy of therapy with clarithromycin, 500 mg twice daily for 6 weeks, in 21 patients with corticosteroid-dependent asthma (i.e., patients had been receiving $\geq$ 5 mg prednisone for $\geq$ 6 months prior to study enrollment).

After 6 weeks of clarithromycin therapy patients were able to tolerate a significant reduction in mean (SD) prednisone dosage from baseline (30% [18%]; $P = 0.02$). Pulmonary function, QOL and asthmatic symptoms did not significantly worsen despite the reduction in prednisone dosage for these patients. Diary reported symptoms such as chest discomfort and cough improved significantly during and after clarithromycin therapy and prednisone taper, respectively ($P = 0.031$ and 0.02, respectively). In a subsequent case series reported by these same authors, of three patients administered clarithromycin for one year, two elderly patients were able to discontinue prednisone therapy altogether [48].

Anti-inflammatory bronchial effects have also been reported with roxithromycin use. In a double-blind, placebo-controlled, crossover study [49] of 14 patients with asthma (aspirin-intolerant), a statistically significant decrease in patients' symptoms, serum eosinophil counts, sputum eosinophil levels, and serum and sputum eosinophilic cationic protein levels was noted after 8 weeks of therapy with 150 mg of roxithromycin given twice daily. Kamoi and collaborators [50] evaluated the impact of roxithromycin on neutrophil activation and bronchial hyperreactivity in 10 asthmatic patients who had been treated with 150 mg daily for 3 months, compared to 10 healthy control subjects. They reported a significant reduction ($p < 0.01$) in bronchial hyperreactivity and synthesis of free radicals (i.e., superoxide anion). Most patients required at least 2 months of macrolide therapy before a demonstrable clinical improvement was noted. Shimizu et al. [45] documented a significant reduction in airway hyperresponsiveness to a histamine challenge in 12 children hospitalized with asthma after 4 weeks ($p < 0.05$) and 8 weeks ($p < 0.01$) of therapy with roxithromycin, 150 mg daily. Shimizu et al. also reported that roxithromycin was noted to attenuate acid induced cough and water induced bronchoconstriction in children with asthma [51].

The efficacy of macrolide therapy in patients with asthma may not be based exclusively on their anti-inflammatory effects. Atypical intracellular pathogens (*Chlamydia pneumoniae* and *Mycoplasma pneumoniae*) may play a role in the pathogenesis of reactive airway diseases [52–55], and macrolides possess antimicrobial activity against these pathogens [56–60]. In one study [61], *M. pneumoniae* or *C. pneumoniae* was present in the airways (detected by polymerase chain reaction [PCR]) in more than half of stable patients with chronic asthma. Thus, it is difficult to distinguish between the anti-inflammatory and antimicrobial effects of macrolides compared with the beneficial responses in some patients with asthma.

There has been some discussion that perhaps macrolide therapy improves lung function in asthmatic patients by the eradication of some occult infection. A growing body of data suggests infection with *C. pneumoniae* may in some way be responsible for the pathogenesis of asthma [62–67]. Hahn and Golubjatnikov [65] treated 46 asthmatic patients with either azithromycin, 1 g once weekly, erythromycin, 1 g daily, or doxycycline (Vibramycin; Pfizer Pharmaceuticals; New York, NY), 100 mg twice daily for a median of 4 weeks. After treatment the mean FEV_1 (67.8% of predicted at baseline) increased by 12.5% ($p = 0.003$). Kraft et al. [66] conducted a double-blind study in which 52 patients with chronic stable asthma were randomized to therapy with either clarithromycin, 500 mg, or placebo twice daily for 6 weeks. Clarithromycin therapy was documented to significantly increase mean (± SD) FEV_1 measurements in those asthmatic patients who were PCR-positive for *Chlamydia* or *Mycoplasma* (baseline, 2.50 ± 0.16 L; post treatment, 2.69 ± 0.19 L; $p = 0.05$). In contrast, no improvement was noted in FEV_1 in those patients who were PCR-negative (baseline, 2.59 ± 0.24 L; post treatment, 2.54 ± 0.18 L; $p = 0.85$).

Black and colleagues [67] studied the effects of roxithromycin in patients with asthma who were antibody positive for *C. pneumoniae*. They randomized 232 asthmatic patients to roxithromycin, 150 mg, or placebo twice daily. After 6 weeks of therapy, patients treated with roxithromycin were documented to have a significantly increased nighttime peak expiratory flow (increase from baseline: roxithromycin, 15 L/min; placebo, 3 L/min; $p = 0.02$), but no significant change in the daytime peak expiratory flow (increase from baseline: roxithromycin, 14 L/min; placebo, 8 L/min). These benefits disappeared within 3 months of stopping the medication. Even though there was a trend towards improvement in the symptom score this was not considered significant. In their discussion the authors speculated that macrolide therapy might possibly have briefly mitigated the effects of *C. pneumoniae* infection on the patient's airways, with infection and its sequelae persisting after the discontinuation of macrolide treatment.

Conclusion

Macrolides, in addition to their antimicrobial properties, possess biological response-modifying mechanisms which are beneficial in asthma. Data published over the last few decades clearly indicates that therapy with the 14- and 15-membered ring macrolides improves the signs and symptoms of asthma, most likely as a result of their anti-inflammatory properties. Significantly, long-term administration of these antimicrobials has not been linked with the emergence of any clinically significant microbial resistance.

Overall, these effects of the macrolides seem to be associated with the downregulation of the nonspecific host inflammatory response to injury to tissues within the respiratory tract. Although precise mechanisms by which these effects occur remains to be identified, it may be likely that these agents act in at numerous points along the inflammatory cascade. In summary, the immunomodulatory/anti-inflammatory effects of the macrolides may improve both the symptoms and function in inflammatory respiratory diseases such as asthma. However, further large-scale studies are necessary to evaluate if macrolide antibiotics are beneficial and safe in asthmatic patients in both short- and long-term treatment and ultimately, improving quality of life.

References

1 Kadota J, Mukae H, Ishii H et al (2003) Long-term efficacy and safety of clarithromycin treatment in patients with diffuse panbronchiolitis. *Resp Med* 97: 844–50

2 National Asthma Education and Prevention Program. *Expert panel report 2: guidelines for the diagnosis and management of asthma*. Bethesda, MD: National Institutes of Health, US Department of Health and Human Services, 1997; NIH Publication No. 97–4051

3 Tamaoki J (2004) The effects of macrolides on inflammatory cells. *Chest* 125: 41S–51S

4 Gotfried MH (2004) Macrolides for the treatment of chronic sinusitis, asthma, and COPD. *Chest* 125: 52S–61S

5 Khair OA, Devalia JL, Abdelaziz et al (1995) Effect of erythromycin on *Haemophilus influenzae* endotoxin-induced release of IL-6, IL-8, and sICAM-1 by cultured human bronchial epithelial cells. *Eur Respir J* 8: 1451–7

6 Takizawa H, Desaki M, Ohtoshi M et al (1995) Erythromycin suppresses interleukin 6 expression by human bronchial epithelial cells: a potential mechanism of its anti-inflammatory action. *Biochem Biophys Res Commun* 210: 781–6

7 Kawasaki S, Takizawa H, Takayuki O, et al (1998) Roxithromycin inhibits cytokine production by and neutrophil attachment to human bronchial epithelial cells *in vitro*. *Antimicrob Agents Chemother* 42: 1499–502

8 Schultz M, Speelman P, Zaat S, et al (1998) Erythromycin inhibits tumor necrosis factor alpha killed and interleukin 6 production induced by heat-killed *Streptococcus pneumoniae* in whole blood. *Antimicrob Agents Chemother* 42: 1605–9

9 Konno S, Asano K Kurokawa M, et al (1994) Antiasthmatic activity of a macrolide antibiotic, roxithromycin: Analysis of possible mechanisms *in vitro* and *in vivo*. *Int Arch Allergy Immunol* 105: 308–16

10 Aoki Y, Kao PN (1999) Erythromycin inhibits transcriptional activation of NF-κB, but not NFAT, through calcineurin-independent signaling in T cells. *Antimicrob Agents Chemother* 43: 2678–84

11 Ichiyama T, Nishikawa M, Yoshitomi T et al (2001) Clarithromycin inhibits NF-κB activation in human peripheral blood mononuclear cells and pulmonary epithelial cells. *Antimicrob Agents Chemother* 45: 44–7

12 Abe S, Nakamura H, Inoue S et al (2000) Interleukin-8 gene repression by clarithromycin is mediated by the activator protein-1 binding site in human bronchial epithelial cells. *Am J Respir Cell Mol Biol* 22: 51–60

13 Desaki M, Takizawa H, Ohtoshi T et al (2000) Erythromycin suppresses nuclear factor-κB and activator protein-1 activation in human bronchial epithelial cells. *Biochem Biophys Res Comm* 267: 124–8

14 Kikuchi T, Hagiwara K Honda Y, et al (2002) Clarithromycin suppresses lipopolysaccharide-induced interleukin-8 production by human monocytes through AP-1 and NF-κB transcription factors. *J Antimicrob Chemother* 49: 745–55

15 Baggiolini M, Wlz A, Kunkel SL (1989) Neutrophil-activating peptide-1/interleukin-8, a novel cytokine that activates neutrophils. *J Clin Invest* 84: 1045–9
16 Standiford TJ, Kunkel SL, Basha MA et al (1990) Interleukin-8 gene expression by a pulmonary epithelial cell line: a model for cytokine networks in the lung. *J Clin Invest* 86: 1945–53
17 Shibata Y, Nakamura H, Kato S et al (1996) Cellular detachment and deformation induce IL-8 gene expression in human bronchial epithelial cells. *J Immunol* 156: 772–7
18 Fujii T, Kadota JI, Morikawa T et al (1996) Inhibitory effect of erythromycin on interleukin 8 production by 1 alpha, 25-dihydroxyvitamin D3-stimulated THP-1 cells. *Antimicrob Agents Chemother* 40: 1548–51
19 Tamaoki J, Takeyama K, Yamawaki I et al (1997) Lipopolysaccharide-induced goblet cell hypersecretion in the guinea pig trachea: inhibition by macrolides. *Am J Physiol* 272: L15–L19
20 Miyajima M, Suga M, Nakagawa K et al (1999) Effect of erythromycin on experimental extrinsic allergic alveolitis. *Clin Exp Allergy* 29: 253–61
21 Fujii Y, Goldberg P, Hussain SNA (1998) Contribution of macrophages to pulmonary nitric oxide production in septic shock. *Am J Respir Crit Care Med* 157: 1645–51
22 Kennedy NJ, Duncan AW (1996). Acute meningococcaemia: recent advances in management (with particular reference to children. *Anaesth Intensive Care* 24: 197–216
23 Tamaoki J, Kondo M, Kohri K et al (1999) Macrolide antibiotics protect against immune-complex-induced lung injury in rats: role of nitric oxide from alveolar macrophages. *J Immunol* 163: 2915–29
24 Labro MT, El Benna J, Babin-Chevaye C (1989) Comparison of the *in vitro* effect of several macrolides on the oxidative burst of human neutrophils. *J Antimicrob Chemother* 24: 561–72
25 Hand WL, Hand D, King-Thompson N (1990). Antibiotic inhibition of the respiratory burst response in human polymorphonuclear leukocytes. *Antimicrob Agents Chemother* 34: 863–70
26 Anderson R (1989) Erythromycin and roxithromycin potentiate human neutrophil locomotion *in vitro* by inhibition of leukoattractant activated superoxide generation and autooxidation. *J Infect Dis* 5: 966–72
27 Theron AJ, Feldman C, Anderson R (2000) Investigation of the anti-inflammatory and membrane-stabilizing potential of spiramycin *in vitro*. *J Antimicrob Chemother* 46: 263–71
28 Takizawa H, Desaki M, Ohitoshi T et al (1998) Erythromycin and clarithromycin attenuate cytokine-induced endothelin-1 expression in human bronchial epithelial cells. *Eur Resp J* 12: 57–63
29 Scaglione F, Rossoni G (1998) Comparative anti-inflammatory effects of roxithromycin, azithromycin, and clarithromycin. *J Antimicrob Chemother* 41 (Suppl B): 47–50
30 Agen C, Danesi R, Blandizzi C et al (1993) Macrolide antibiotics as anti-inflammatory agents: roxithromycin in an unexpected role. *Agents Actions* 38: 85–90
31 Kaplan MA, Goldin M (1959) The use of triacetyloleandomycin in chronic infectious

asthma. In: Welsh H, Marti-Ibanez F (eds): *Antibiotic Annual 1958–1959*. Interscience Publishers, New York, NY, 273–6
32 Zieger RS, Schatz M, Sprerling W et al (1980) Efficacy of troleandomycin in outpatients with severe, corticosteroid-dependent asthma. *J Allergy Clin Immunol* 66: 438–46
33 Wald JA, Friedman BF, Farr RS (1986) An improved protocol for the use of troleandomycin (TAO) in the treatment of steroid-requiring asthma. *J Allergy Clin Immunol* 78: 36–43
34 Ball BD, Hill MR, Brenner M et al (1990) Effect of low-dose troleandomycin on glucocorticoid pharmacokinetics and airway hyperresponsiveness in severely asthmatic children. *Ann Allergy* 65: 37–45
35 Flotte TR, Loughlin GM (1991) Benefits and complications of troleandomycin (TAO) in young children with steroid-dependent asthma. *Pediatr Pulmonol* 10: 178–82
36 Kamada AK, Hill MR, Ikle DN et al (1993) Efficacy and safety of low-dose troleandomycin therapy in children with severe, steroid-requiring asthma. *J Allergy Clin Immunol* 91: 873–82
37 Eitches RW, Rachelefsky GS, Katz RM et al (1985) Methylprednisolone and troleandomycin in treatment of steroid-dependent asthmatic children. *Am J Dis Child* 139: 264–8
38 Siracusa A, Brugnami G, Fiordi T et al (1993) Troleandomycin in the treatment of difficult asthma. *J Allergy Clin Immunol* 92: 677–82
39 Spector S, Katz F, Farr R (1974) Troleandomycin: effectiveness in steroid dependent asthma and bronchitis. *J Allergy Clin Immunol* 54: 367–79
40 Weinberger M, Hudgel D, Spector S, Chisey C (1977) Inhibition of theophylline clearance by troleandomycin. *J Allergy Clin Immunol* 59: 228 31
41 Rosenberg SM, Gerhard H, Grunstein MM (1991) Use of TAO without methylprednisolone in the treatment of severe asthma. *Chest* 100: 849 50
42 Spahn JD, Fost DA, Covar R et al (2001) Clarithromycin potentiates glucocorticoid responsiveness in patients with asthma: results of a pilot study. *Ann Allergy Asthma Immunol* 87: 501–5
43 Miyatake H, Taki F, Taniguchi H et al (1991) Erythromycin reduces the severity of bronchial hyperresponsiveness in asthma. *Chest* 99: 670–3
44 Amayasu H, Yoshida S, Ebana S et al (2000) Clarithromycin suppresses bronchial hyperresponsiveness associated with eosinophilic inflammation in patients with asthma. *Ann Allergy Asthma Immunol* 84: 594–8
45 Shimizu T, Kato M, Mochizuki H et al (1994) Roxithromycin reduces the degree of bronchial hyperresponsiveness in children with asthma. *Chest* 106: 458–61
46 Ekici A, Ekici M, Erdemoglu AK (2002) Effect of azithromycin on the severity of bronchial hyperresponsiveness in patients with mild asthma. *J Asthma* 39(2): 181–5
47 Gotfried MH, Jung R, Messick C et al (2004) Effects of six week clarithromycin therapy in corticosteroid-dependent asthma: A randomized double-blind, placebo-controlled pilot study. *Curr Ther Res* 65: 1–12
48 Garey KW, Rubinstein I, Gotfried MH et al (2000) Long-term clarithromycin decreases

prednisone requirements in elderly patients with prednisone-dependent asthma. *Chest* 118: 1826–7
49 Shoji T, Yoshida S, Sakamoto H et al (1999) Anti-inflammatory effect of roxithromycin in patients with aspirin-intolerant asthma. *Clin Exp Allergy* 29: 950–6
50 Kamoi H, Kurihara N, Fujiwara H et al (1995) The macrolide antibacterial roxithromycin reduces bronchial hyperresponsiveness and superoxide anion production by polymorphonuclear leukocytes in patients with asthma. *J Asthma* 32: 191–7
51 Shimizu T, Kato M, Mochizuki H et al (1997) Roxithromycin attenuates acid-induced cough and water-induced bronchoconstriction in children with asthma. *J Asthma* 34: 211–17
52 Emre U, Roblin PM, Gelling M et al (1994) The association of *Chlamydia pneumoniae* infection and reactive airway disease in children. *Arch Pediatr Adolesc Med* 148: 727–32
53 Hahn DL, Bukstein D, Luskin A et al (1998) Evidence for *Chlamydia pneumoniae* infection in steroid-dependent asthma. *Ann Allergy Asthma Immunol* 80: 45–9
54 Kraft M, Cassell GH, Henson JE et al (1998) Detection of *Mycoplasma pneumoniae* in the airways of adults with chronic asthma. *Am J Respir Crit Care Med* 158: 998–1001
55 Black PN, Scicchitano R, Jenkins CR et al (2000) Serological evidence of infection with *Chlamydia pneumoniae* is related to the severity of asthma. *Eur Respir J* 15: 254–9
56 Waites KB, Cassell GH, Canupp KC et al (1988) *In vitro* susceptibilities of mycoplasmas and ureaplasmas to new macrolides and aryl-fluoroquinolones. *Antimicrob Agents Chemother* 32: 1500–2
57 Fenelon LE, Mumtaz G, Ridgway GL (1990) The *in vitro* antibiotic susceptibility of *Chlamydia pneumoniae*. *J Antimicrob Chemother* 26: 763–7
58 Critchley IA, Jones ME, Heinze PD et al (2002) *In vitro* activity of levofloxacin against contemporary clinical isolates of *Legionella pneumophila*, *Mycoplasma pneumoniae*, and *Chlamydia pneumoniae* from North America and Europe. *Clin Microbiol Infect* 8: 214–21
59 Hammerschlag MR, Qumei KK, Roblin PM (1992) *In vitro* activities of azithromycin, clarithromycin, L-ofloxacin, and other antibiotics against *Chlamydia pneumoniae*. *Antimicrob Agents Chemother* 36: 1573–4
60 Renaudin H, Bebear C (1990) Comparative *in vitro* activity of azithromycin, clarithromycin, erythromycin and lomefloxacin against *Mycoplasma pneumoniae*, *Mycoplasma hominis* and *Ureaplasma urealyticum*. *Eur J Clin Microbiol Infect Dis* 9: 838–41
61 Kraft M, Cassell GH, Pak J et al (2002) Mycoplasma pneumoniae and *Chlamydia pneumoniae* in asthma: effect of clarithromycin. *Chest* 121: 1782–8.
62 Cook PJ, Davies P, Tunnicliffe W et al (1998) *Chlamydia pnuemoniae* and asthma. *Thorax* 53: 254–9
63 Hahn DL, Dodge RW, Golubjatnikov R (1991) Association of *Chlamydia pneumoniae* (strain TWAR) infection with wheezing, asthmatic bronchitis and adult onset asthma. *J Am Med Assoc* 266: 255–30

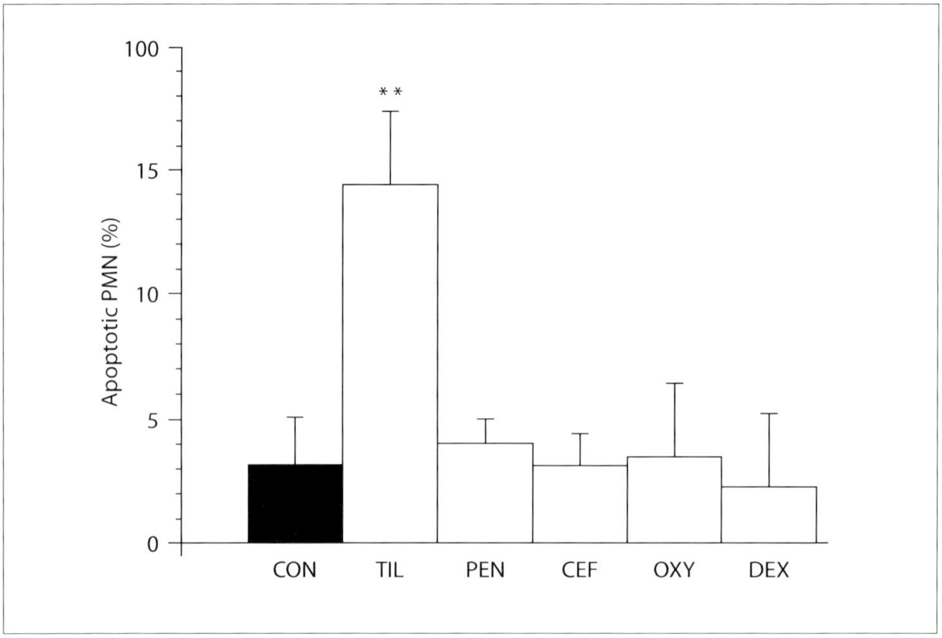

Figure 1
*Percentage of apoptotic peripheral bovine PMNs incubated for 2 h with either 1 × PBS (CON), 0.5 µg of tilmicosin per ml (TIL), 0.5 µg of penicillin per ml (PEN), 0.5 µg of ceftiofur per ml (CEF), 0.5 µg of oxytetracycline per ml (OXY), or 10^8 M dexamethasone (DEX). Values are means ± standard errors of the means; n = 5 per experimental group. **,* P *< 0.01* versus *control.*

presence or absence of live *P. haemolytica*, while other antibiotics, including penicillin, ceftiofur, and oxytetracycline, did not induce neutrophil apoptosis (Fig. 1).

These experimental results suggest that, though some macrolides induce neutrophil apoptosis, factors such as the presence or absence of bacilli may affect inflammation. Macrolides decrease the lifespan of activated neutrophils by accelerating apoptosis and phagocytosis of apoptotic neutrophils by macrophages, in contrast to steroids, which prolong the survival of neutrophils. The apoptosis of neutrophils induced by macrolides appears to markedly affect chronic inflammation.

Inhibition of radical formation and neutrophil migration into tissue

Tamaoki and colleagues reported that the macrolide compounds erythromycin and josamycin, but neither amoxicillin nor cephaclor, inhibited IgG immune-complex

(IgG-ICx)-induced lung injury in a rat model [4]. They found that the inhibitory effects of macrolides on IgG–ICx-induced lung injury were probably associated with reduction of cytokine release and induction of nitric oxide synthase. Erythromycin also inhibited lipopolysaccharide (LPS)-induced acute lung injury, by decreasing vascular leakage (6.7 ± 1.2 to 1.4 ± 0.3% ($p < 0.01$)), neutrophil recruitment (365 ± 51 to 149 ± 30 cells/mm^2 ($p < 0.01$)), and the wet/dry ratio of the lung (6.76 ± 0.30 to 5.39 ± 0.21 ($p < 0.01$)) [5]. Pretreatment with erythromycin was more effective than concurrent or post-treatment with it. The priming effect of erythromycin may stabilize the hyperresponsiveness of mice infected with influenza. In a model of influenza virus (A/Kumamoto/Y5/67 (H_2N_2)) infection, concurrent administration of erythromycin (from days 1 to 6 after infection) significantly improved the rate of survival of infected mice. The rate of survival of virus-infected mice at day 20 after infection increased in dose-dependent fashion with administration of erythromycin (control 14%, erythromycin 1.0 mg/kg/d 42% and 3.3 mg/kg/d 57%) [6] (Fig. 2). Erythromycin inhibited induction of IFN-γ, a key molecule promoting lymphocyte alveolitis in influenza virus-induced lung injury. Erythromycin may thus have significant therapeutic value for various types of acute inflammation such as influenza-induced lung injury.

Macrolide treatment of drug-induced lung injury

We have found that macrolides improve drug-induced acute lung injury and subsequent fibrosis in a rat model. In this model neutrophils migrated into the airways and released injurious substances, such as oxygen radicals and proteases, resulting in epithelial damage and fibroblast proliferation. Erythromycin inhibited migration of neutrophils into the airways, resulting in significant decrease in levels of neutrophil-derived elastase [7]. The effect of erythromycin was stronger when it was administered prophylactically than after bleomycin instillation.

In addition, in a mouse model, bleomycin-induced pulmonary fibrosis was attenuated by treatment with 14-membered ring macrolides including erythromycin, clarithromycin, and roxithromycin. Treatment with 14-membered ring macrolides inhibited induction of mRNAs of some adhesion molecules such as vascular cell adhesion molecule-1 (VCAM-1) and intercellular adhesion molecule-1 (ICAM-1), but not that of mRNAs of selectins [8] (Fig. 3). These findings indicate that attenuation of inflammatory cell migration into the airspace by 14-membered ring macrolides, especially that of neutrophils and macrophages, inhibited lung injury and subsequent fibrosis. Fourteen-membered ring macrolides inhibited neutrophil migration and release of injurious substances from them, resulting in attenuation of epithelial destruction. In another study, roxithromycin directly inhibited the proliferation of fibroblasts derived from nasal polyps [9]. We expect prophylactic administration of 14-membered ring macrolides to be clinically effective in preventing

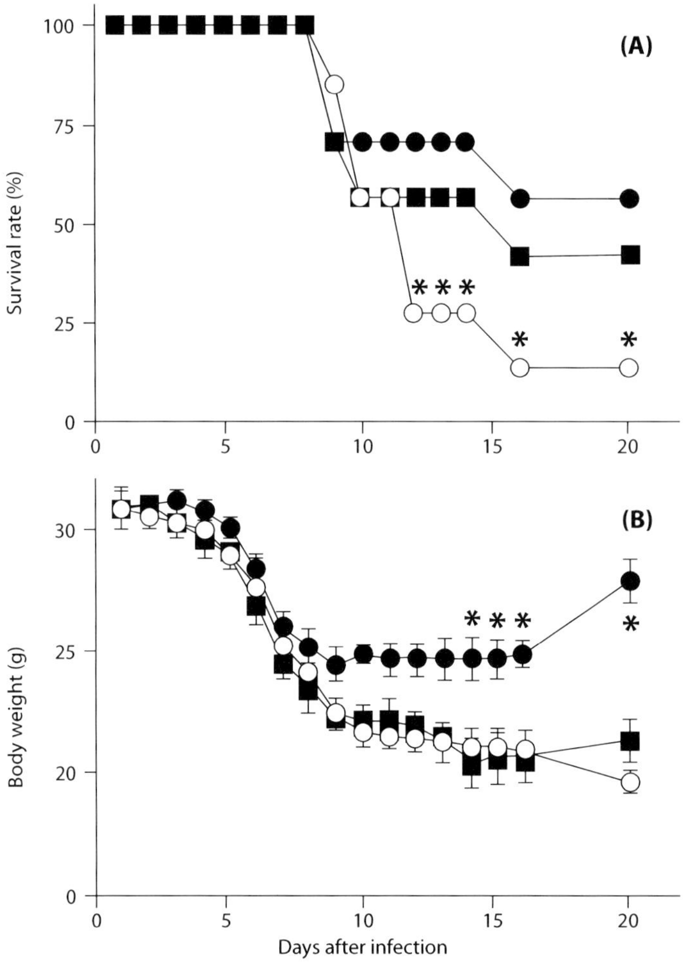

Figure 2

Effect of EM on survival rate (A) and body weight (B) of mice infected with influenza virus. At each time point after influenza virus infection [1.5 × LD50 of influenza virus (A/Kumamoto/Y5/H_2N_2)], the survival rate of infected mice was evaluated and the body weight of mice was measured. The mice were given EM intraperitoneally (solid squares: 1.0 mg/kg/d, solid circles: 3.3 mg/kg/d in saline/0.5% DMSO) every 24 h from Day 1 to Day 6 after virus infection. The control group (open circles) was injected intraperitoneally with 0.5 ml saline/0.5% DMSO. Fourteen mice were used in each experimental group: $p < 0.05$, control versus *EM-treated groups.*

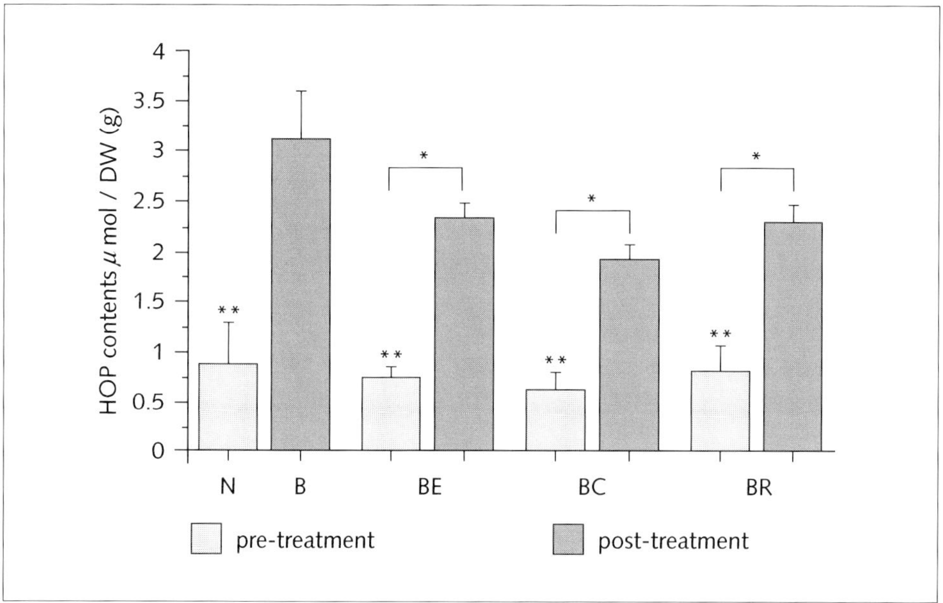

Figure 3
*Comparison of hydroxyproline contents. Significantly different from bleomycin alone-treated group by the Mann-Whitney U test, **$p < 0.01$; significantly different from 14-MRML-pretreated groups by the Mann-Whitney U test, $p < 0.01$. Values are means; bars = SD, $n = 10$; HOP, hydroxyproline; DW, dry lung weight; N, normal saline-treated group; B, bleomycin-alone-treated group; BE, bleomycin- and erythromycin-treated group; BC, bleomycin- and clarithromycin-treated group; BR, bleomycin and roxithromycin-treated group.*

acute lung injury, acute exacerbation of interstitial pneumonia, and progression of fibrosis. This expectation is supported by similar experimental findings [10].

Recent developments concerning the anti-inflammatory effects of ketolides

New ketolide compounds, such as HMR 3647, are active against intracellular pathogens. Ketolides include a 3-keto group instead of an L-cladinose, a neutral sugar characteristic of erythromycin-A derivatives. Recently, HMR3647, which is now available for treatment of respiratory infectious diseases, was reported to exhibit anti-inflammatory effects. The anti-inflammatory activities of ketolides are mainly due to time- and concentration-dependent inhibition of the production of superoxide anions [11]. HMR3647 is avidly uptaken and concentrated by polymorphonuclear cells (PMNs), with cellular-to-extracellular concentration ratios of 31 ± 4.2 at 5 min and up to 348 ± 27.1 at 3 h. HMR3004, a derivative of HMR3647, also inhibited oxidant production by PMNs [12]. HMR3004 exhibited membrane-stabilizing potential, as well as effects on the production of superoxide by human neutrophils activated by several stimulants. Labro and colleagues compared the anti-inflammatory effects of new ketolides, HMR3647 and HMR3004, with those of roxithromycin (RXM) [13]. TNF-α and GM-CSF each decreased the inhibitory effect of HMR3647 on oxidant production by PMNs. The 50% inhibitory concentrations of HMR3647 were in the same range for control and cytokine-treated cells. These findings suggest that HMR3647 acts downstream of the priming effect of cytokines [13]. In contrast, the decrease in production of oxidants induced by RXM and HMR3004 was not changed (or an increase in production of oxidants was observed) with treatment of cells with TNF-α or GM-CSF. The authors hypothesized that the effects of TNF-α and GM-CSF may be associated with protein kinase A- and tyrosine kinase-dependent phosphorylation, which is necessary for optimal uptake of macrolides into cells. In other words, accumulation of macrolides in cells regulated by protein kinases modulates the inhibitory effect of macrolides on oxidant production. The quinoline side chain of HMR3004 plays a key role in inhibition of oxidant synthesis.

In another study, HMR3004 significantly prevented recruitment of neutrophils and monocytes into lung and attenuated IL-6 release and nitric oxide production in lung tissue infected with *Streptococcus pneumoniae*. HMR3004 protected interstitial tissue against edema and led to rapid and profound modifications of the host response in lungs, which may protect mice from deleterious inflammatory reactions by controlling bacterial invasion. Both anti-inflammatory and antimicrobial effects of ketolides are exhibited in bacterially infected lung tissue [14].

Summary

Lung injury is in many cases ameliorated by treatment with macrolides. The principal mechanism of action of macrolides is inhibition of neutrophil activity. The effects of macrolides are often compared with those of corticosteroids. Many investigations have found significant inhibitory effects of macrolides on the activity of neutrophils associated with inflammation, but the degree of such inhibition used to be only partial, in contrast to inhibition by corticosteroids, which decrease most cytokine activities to baseline level. It thus appears that macrolides may stabilize over-responsiveness of several factors in inflammation to physiological levels, but that corticosteroids have immunosuppressive effects and often result in immunocompromised status associated with fungal infections, *Pneumocystis carinii pneumoniae*, and cytomegalovirus infection. More detailed investigations are needed to clarify the differences between macrolides and corticosteroids in effects on lung

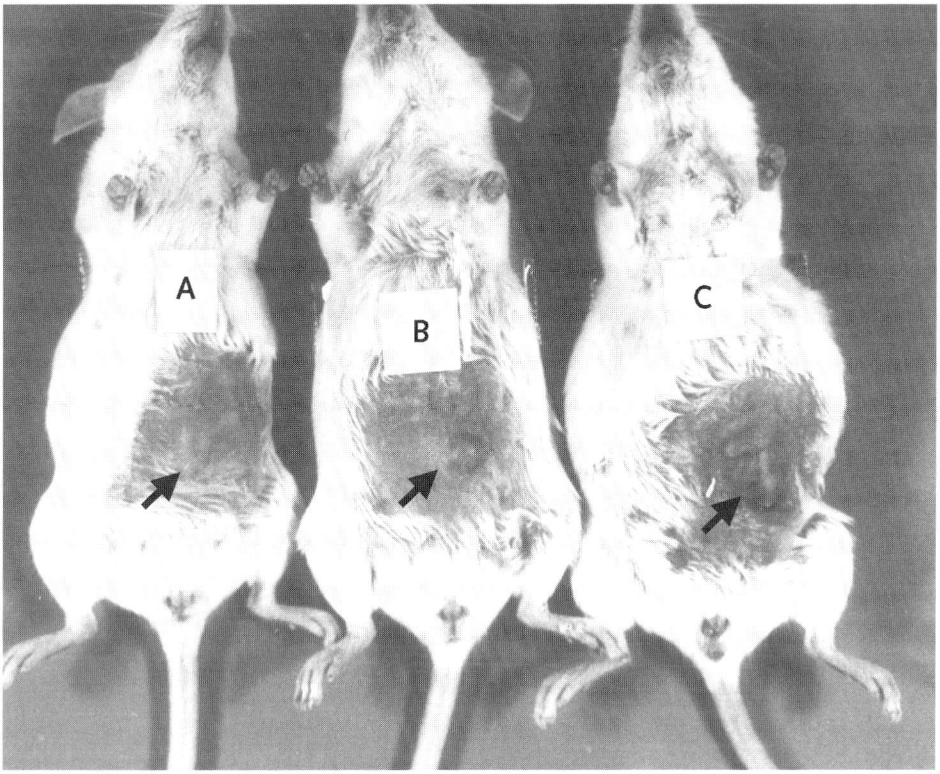

Figure 1
Antitumor effect of EM in ddY mice injected subcutaneously with 5 × 10^6 cells of Ehrlich ascites carcinoma
(A) no apparent tumor formation in mice receiving 5 mg/kg of EM at 60 days after inoculation; (B) apparent tumor formation at 7 days after inoculation and further growth of the tumor at 30 days in mice receiving vehicle treatment at 30 days after inoculation.
(from Chemotherapy *(1995) 41: 59–69, with permission of S. Karger AG, Basel)*

occurred following the increase in NK activity of spleen cells [10]. These facts suggested that NK cells accounted for the increased production of IFN-γ in EM-treated mice. Such responses in tumor-bearing mice were more prominent when they received chemotherapy than when they received vehicle therapy. Treatment with CAM for 2–4 weeks following chemotherapy significantly suppressed the growth of Lewis tumor (3LL) cells in the lung, and afforded long-term mouse survival [10]. Similar to EM therapy, long-term treatment of CAM increased the number of IL-4-producing cells in the spleen of 3LL-bearing mice, which accounted for the induc-

Table 1 - Effect of CAM treatment following chemotherapy on NK activity in the spleen of tumor-bearing mice.

Treatment	NK activity, %			
	Day 1	Day 7	Day 14	Day 21
Untreated mice	5.7 ± 1.4	8.6 ± 2.3	12.3 ± 4.5	–
Mice receiving chemotherapy (day 7)				
Without CAM treament	–	–	6.3 ± 2.2	19.4 ± 4.8
CAM treatment from day 7	–	–	15.7 ± 5.9	23.6 ± 6.2
CAM treatment from day 14	–	–	–	43.6 ± 7.2

Significance brackets printed beside the table: Day 14: untreated vs. without CAM, a; without CAM vs. CAM from day 7, a; untreated vs. CAM from day 7, b. Day 21: without CAM vs. CAM from day 7, b; CAM from day 7 vs. CAM from day 14, a; without CAM vs. CAM from day 14, a.

C57Bl/6 mice bearing Lewis lung carcinoma were given chemotherapy 7 days after tumor inoculation (day 0), and then received CAM treatment for 7 (until day 7) or 14 days (until day 14) just after chemotherapy.
Each cytotoxic test was done in triplicate. Data were obtained from 3 different experiments and expressed as the mean ± SD of 9 determinants.
–, not tested; a, $p<0.05$; b, not significant.

tion and activation of cytotoxic macrophages (Tab. 2) [10]. This indicated that long-term treatment of 14-membered ring macrolides stimulated both Th1 and Th2 responses in tumor-bearing mice. In addition, the number of $CD8^+$ T cells cytotoxic to 3LL cells increased in the lung of the tumor-bearing C57BL/6 mice [10]. In contrast to the enhanced production of IFN-γ, IL-4 and IL-12, the long-term therapy decreased serum levels of TNF-α and IL-6 in tumor-bearing mice; accordingly, these animals lost body weight to a lesser extent compared with vehicle controls. Taken together, the long-term treatment of 14-membered ring macrolides may possibly induce the well-balanced resistance (namely, the balance between Th1 and Th2 responses) to the growth of tumor cells as well as to prevent nutritional decline. More importantly, these beneficial effects resulting from the long-term macrolide therapy are more active in mice receiving anti-cancer therapy than those receiving vehicle treatments.

There were no double-blind trials on the long-term macrolide therapy as an adjuvant treatment for anti-cancer therapy. However, Mikasa et al. [11] performed a prospective randomized trial; treatment with CAM (400 mg/day) following prior basic anti-cancer therapy was compared to the same basic therapy alone in patients

Table 2 - Effect of CAM treatment following chemotherapy on the number of cytokine-producing T cell in the spleen of tumor-bearing mice.

Treatment	Cytokine-producing T cell number/106 cells					
	day 7		day 14		day 21	
	IFN-γ	IL-4	IFN-γ	IL-4	IFN-γ	IL-4
Untreated mice	228 ± 64	21 ± 10	206 ± 53	42 ± 7	–	–
			a ┐	a ┐		
Mice receiving chemotherapy (day 7)						
Without CAM treament	–	–	86 ± 33 ┤ c	19 ± 6 ┤ c	108 ± 41	32 ± 13
			a ┘	a ┘	a ┐	a ┐
CAM treatment from day 7	–	–	247 ± 42	67 ± 14	423 ± 52 ┤ b	132 ± 29 ┤ b
					a ┘	a ┘
CAM treatment from day 14	–	–	–	–	895 ± 93	256 ± 31

C57Bl/6 mice bearing Lewis lung carcinoma were given chemotherapy 7 days after tumor inoculation (day 0), and then received CAM treatment for one week from day 7 (until day 14) or for two weeks (until day 21).
Data were obtained from 2 separate experiments with 3 mice each and expressed as the mean ± SD of 6 determinants.
–, not tested; a, $p<0.05$; b, $p<0.01$; c, not significant.

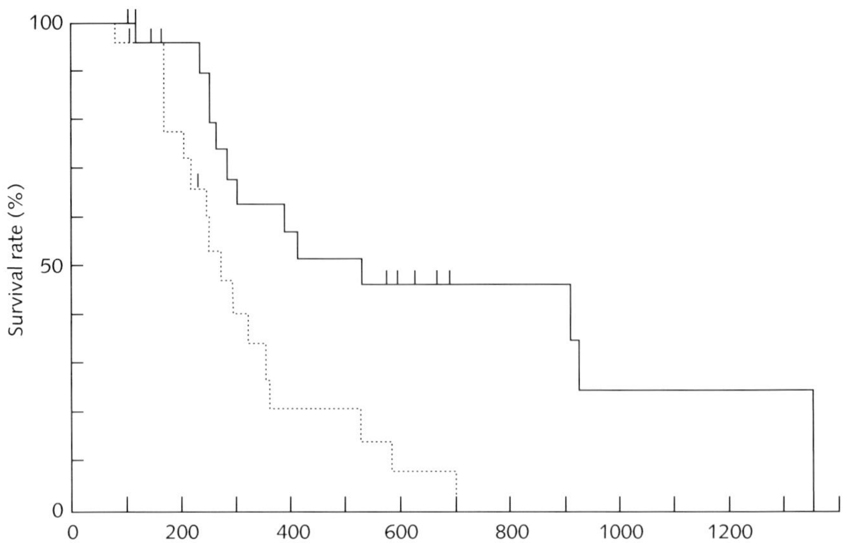

Figure 2
Estimates of survival according to treatment using the Kaplan-Meier methods for patients with unresectable NSC lung cancer
Vertical bars indicate patients still alive. solid line: CAM group (n = 22), dotted line: non-CAM group (n = 20), p = 0.0132 by the generalized Wilcoxon method, p = 0.0032 by the log-rank test.
(from Chemotherapy *(1997) 43: 288–96, with permission of S. Karger AG, Basel)*

with advanced NSC lung cancer (stages IIIA, IIIB and IV). The endpoint of this trial was survival: the median survival time was calculated by the method of Kaplan and Meier, and the statistical significance of the difference in median survival time between the CAM group and the control group was analyzed by the generalized Wilcoxon method. They showed that the long-term treatment with CAM following radiation, chemotherapy or their combination therapy significantly increased the median survival time for NSC lung cancer patients, which was estimated by the method of Kaplan and Meier. The median survival for the CAM-treated group was 535 days and that for the non-CAM group was 277 days ($p = 0.0132$ by the generalized Wilcoxon, $p = 0.0032$ by the log-rank test) (Fig. 2). Furthermore, multivariate analysis of prognostic factors by the Cox proportional hazard model demonstrated that only treatment with CAM was predictive of longer survival for NSC lung cancer ($p = 0.0181$; hazard ratio = 0.23).

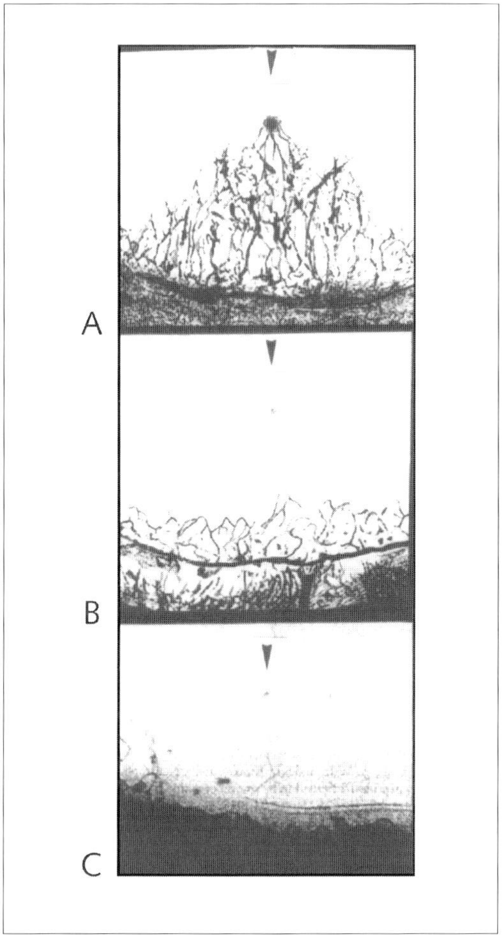

Figure 3
Suppression of neovascularization by CAM in the rabbit cornea
Injection of human lung cancer cell (A549) extract into a rabbit corneal pocket induced vascularization in the cornea (A), while co-inoculation of monoclonal antibody to IL-8 almost completely abolished its potency (C). Extract of the cells cultured with CAM (2-5 μg/ml) decreased its potency (B).

Following their report, many investigators attempted to determine whether or not CAM had a direct action on the growth of tumor cells *in vitro* or *in vivo*. Sawaki et al. [12] first reported that CAM was able to suppress the tumor-induced angiogenesis (Fig. 3), the degree of which was apparently related to its ability to suppress IL-8 production. Subsequently, Yatsunami et al. [13] demonstrated the

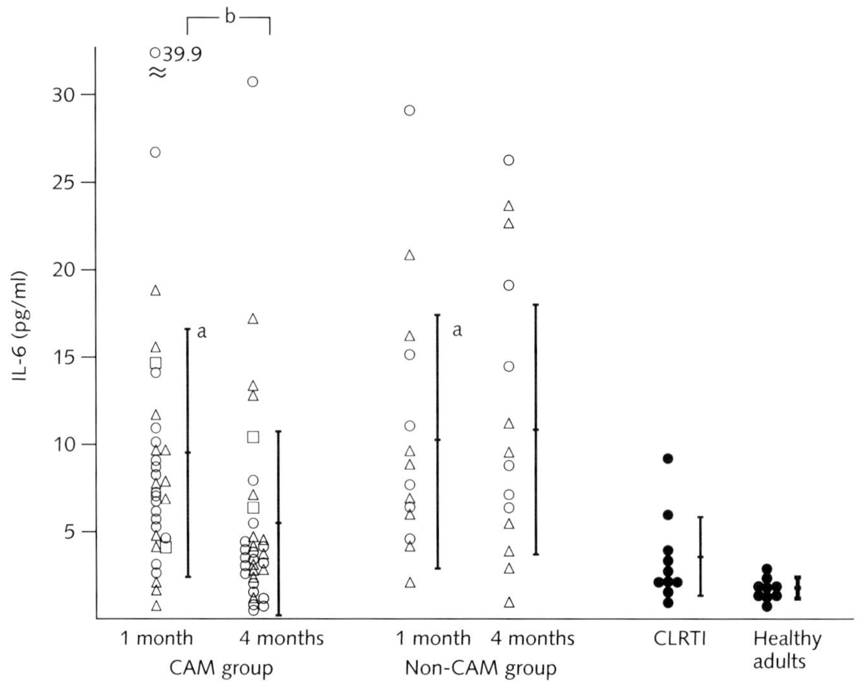

Figure 4
Changes in serum IL-6 levels before (1 month after basic cancer therapy) and after CAM treatment (4 months after basic cancer therapy)
At 3 months of CAM treatment, serum levels of IL-6 were significantly decreased. a: $p < 0.05$ compared with IL-6 levels of patients with chronic lower respiratory tract infections (CLRTI) and healthy adults, b: $p < 0.05$ compared with IL-6 levels before CAM treatment. ○ = adenocarcinoma; □ = large cells carcinoma; △ = squamous cell carcinoma; ● = controls.
(from Chemotherapy *(1995) 41: 59–69, with permission of S. Karger AG, Basel)*

potential of CAM as an inhibitor of tumor-induced angiogenesis through its suppressive effects on endothelial cell tube formation. The report by the same group [14] demonstrated the inhibitory effects of roxithromycin (RXM), a 14-membered ring semi-synthetic macrolide, on tumor angiogenesis and growth of mouse B16 melanoma cells; this activity was assumed to suppress the metastasis of implanted melanoma cells in the tumor-bearing mice. Nevertheless, CAM alone was unable to control the spread of NSC lung cancer in SCID mice [15], despite the fact that CAM augmented the IL-2-induced killer (LAK) activity *in vitro* [15], and also that CAM treatment decreased metastatic development in patients with NSC lung cancer [11]. The inability to control the spread of NSC lung cancer in SCID mice seemed to be accounted for by their severe innate T-cell dysfunction, and suggested that at least some of the macrolide effect was due to inadvertently boosting host immunity. Teramoto et al. [16] first demonstrated the ability of CAM to induce high levels of IL-12 and IFN-γ expression in the vicinity of tumor lesion surgically resected from the lung of patients with NSC lung cancer; IL-12 mRNA was not expressed in NSC lung cancer patients receiving drugs other than 14-membered ring macrolides. More importantly, there was a statistically significant correlation between the decrease in serum IL-6 and longer survival time in CAM-treated patients with NSC lung cancer (Fig. 4) [17], which may suggest that nutritional status is an important factor for the full expression of CAM's adjuvant effect on standard anticancer therapy.

As mentioned previously [9–11], treatment with 14-membered ring macrolides suppressed metastasis of implanted tumor cells in both mouse models and patients. In this regard, a recent report by Sasaki et al. [18] has shown that EM and CAM have the ability to modulate the growth factor-induced expression of heparanase mRNA on human lung cancer cells *in vitro*. Heparan sulfate proteoglycans (HSPGs) are the major component of extracellular matrix protein, and also important structures in basement membranes. This type of cell-surface proteoglycans is involved in cell adhesion, migration, proliferation, and angiogenesis [19, 20]. Initial process of tumor cell metastasis may be attributable to the action of several types of proteolytic enzymes, including matrix metalloproteinases, serine and cysteine proteases, and heparanases [21–23]. Moreover, overexpression of heparanase in tumor cells is shown to confer a high metastatic potential in mouse tumor models [24, 25]. Recently, Kita et al. [26] have demonstrated that RXM suppressed syndecan-1 shedding, upon microbial adhesion, from human airway epithelia: syndecan-1 is the major constituent of HSPGs on human epithelial cells. The suppression of syndecan-1 shedding by RXM was dependent on its ability to inhibit the activation of matrix metalloproteinase-7 (MMP-7: Matrilysin) [26]. Furthermore, recent clinical investigations [27, 28] have shown that high serum syndecan-1 levels at diagnosis were associated with poor outcome prognosis in lung cancer. Taken together, inhibition of tumor metastasis by 14-membered ring macrolides is most probably related to their suppressive activity in several classes of proteolytic enzymes.

Finally, it is worth noting that the long-term administration of 14-membered ring macrolides has a high inhibitory potential in the development of lung tumor initiated by *N*-nitrosobis (2-hydroxypropyl) amine in rats [29]. This novel effect is assumed to result from the anti-inflammatory effect of the macrolides, based on the analyses of microbial culture and histological examination in the airway during tumor development. In connection with this effect, inhibition of Martilysin by these macrolides may be crucial to the anti-inflammatory action of 14-membered ring macrolides, since this proteinase plays a major role in the formation of transepithelial gradient of chemokines in the respiratory tract [30]. Persistent microbial infection is defined as one of etiological factors for chronic inflammatory diseases such as atherosclerosis, and gastric ulcer, and also some neoplastic diseases. Further, a double-blind placebo-controlled randomized trial [31] has shown that prophylactic use of RXM in combination with ciprofloxacin reduced the incidence of febrile leucopenia during the standard chemotherapy in the treatment of lung cancer. This effect was primarily accounted for by the prophylactic capacity of both antibiotics in preventing infection; however it could not be ruled out that RXM exhibited the overall anti-inflammatory effects. The group of 14-membered ring macrolides therefore might weigh up the potential benefits of prophylactic use in patients at risk of cancer development initiated by persistent microbial infection and also for chemotherapy-induced febrile leucopenia, since long-term administration of the drugs has never brought any life-threatening side effects.

In conclusion, long-term administration of 14-membered ring macrolides as an adjuvant treatment for anticancer therapy may possibly benefit patients with NSC lung cancer by increasing the overall tumor resistance and by improving nutritional status. Beneficial effects of 14-membered ring macrolides as an adjuvant drug therefore seem to be attributed to its biological response modifier (BRM) activity, which is characterized by the ability to induce the well-balanced immunological response between Th1 and Th2, by the anti-inflammatory activity including the suppression of proinflammatory cytokine/chemokine production, and by the regulatory effect of proteinase activation.

Macrolides *versus* arthritis

Until now, most investigators have been interested in infection-induced arthritis as well as rheumatoid arthritis (RA) for the target of macrolide treatment. The first use of macrolide compounds was experimentally in the treatment of arthritis induced by *Borrelia burgdorferi* in hamsters, which mimicked Lyme disease in humans [32]. CAM was effective in preventing the onset of *B. burgdorferi*-induced arthritis as determined by several parameters of paw swelling; this effect was due to its direct antimicrobial activity that was at least 1 log more potent than tetracycline against clinical isolates of the pathogen. Even after the onset of arthritis, CAM therapy was able to reduce the degree of swelling and shorten recovery time. The latter effect may possibly result from anti-inflammatory action of this macrolide.

The next experimental evidence for the availability of macrolides in the infection-induced arthritis was obtained from experimental therapy with azithromycin (AZM) for septic arthritis caused by group B streptococci (GBS) [33]. A single intraperitoneal dose of AZM (100 mg/kg) strongly reduced the incidence or articular lesions in CD-1 mice infected with 10^7 cfu of type IV GBS, the level of which was much higher than that afforded by EM or penicillin G. Three repeated intraperitoneal injections of AZM (50 mg/kg at 12 h intervals) resulted in the complete prevention of arthritis in infected mice. Moreover, AZM was able to cure about 70% of infected mice when administered on days 7, 8, and 9 of infection. These effects of this macrolide may be accounted for by its higher ability of bacterial killing, longer half-life, and higher affinity for the joints. The fact that delayed initiation of AZM treatment was still effective in this infection model might imply the capacity of this macrolide to regulate the inflammatory process in the affected joints. In this regard, AZM was shown to decrease extracellular release of lysosomal enzymes in the synovial fluid of the injected hind paw of rats that had received a single subplantar injection of Freund's complete adjuvant (adjuvant arthritis) [34]. This effect was mainly due to reduced exocytosis of lysosomal enzymes, β-glucuronidase and β-*N*-acetylglucosaminidase from polymorphonuclear leukocytes (PMN) accumulating at the affected lesion. In this regard, 14-membered ring macrolides have the ability to reduce the increment in vascular permeability during inflammatory processes [35], in addition to the capacity for suppressing the production of PMN chemokines [36, 37] and proinflammatory cytokines [6, 35, 38]. Thus, the overall efficacy of AZM on adjuvant arthritis is likely to result from the synergy of these individual anti-inflammatory actions.

Furthermore, CAM was able to suppress the expression of HLA-DR and co-stimulatory molecules by IFN-γ-stimulated synoviocytes, which in turn might inhibit the antigen-specific T cell proliferation induced by synoviocytes [39]. Recently, a small pilot study, an open uncontrolled trial, demonstrated that according to the American College of Rheumatology (ACR) criteria, CAM was very beneficial in RA patients who had not responded to disease-modifying antirheumatic drugs [40]. In the study, patients were treated with CAM at the dose of 500 mg twice per day for the first 10 days, followed by a daily maintenance dose of 250 mg twice per day. Regression of RA symptoms and reduction in plasma levels of prostaglandin E2 and soluble phospholipase A2 were closely related to plasma levels of CAM. However, they did not measure IL-6 in the circulation or joint fluid, despite the fact that IL-6 is defined as one of critical factors for the progression of RA [41, 42]. In this connection, Sakamoto et al. [17] demonstrated that reduction in plasma IL-6 levels by CAM administration was closely related to longer survival for NSC lung cancer patients (Fig. 4). We found that CAM administration increased survival for Balb/c mice intravenously inoculated with murine myeloma cells (unpublished data); this effect was closely related with the decrease in serum levels of IL-6. In fact, IL-6 was shown to promote the growth of myeloma cells in association with CD45 induction [43]. These facts strongly suggested that the efficacy of the macrolide therapy for controlling chronic inflammatory process in the airway and the joint might be in part dependent on its ability to suppress IL-6 production.

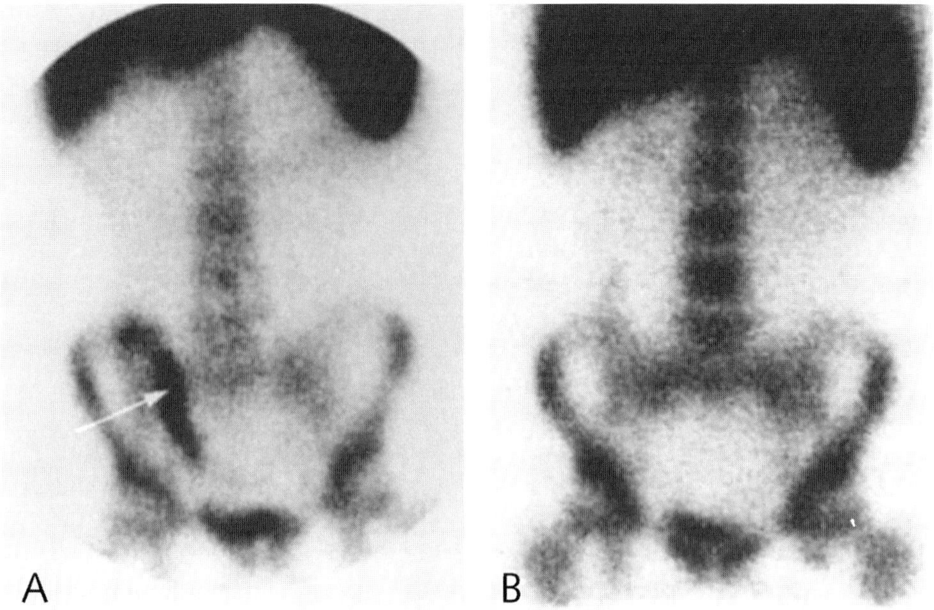

Figure 5
White cell scan of a 22 year old female patient (A) before treatment with rifabutin and CAM showing (arrow) active inflammation in the diseased bowel, and (B) 22 months after treatment showing complete abolition of abnormal technetium-99 uptake.
(from Journal of Antimicrobial Chemotherapy *(1997) 39: 393–400, with permission of the British Society for Antimicrobial Chemotherapy)*

Potential therapy for inflammatory bowel diseases

Among inflammatory bowel diseases (IBD), Crohn's disease was the original target to be treated with macrolides. While the exact cause of Crohn's disease remains controversial, a variety of microbes, including normal intestinal bacteria and yeasts, can be recovered from mesenteric lymph nodes [44, 45]. It therefore appears likely that microbes invade the mucosa in patients of Crohn's disease, possibly resulting from a breakdown in the normal intestinal barrier.

Recently, evidence for the involvement of *Mycobacterium paratuberculosis* in the pathogenesis of this disease has been accumulating from both long-term culture [46] and polymerase chain reaction (PCR) tests [47]. Although the role of viable mycobacteria in its pathogenesis remains unproven, *M. paratuberculosis* is a specific chronic pathogen in the intestine of many animal species, including primates.

Recent evidence implies that this pathogen could be conveyed to humans in pasteurized cows' milk [48]. Based on these facts, small open studies using antimycobacterial agents were carried out, and such treatments appeared promising but not to a satisfactory extent [49], probably due to the fact that *M. paratuberculosis* was generally resistant to standard antimycobacterial drugs. Then, Gui et al. [50] performed the combination therapy of rifabutin and CAM or AZM for 46 patients with severe Crohn's disease, since these three antibiotics had enhanced activity against *M. paratuberculosis*. Two-year outcomes analysis of active Crohn's disease treated with rifabutin and one of these other macrolides demonstrated that this treatment brought a substantial clinical improvement in the disease (Fig. 5). A reduction in the Harvey-Bradshaw Crohn's disease activity index occurred after 6 months' treatment ($p = 0.004$, paired Wilcoxon test) and was maintained at 24 months ($p = 0.001$ *versus* pretreatment). An improvement in inflammatory parameters, including reduction in erythromycin sedimentation rate ($p = 0.009$) and C-reactive protein ($p = 0.03$) at 18 months compared with pretreatment levels, was obtained. Further, this treatment increased serum levels of albumin at 12 months ($p = 0.04$). These clinical parameter changes were consistent with the data obtained by the long-term treatment of CAM for patients with NSC lung cancer [17]. These clinical benefits therefore reflected the BRM activity of 14-membered ring macrolides.

In IBD, the normal healing process during restitution can be disrupted by the inflammation; this may be due to loss of growth factors, surface adhesion molecules or both, leading to a reduced rate of healing [51]. The expression of syndecan-1, as the predominant epithelial syndecan, was markedly reduced in reparative epithelium from IBD patients [52]. Syndecan-1 shedding is thought to reflect the activation of cellular proteases such as MMP-7, capable of cleaving ectodomains of this syndecan from the cell surface, which in turn may result in the formation of transepithelial CXC chemokines gradient [30]. Since IBD is characterized by intestinal permeability changes and large numbers of neutrophils trafficking through the epithelium [53], an increased rate of syndecan-1 shedding may initiate subsequent inflammatory processes including activation of proteases and production of CXC chemokines [26]. Contrary to the function as an initiator of inflammation, syndecan-1 could contribute to the healing process through its function as a co-receptor for epidermal growth factor-2 (EGF-2). Since 14-membered ring macrolides have the capacity for suppressing MMP-7 activation and CXC chemokine production [26], as well as regulating the epithelial and endothelial permeability [35], they would possibly be expected to reduce the inflammatory process in the intestinal epithelium. In contrast, it is postulated that syndecan-1 participates in cell adhesion mainly in the foveolar epithelium of digestive tracts, and plays a role in the healing of ulcerative lesions by interacting with heparin binding growth factors, based on the study by *in situ* hybridization and immunohistochemistry in stomachs of patients with active ulcers as well as those undergoing early scar formation [54].

The group of 14-membered ring macrolides exhibits the suppressive effect on proinflammatory cytokine/chemokine production at damaged tissue, but not on healthy tissue [35]. Such dual effects of the macrolides might be expected to function even in the expression of syndecan-1, depending on the degree of inflammation in the damaged tissue. The binding activity of syndecan-1 to heparin-binding growth factors such as FGF-1 and -2 may facilitate the healing process of damaged tissue [52]. Thus, the regulatory action of 14-membered ring macrolides on syndecan-1 shedding [26] may reduce inflammatory damage but also enhance healing process in IBD.

Conclusion

Recently, tetracyclines have been shown to exert a number of anti-inflammatory and immunomodulatory activities; they specifically inhibit activated B cell function [55], decrease NO synthase [56, 57] and phospholipase A2 [58] in activated macrophages, and also modulate secretion of soluble factors such as TNF-α [59] and Fas ligand [60]. Furthermore, tetracyclines can specifically inhibit the expression, activation from proenzyme precursor, as well as the enzyme activity of MMPs (reviewed in [61, 62]); this function is in part responsible for the anti-inflammatory activity of tetracyclines. These regulatory activities of tetracyclines are very similar to those of 14-membered ring macrolides. However, it has never been confirmed whether the long-term therapy of tetracyclines can be safe and effective in patients with chronic inflammatory diseases.

At the present time, the precise mechanisms underlying non-antimicrobial activity of the macrolides have not been clarified, but recent development of molecular study has suggested that macrolides are able to modify the function of various molecules responsible for intracellular signaling. Intracellular uptake of drugs generally results in profound modifications of host cell metabolism and functions. It is plausible that macrolides confer such effects on intracellular molecules, since macrolides are usually highly concentrated in host cells. Against arthritis and IBD, regulatory activity of matrix metalloproteases appears to be critical for macrolides to express anti-inflammatory effects: activation of the proteases must be a prerequisite to syndecan shedding as well as formation of transepithelial CXC chemokines gradient. Furthermore, the healing process of damaged tissue requires enhanced syndecan expression. Against lung cancer, suppressive activity of heparanase is also very important for macrolides to exhibit an anti-tumor effect, since heparansulphate of syndecan ectodomains functions as a co-receptor for several growth factors and chemokines. In addition to these effects due to modifications of cell-molecular function, long-term administration of macrolides apparently alters T cell-mediated responses. It is not yet explained whether such alteration is a result from the direct action of macrolides on T cells or from the consequence of the above-described effects. Further study on the interaction between lymphocytes and macrolides would help to make the long-term administration of 14-membered ring macrolides more reliable and safer for clinical use in the treatment of chronic inflammatory diseases and lung cancer.

References

1 Labro MT (1993) Effects of macrolides on host natural defences. In: AJ Bryskier, JP Butzler, HC Neu, PM Tulkens (eds): *Macrolides*. Arnette Blackwell, Paris, 389–408

2 Desaki M, Takizawa H, Ohtoshi T, Kasama T, Kobayashi K, Sunazuka T, Omura S, Yamamoto K, Ito K (2000) Erythromycin suppresses nuclear factor-κB and activator protein-1 activation in human bronchial epithelial cells. *Biochem Biophys Res Commun* 267: 124–8

3 Yamanaka A, Saiki S, Tamura S, Saito K (1969) Problems in chronic obstructive bronchial disease, with special reference to diffuse panbronchiolitis. *Naika* 23: 442–51

4 Kobayashi H (1995) Airway biofilm disease: clinical manifestations and therapeutic possibilities using macrolides. *J Infect Chemother* 1: 1–15

5 Sawaki M, Mikami R, Mikasa K, Kunimatsu M, Ito S, Narita N (1986) The long-term chemotherapy with erythromycin in chronic lower respiratory tract infection – first report: comparison with amoxicillin. *Kansenshogaku Zasshi* 60: 37–44

6 Khair OA, Devalia JL, Abdelaziz MM, Sapsford RJ, Davis RJ (1995) Effect of erythromycin on *Haemophilus influenzae* endotoxin-induced release of IL-6, IL-8 and sICAM-1 by cultured human bronchial epithelial cells. *Eur Respir J* 8: 1451–7

7 Ueda K, Mikasa K, Hamada K, Sakamoto M, Konishi M, Maeda K, Majima T, Kita E, Narita N (1999) Effects of clarithromycin on expression of cytokine mRNA in spleen cells of mice bearing Lewis lung carcinoma cells. *Haigan* 39: 117–24

8 Mikasa K, Sawaki M, Konishi M, Egawa S, Yoneda T, Yagyu Y, Fujimura M, Hamada K, Kunimatsu M, Narita N (1989) The effect of erythromycin treatment of natural killer (NK) cell activity in patients with chronic lower respiratory tract infections. *Kansenshogaku Zasshi* 63: 811–15

9 Hamada K, Kita E, Sawaki M, Mikasa K, Narita N (1995) Antitumor effect of erythromycin in mice. *Chemotherapy* 41: 59–69

10 Hamada K, Mikasa K, Yunou Y, Kurioka T, Majima T, Narita E (2000) Adjuvant effect of clarithromycin on chemotherapy for murine lung cancer. *Chemotherapy* 46: 49–61

11 Mikasa K, Sawaki M, Kita E, Hamada K, Teramoto S, Sakamoto M, Maeda K, Konishi M, Narita N (1997) Significant survival benefit to patients with advanced non-small-cell lung cancer from treatment with clarithromycin. *Chemotherapy* 43: 288–96

12 Sawaki M, Kita E, Mikasa K, Narita N (1995) Clarithromycin as a potent ant-angiogenesis agent: possible application for the antitumor agent. *Can J Infect Dis* 6 (Suppl C): 213

13 Yatsunami J, Tsuruta N, Wakamatsu K. Hara N, Hayashi S (1997) Clarithromycin is a potent inhibitor of tumor-induced angiogenesis. *Res Exp Med* 197: 189–97
14 Yatsunami J, Tsuruta N, Fukuno Y, Kawashima M, Taniguchi S, Hayashi S (1999) Inhibitory effects of roxithromycin on tumor angiogenesis, growth and metastasis of mouse B16 melanoma cells. *Clin Exp Metastasis* 17: 119–24
15 Parajuli P, Yano S, Hanibuchi M, Nokihara H, Shinohara T, Sone S (1998) Effect of clarithromycin on the distant metastases of human lung cancer cells in SCID mice. *J Med Invest* 44: 205–10
16 Teramoto S, Kita E, Mikasa K, Hamada K, Konishi M, Maeda K, Sakamoto M, Tsujimoto M, Mori K, Sawaki M et al (1998) Effect of clarithromycin administration on interferon-gamma and interleukin 12 mRNA expression in the tumor tissue of non-small-cell lung cancer. *Jpn J Antibiot* 51 (Suppl): 53–5
17 Sakamoto M, Mikasa K, Majima T, Hamada K, Konishi M, Maeda K, Kita E, Narita N (2001) Anti-cachectic effect of clarithromycin for patients with unresectable non-small cell lung cancer. *Chemotherapy* 47: 444–51
18 Sasaki M, Ito T, Fukui S, Izumiyama N, Kashima M, Sano M, Fujiwara Y, Miura H (2001) Effect of 14-membered ring macrolides on heparanase mRNA expression in lung cancer cells. *Jpn J Antibiot* (Suppl): 54: C97–100
19 Sasaki M, Kashima M, Ito T, Watanabe A, Sano M, Kagaya M, Shioya T, Miura M (2000) Effect of heparin and related glycosaminoglycan on PDGF-induced lung fibroblast proliferation, chemotactic response and matrix metalloproteinases activity. *Mediators Inflamm* 9: 85–91
20 Lapierre F, Holme K, Lam L, Tressler RJ, Storm N, Wee J, Stack RJ, Castellot J, Tyrrell D (1996) Chemical modifications of heparin that diminish its anticoagulant but preserve its heparanase-inhibitory, angiostatic, anti-tumor and anti-metastatic properties. *Glycobiol* 6: 355–66
21 Nakajima M, Irimura T, Nicolson GL (1988) Heparanases and tumor metastasis. *J Biol Chem* 36: 157–67
22 Vaday GG, Lider O (2000) Extracellular matrix miotics, cytokines, and enzymes: dynamic effects on immune cell behavior and inflammation. *J Leukoc Biol* 67: 149–59
23 Nakajima M, Irimura T, Di Ferrante N, Nicolson GL (1984) Metastatic melanoma cell heparanase. Characterization of heparan sulphate degradation fragments produced by B16 melanoma endoglucuronidase. *J Biol Chem* 259: 2283–90
24 Vlodavsky I, Friedman Y, Elkin M, Aingorn H, Atzmon R, Ishai-Michaeli R, Bitan M, Pappo O, Peretz T, Michal I et al (1999) Mammalian heparanase: gene cloning, expression and function in tumor progression and metastasis. *Nat Med* 5: 793–802
25 Hulett MD, Freeman C, Hamdorf BJ, Baker RT, Harris MJ, Parish CR (1999) Cloning of mammalian heparanase, an important enzyme in tumor invasion and metastasis. *Nat Med* 5: 803–9
26 Kita E, Mikasa K, Kasahara K (2003) Syndecan shedding from epithelial cells affects host defense against respiratory infection. *International Congress Series* 1257C: 21–5

27 Joensuu H, Anttonen A, Eriksson M, Mäkitaro R, Alfthan H, Kinnula V, Leppä S (2002) Soluble syndecan-1 and serum basic fibroblast growth factor are new prognostic factors in lung cancer. *Cancer Res* 62: 5210–17

28 Anttonen A, Leppa S, Ruotsalainen T, Alfthan H, Mattson K, Joensuu H (2003) Pretreatment serum syndecan-1 levels and outcome in small cell lung cancer patients treated with platinum-based chemotherapy. *Lung Cancer* 41: 171–7

29 Tsutsumi M, Kitada H, Shiraiwa K, Takahama M, Tsujiuchi T, Sakitani H, Sasaki Y, Murakawa K, Yoshimoto M, Konishi Y (2000) Inhibitory effects of combined administration of antibiotics and anti-inflammatory drugs on lung tumor development initiated by N-nitrosobis (2-hydroxypropyl) amine in rats. *Carcinogenesis* 21: 251–6

30 Li Q, Park PW, Wilson CL, Parks WC (2002) Matrilysin shedding of syndecan-1 regulates chemokine mobilization and transepithelial efflux of neutrophils in acute lung injury. *Cell* 111: 635–46

31 Tjan-Heijnen VCG, Postmus PE, Ardizzoni A, Manegold CH, Burghouts J, van Meerbeeck J, Gans S, Mollers M, Buchholz E, Biesma B et al (2001) Reduction of chemotherapy-induced febrile leucopenia by prophylactic use of ciprofloxacin and roxithromycin in small-cell lung cancer patients: an EORTC double-blind placebo-controlled phase III study. *Ann Oncol* 12: 1359–68

32 Alder J, Mitten M, Jarvis K, Gupta P, Clement J (1993) Efficacy of clarithromycin for treatment of experimental Lyme disease *in vivo*. *Antimicrob Agents Chemother* 37: 1329–33

33 Tissi L, von Hunolstein C, Mosci P, Campanelli C, Bistoni F, Orefici G (1995) *In vivo* efficacy of azithromycin in treatment of systemic infection and septic arthritis induced by type IV group B *Streptococcus* strain in mice: comparative study with erythromycin and penicillin G. *Antimicrob Agents Chemother* 39: 1938–47

34 Carevic O, Djokic S (1988) Comparative studies on the effects of erythromycin A and azithromycin upon extracellular release of lysosomal enzymes in inflammatory processes. *Agents Actions* 25: 124–31

35 Mikasa K, Kita E, Sawaki M, Kunimatsu M, Hamada K, Konishi M, Kashiba S, Narita N (1992) The anti-inflammatory effect of erythromycin in zymosan-induced peritonitis of mice. *J Antimicrob Chemother* 30: 339–48

36 Kadota J, Sakito O, Kohno S, Sawa H, Mukae H, Oda H, Kawakami K, Fukushima K, Hiratani K, Hara K (1993) A mechanism of erythromycin treatment in patients with diffuse panbronchiolitis. *Am Rev Respir Dis* 147: 153–9

37 Takizawa H, Desaki M, Ohtoshi T, Kawasaki S, Kohyama T, Sato M, Tanaka M, Kasama T, Kobayashi K, Nakajima J et al (1997) Erythromycin modulates IL-8 expression in normal and inflamed human bronchial epithelial cells. *Am J Respir Crit Care Med* 156: 266–71

38 Takizawa H, Desaki M, Ohtoshi T, Kikutani S, Okazaki H, Sato M, Tanaka M, Akiyama N, Shoji S, Hiramatsu K et al (1995) Erythromycin suppresses interleukin 6 expression by human bronchial epithelial cells: A potential mechanism of its anti-inflammatory action. *Biochem Biophys Res Commun* 210: 781–6

39 Matsuoka N, Eguchi K, Kawakami A, Tsuboi M, Kawabe Y, Aoyagi T, Nagataki S (1996) Inhibitory effect of clarithromycin on costimulatory molecule expression and cytokine production by synovial fibroblast-like cells. *Clin Exp Immunol* 104: 501–8
40 Saviola G, Abdi Ali L, Rossini P, Campostrini L, Coppini A, Gori M, Ianaro A, Bucci M, de Nucci G, Cirino G (2002) Clarithromycin in rheumatoid arthritis patients not responsive to disease-modifying antirheumatic drugs: an open, uncontrolled pilot study. *Clin Exp Rheumatol* 20: 373–8
41 Nishimoto N, Sasai M, Shima Y, Nakagawa M, Matsumoto T, Shirai T, Kishimoto T, Yoshizaki K (2000) Improvement in Castleman's disease by humanized anti-interleukin-6 receptor antibody therapy. *Blood* 95: 56–61
42 Yoshizaki K, Nishimoto N, Mihara M, Kishimoto T (1998) Therapy of rheumatoid arthritis by blocking IL-6 signal transduction with a humanized anti-IL-6 receptor antibody. *Springer Semin Immunopathol* 20: 247–59
43 Mahmoud MS, Ishikawa H, Fujii R, Kawano MM (1988) Induction of CD45 Expression and Proliferation in U-266 myeloma cell line by interleukin-6 (IL-6). *Blood* 92: 3887–97
44 Liu Y, van Kruiningen HJ, West AB, Cartun RW, Cortot A, Colombel JF (1995) Immunocytochemical evidence of *Listeria*, *Escherichia coli*, and *Streptococcus antigens* in Crohn's disease. *Gastroenterology* 108: 1396–404
45 Cartun RW, Van Kruiningen HJ, Pedersen CA, Berman MM (1993) An immunocytochemical search for infectious agents in Crohn's disease. *Mod Pathol* 6: 212–19
46 Hermon-Tayler J, Barnes N, Clarke C, Finlayson C (1998) *Mycobacterium paratuberculosis* cervical lymphadenitis, followed five years later by terminal ileitis similar to Crohn's disease. *Br Med J* 316: 449–53
47 Dell'Isola B, Poyart C, Goulet O, Mougenot JF, Sadoun-Journo E, Brousse N, Schmitz J, Ricour C, Berche P (1994) Detection of *Mycobacterium paratuberculosis* by polymerase chain reaction in children with Crohn's disease. *J Infect Dis* 169: 449–51
48 Millar D, Ford J, Sanderson J, Withey S, Tizard M, Doran T, Hermon-Taylor J (1996) IS900 PCR to detect *Mycobacterium paratuberculosis* in retail supplies of whole pasteurized cows' milk in England and Wales. *Appl Environ Microbiol* 62: 3446–52
49 Swift GL, Srivastava ED, Stone R, Pullan RD, Newcombe RG, Rhodes J, Wilkinson S, Rhodes P, Roberts G, Lawrie BW (1994) Controlled trial of anti-tuberculous chemotherapy for two years in Crohn's disease. *Gut* 35: 363–8
50 Gui GP, Thomas PR, Tizard ML, Lake J, Sanderson JD, Hermon-Taylor J (1997) Two-year-outcomes analysis of Crohn's disease treated with rifabutin and macrolide antibiotics. *J Antimicrob Chemother* 39: 393–400
51 Day R, Forbes A (1999) Heparin, cell adhesion, and pathogenesis of inflammatory bowel disease. *Lancet* 354: 62–5
52 Day R, Ilyas M, Daszak P, Talbot I, Forbes A (1999) Expression of syndecan-1 in inflammatory bowel disease and a possible mechanism of heparin therapy. *Dig Dis Sci* 44: 2508–15
53 Colgan SP, Comerford KM, Lawrence DW (2002) Epithelial cell-neutrophil interactions

in the alimentary tract: a complex dialog in mucosal surveillance and inflammation. *The Scientific World Journal* 2: 76–88

54 Tanabe H, Yokota K, Kohgo Y (1999) Localization of syndecan-1 in human gastric mucosa associated with ulceration. *J Pathol* 187: 338–44

55 Kuzin II, Snyder JE, Ugine GD, Wu D, Lee S, Bushnell T Jr, Insel RA, Young MF, Bottaro A (2001) Tetracyclines inhibit activated B cell function. *Int Immunol* 12: 921–931

56 Amin AR, Attur MG, Thakker GD, Patel PD, Vyas PR, Patel RN, Patel IR, Abramson SB (1996) A novel mechanism of action of tetracyclines: effects on nitric oxide synthases. *Proc Natl Acad Sci USA* 93: 14014–19

57 Amin AR, Patel RN, Thakker GD, Lowenstein CJ, Attur MG, Abramson SB (1997) Post-transcriptional regulation of inducible nitric oxide synthase mRNA in murine macrophages by doxycycline and chemically modified tetracyclines. *FEBS Lett* 410: 259–64

58 Pruzanski W, Greenwald RA, Street IP, Laliberte F, Stefanski E, Vadas P (1992) Inhibition of enzymatic activity of phospholipases A2 by minocycline and doxycycline. *Biochem Pharmacol* 44: 1165–70

59 Shapira L, Soskolne WA, Houri Y, Barak V, Halabi A, Stabholz A (1996) Protection against endotoxic shock and lipopolysaccharide-induced local inflammation by tetracycline: correlation with inhibition of cytokine secretion. *Infect Immun* 64: 825–8

60 Liu J, Kuszynski CA, Baxter BT (1999) Doxycycline induces Fas/Fas ligand-mediated apoptosis in Jurkat T lymphocytes. *Biochem Biophys Res Commun* 260: 562–7

61 Vernillo AT, Rifkin BR (1998) Effects of tetracyclines on bone metabolism. *Adv Dental Res* 12: 56–62

62 Golub LM, Lee HM, Ryan ME, Giannobile WV, Payne J, Sorsa T (1998) Tetracyclines inhibit connective tissue breakdown by multiple non-antimicrobial mechanisms. *Adv Dental Res* 12: 12–6

Anti-inflammatory properties of antibiotics other than macrolides

Bruce K. Rubin[1], *Markus O. Henke*[2] *and Axel Dalhoff*[3]

[1]Department of Pediatrics, School of Medicine, Wake Forest University, Medical Center Boulevard, Winston-Salem, NC 27157-1081, USA; [2]Department of Pulmonary Medicine, Universität Marburg, Baldingerstrasse 1, 35043 Marburg, Germany; [3]Bayer AG, Aprather Weg, 42096 Wuppertal, Germany

Introduction

The interactions of bacterial pathogens with the host comprise several steps.

1. Bacteria adhere to and colonise epithelial surfaces followed by penetration into and dissemination within the macroorganism. Usually, the early stages of infection are passed through without antibacterial therapy; once a bacterial infection has been diagnosed most infectious diseases are treated with antibiotics. Antimicrobial agents interfere with the bacterium and infection but these can modify the interaction between bacteria and the host cells [1].
2. The infection is a potent activator of the immune response causing inflammatory responses and triggering the cytokine network. Bacterial exoenzymes, exotoxins, like polysaccharide, lipoteichoic acid and teichoic acid, peptidoglycan and even bacterial DNA released by bacteria affect the immune system and induce the release of cytokines. Bacterial products may also induce B cell proliferation or activate the complement pathways. Antimicrobial agents may not only reduce bacterial numbers but also these bacterial products. Consequently, antibiotics modulate the release of proinflammatory bacterial compounds. Both the direct antibacterial effect and the differential interaction of various antibiotics with the release of bacterial products have a significant effect on treatment outcome (for summaries see [2, 3]).
3. The indigenous microflora plays a role in maintaining a healthy immune system. The depletion of some components of the intestinal flora affects the maintenance of a healthy immune system [4]. Also, bacterial translocation may significantly influence antibody response. Although the immune responses in the intestine are of immense importance and powerful immunological forces are present, the mechanisms are poorly understood [5, 6].
4. Antibacterial agents may indirectly interfere with the host bacteria relationship. They may enhance phagocytosis and/or may make bacteria more vulnerable to

Antibiotics as Anti-Inflammatory and Immunomodulatory Agents, edited by Bruce K. Rubin and Jun Tamaoki

intraleukocytic killing by altering the morphology and structure of bacterial surface [7, 8].

5. Antibacterial agents may indirectly interfere with phagocytic efficacy. Quinolones and macrolides in particular, readily penetrate into phagocytes, are accumulated intracellularly and exhibit intracellular bacterial activities despite the acidic intracellular pH [9–11].
6. Antibiotics may cause immunopharmacological effects by being antigenic. Penicillins, cephalosporins and rifampin are examples of antigenic antibiotics [12, 13].
7. Antibacterial agents may directly interact with the immune system.

Almost all drug classes exert effects on the specific immune system and complement activation. Aminoglycosides, ansamycins, benzylpyrimidinones, β-lactams, cyclines, fosfomycin gyrase B-inhibitors, lincosamides, peptides, sulfonamides, and in particular macrolides as well as quinolones either increase or decrease phagocyte functions. For many of these agents the underlying mechanisms of immunomodulation are not well defined and some events are strongly dependent on the methods applied. By contrast, the immunomodulation by macrolides [14] and quinolones [15] is much better described phenotypically.

The main *in vitro* and *in vivo* effects of the various families of antibacterial agents were reviewed by Labro [16], Nau [2] and Dalhoff [17] and are summarized and updated in this review.

Aminoglycosides

Aminoglycosides interfere with bacterial protein synthesis by acting on the 30S ribosomal subunit.

There are controversial data on the inhibitory effect of aminoglycosides at therapeutic concentrations on polymorphonuclear leukocyte (PMN) chemotaxis, oxidative metabolism, and yeast killing [18, 19]. Various mechanisms have been advanced, including binding to negatively-charged membrane phospholipids, leading to membrane disturbances, specific binding to inositol biphosphate resulting in inhibition of phospholipase C, and inhibition of protein kinase C (PKC).

The intraphagocytic activity of streptomycin on intracellular *Escherichia coli* has been suggested to rely on stimulation of O_2-dependent cellular microbactericidal mechanisms in macrophages, although drug uptake was not studied in this model [20].

Neomycin showed concentration dependant inhibitory or stimulatory effects on the generation of leukotrienes by PMNs, depolymerization of actin and GTPase activity of crude membrane fractions. These drugs (particularly neomycin) appear to be useful tools studying transmembrane signaling pathways [18].

Gentamicin and other aminoglycosides inhibit protein synthesis and may induce cell lysis, which might then increase release of endotoxins (lipopolysaccharide [LPS]). For this reason, aminoglycosides may not be ideal candidates for the reduction of proinflammatory bacterial compounds [21, 22]. Gentamicin may also correct the function of the cystic fibrosis transmembrane conductance regulator (CFTR) when there is a stop mutation. By suppressing premature termination codons, these aminoglycosides permit mRNA to "read through" increasing the expression of CFTR in target epithelia [124]. This could have a secondary effect on the chronic bacterial infection and inflammation that is characteristic of CF airway disease.

Ansamycins

Antibacterial ansamycins (rifamycins) are mainly effective against Mycobacteria and alter RNA biosynthesis by interfering with RNA polymerase activity. Rifampin, the most important representative of this group, impairs various PMN functions such as chemotaxis and oxidative burst. The compound has been claimed to bind at glucocorticoid receptors leading to pharmacological glucocorticoid-like effects such as host immunosuppression, thereby acting as an immunosuppressant. Recent studies in human alveolar and neuroblastoma cells and in mouse hippocampal cells, however, have found no evidence of activation of the glucocorticoid receptors by rifampin [23, 24].

Immunosuppression, including inhibition of T cell activity, reduced humoral and cell-mediated immunity has long been noted during rifampin therapy [25, 26]. A possible effect offsetting drug-induced immune supression was shown in a study in which rifampin increased GM-CSF- and IL-4-induced expression of CD1b (a human antigen-presenting molecule belonging to the nonclassical MHC-independent system involved in the presentation of nonpeptide antigens) thereby favoring the lipid/glycolipid antigen presentation mediated by CD1b on peripheral blood monocytes [27].

In vivo administration of geldanamycin attenuates lung inflammation and acute lung injury in animal models, thereby suggesting that geldanamycin also has anti-inflammatory effects. Supporting this *in vivo* effect, geldanamycin inhibits the TNF-α-mediated IL-8 gene expression possibly through inhibition of NF-κB activation [28]. Another explanation might be that geldanamycin inhibits the production of TNF-α, IL-6, and IL-1β in activated macrophages possibly through heat shock protein (HSP) 90 which is also critical in the intracellular signaling pathways promoting inflammatory cytokine production [29]. Furthermore ansamycin antibiotics inhibit function of HSP90, causing selective degradation of several intracellular proteins regulating such processes as proliferation, cell cycle regulation, and prosurvival signaling cascades. HSP90 has been identified previously as a molecular target for anticancer agents [30].

Benzylpyrimidines (trimethoprim and analogs)

Benzylpyrimidines include trimethoprim (TMP), tetroxoprim, epiroprim, and brodimoprim, which all inhibit dihydrofolate reductase. TMP is generally used in combination with another antifolate drug (sulfamethoxazole). In most studies, TMP, alone or in combination, had an inhibitory effect on PMN functions. In one study the PMN chemotaxis and chemiluminescence were increased, and this effect was also observed with defective functions [31]. The liposolubility of brodimoprim was about three times higher than that of TMP and also had a greater cellular uptake. Brodimoprim did not decrease either phagocytosis or phagocyte-mediated bactericidal activity, nor did it affect oxidative burst activity, whereas TMP influences the oxidative burst [32]. It is speculated that the underlying mechanism of TMP-induced inhibition of PMN oxidative metabolism is probably an inhibitory effect of TMP on the PLD-phosphatidate phosphohydrolase pathway, leading to decreased generation of diradylglycerol, leading to activation of the $O_{2(-)}$-generating respiratory burst. However, the concentration which impaired the PMN oxidative burst by about 50% was far higher than therapeutic concentrations [33].

β-lactams

β-lactam antibiotics represent more than half of all antimicrobial drugs used therapeutically. Structurally, they comprise of five groups of compounds: penams (penicillins and β-lactamase inhibitors), penems (faropenem), carbapenems (imipenem, meropenem), cephems (cephalosporins, cephamycins, oxacephens, and carbacephems), and monobactams (aztreonam, etc.). All groups have a common antibacterial mechanism involving inhibition of various enzymes (such as penicillin-binding protein) involved in the synthesis of peptidoglycan. Many data are available on the *in vitro* effects of these drugs on phagocyte functions and specific immune effectors, but no class- or subgroup-related effect has been demonstrated. β-lactam-induced modulation of immune responses does not appear to be of major clinical relevance, with the possible exception of cefodizime.

Cefodizime a 2-amino-5-thiazolyl cephalosporin has been investigated *in vitro*, *ex vivo*, and *in vivo* in humans and animals (both healthy and immunocompromised). Overviews have summarized the main immunomodulatory properties of cefodizime [34–36]. It enhances the immune function of natural killer cells and phagocytic activity of monocytes and macropahges in immunocompromised animals.

Experimental models using immunocompromised animals confirm the efficacy of cefodizime. In a mouse pneumonia model, it upregulated the early *Klebsiella pneumoniae* induced secretion of TNF-α and the number and phagocytic efficacy of alveolar macrophages [8]. Prophylactic administration of cefodizime increased the survival of some mouse strains after infection with *Toxoplasma gondii* or *Can-*

dida albicans [34, 37]. In contrast, cefodizime inhibited the LPS-stimulated release of TNF-α and IL-1 from human monocytes [38] and TNF-α and IL-6 secretion into the bronchoalveolar lavage fluid after intranasal challenge with heat-killed *pneumococci* [39]. In healthy rats, an intravenous bolus of 30 mg per kg of ceftazidime led to a substantial increase in serum IL-6 and TNF-α concentrations [40].

Ex vivo studies demonstrated a strain- and concentration-dependent responsiveness of the immune system to cefodizime with regard to delayed-type hypersensitivity, antibody production, and lymphocyte proliferation. In healthy humans given cefodizime the immune system was not affected, whereas in immunocompromised individuals (with immune systems suppressed by cancer, hemodialysis, old age, surgical stress, etc.) there appears to be enhanced immune system function after cefodizime administration. In particular, cefodizime administration increased phagocytic functions. When placebo or comparator antibiotics were given, the beneficial effect was seen only in the cefodizime-treated group. The chemical structure responsible for the immunomodulatory properties was identified as the thio-thiazolyl moiety at position 3 of the cephem ring [41], but the cellular mechanism responsible for the immunomodulatory properties remains unclear.

In vitro, cefodizime stimulates the proliferative response of lymphocytes, increases the phagocytotic and bactericidal activity of PMNs, and downmodulates the production of proinflammatory cytokines by stimulated monocytes. In contrast to all β-lactams, cefodizime was also reported to significantly increase colony formation by granulocyte–monocyte progenitors [42]. Alteration of bacterial virulence in susceptible and resistant bacteria has also been demonstrated with cefodizime.

In a study comparing 15 infected patients receiving cefodizime with a comparable group treated with ceftriaxone, although phagocyte function recovered significantly earlier, the only apparent clinical advantage was earlier defervescence in the cefodizime-treated group [43]. In a study of subjects with multiple myeloma and chronic uremia given cefodizime for 5–7 days, there was increased monocyte and neutrophil chemotaxis and an increased percentage of lymphocyte subgroups [44]. The effect of cefodizime on phagocytosis and candidacidal capacity was generally greater than that of ofloxacin, ciprofloxacin and IFN-2α [45]. Candidacidal capacity only increased significantly with ciprofloxacin at 2 micg/ml, but ciprofloxacin had no effect on phagocytosis. Ofloxacin and IFN-2α had no effect, and combinations of these three antibiotics with IFN-2α showed the same effects as the drugs alone. These results indicate that cefodizime has an enhancing effect on PMN function in patients with chronic renal failure.

Cyclines

Cyclines interfere with bacterial protein synthesis by acting on the 30S ribosomal subunit.

Minocyclin inhibits caspase-1 and caspase-3 expression, thereby delaying mortality in a transgenic mouse model of Huntington disease [48]. In a rat model of ischemic stroke, minocycline reduced the infarct size when started before and up to 4 h after the onset of ischemia [49].

Tetracyclines are potent inhibitors of the matrix metalloproteinase (MMP) family of enzymes, [50]. Doxycycline, a tetracycline derivative, has been used experimentally to inhibit matrix degradation during abdominal aortic aneurysm formation [51, 52], and recent clinical studies have investigated the use of doxycycline to limit aneurysm growth [53, 54]. Tetracyclines also inhibit cell proliferation, cell migration, and synthesis of the extracellular matrix in a variety of cell types studied in culture [55–57].

Few studies have investigated the effect of cyclines on cytokine production: paradoxically, minocycline and tetracycline, increased IL-1β secretion by LPS-stimulated human monocytes [58]. Various mechanisms have been proposed to explain the inhibitory action of cyclines. Structure-activity relationships indicate a parallel increase in lipid solubility (possibly cellular accumulation) and inhibitory properties (for example, doxycycline > chlortetracycline > tetracycline > oxytetracycline) [59, 60]. However, other studies stress the different chemical reactivities of the various molecules under UV exposure.

Clinical relevance

The clinical relevance of the inhibitory properties of cyclines on phagocyte functions is widely acknowledged. Tetracyclines are widely used in the treatment of inflammatory acne. These antibiotics inhibit the proliferation of *Propionibacterium acnes*, but tetracycline also significantly inhibits the release of reactive oxygen species (ROS) from human PMN and reduces the capacity of *Propionibacterium acnes* to produce neutrophil chemotactic factors, providing evidence that it has anti-inflammatory actions [61].

Tetracyclines have been also used in reactive arthritis, i.e., nonpurulent inflammation of a joint following, urogenital, gastrointestinal, or lower respiratory tract infections [62] possibly mediated via inhibiting the nitric synthase activity and nitrosothiols [63–66]. A multicenter double-blind placebo-controlled trial concluded that minocycline was safe and effective in patients with mild-to-moderate rheumatoid arthritis [67, 68], supporting the use of this drug alone or as adjunctive therapy in rheumatic diseases. The anti-inflammatory action of tetracyclines seems related to a non-antibacterial mechanism. In addition, the anti-inflammatory action of tetracycline has been proposed to be of benefit to prevent endotoxic shock by blockade of LPS-induced TNF-α and IL-1β secretion [69].

Fosfomycin

Fosfomycin (1-*cis*-1,2-epoxypropylphosphoric acid) is a natural product with broad-spectrum bactericidal antibiotic. It interferes with bacterial cell wall biosynthesis by inhibiting the pyruvate-uridine diphosphate-*N*-acetylglucosamine transferase.

In a mouse model of gut-derived *P. aeruginosa* sepsis, treatment with an isomer of fosfomycin without antibacterial activity significantly increased the survival rate in comparison to saline-treated mice. It was speculated that the fosfomycin isomer possessed immunomodulatory activity inducing protection against *P. aeruginosa* bacteremia [70].

In a rat air pouche inflamed with carrageenan it was found that the volume, protein amounts and cell counts in the exudates obtained from fosfomycin-treated animals were significantly reduced compared with that from placebo-treated animals. The content of PGE_2, TNF-α, and mRNA for cyclooxygenase-2 were also markedly suppressed in fosfomycin-treated rats. Histological examination showed suppression of the inflammatory response in the pouch tissues from fosfomycin-treated rats [71].

In vitro, fosfomycin has immunomodulatory activity on B- and T-lymphocyte function, and also inhibits histamine release from basophils [72, 73]. It was reported that fosfomycin decreased the rate of synthesis of TNF-α and IL-1 but increased that of IL-6 in phagocytes [74]. In mice injected with LPS, fosfomycin significantly lowered the serum levels of TNF-α and IL-1β, indicating that fosfomycin alters inflammatory cytokine production after LPS stimulation [75]. Fosfomycin also reduced the formation of biofilms, produced by uropathogenic *E. coli* strains [76]. The therapeutic relevance of these effects is under evaluation.

Gyrase B inhibitors

Gyrase B inhibitors impair bacterial DNA replication and consist of novobiocin and coumermycin. Novobiocin interferes with metabolic processes in eukaryotic cells. In particular, it is a potent inhibitor of ADP ribosylation.

At therapeutic concentrations, coumermycin has been reported to impair chemotaxis, superoxide anion production, and intracellular killing of PMNs [77]. It suppresses the production of proinflammatory cytokines (TNF-α, IL-1, and IL-6), and the anti-inflammatory cytokine IL-10, by LPS-stimulated human monocytes [78].

Novobiocin downregulates the surface molecules on monocytes, in particular CD14. The cytosolic protein phosphorylation pattern was altered by novobiocin and other inhibitors of ADP ribosylation, suggesting a role in monocyte signal transductional pathways. A species dependence with novobiocin was shown, since mouse macrophages were far less susceptible to the inhibitory effect of novobiocin on TNF-α production than were human monocytes [78]. Although the drug had hepatoprotective properties *in vivo*, elevated TNF-α levels in mice treated with D-galactosamine were not reduced by novobiocin administration [78].

Novobiocin is also known to enhance anticancer drug sensitivity of cancer cells *in vitro* and *in vivo*. This is probably due to inhibition of DNA repair, topoisomerase II and/or drug-efflux activities but the mechanism remains undetermined [79–82].

Lincosamides

Lincomycin and clindamycin interfere with bacterial protein synthesis at the level of the 50S ribosomal subunit. Clindamycin, although not effective against *E. coli*, suppressed the production of hemolysin in an animal model of haemolytic *E. coli* peritonitis [83]. The ceftazidime-induced release of cytokines (TNF-α, and IL-1β) by *E. coli*-LPS could be suppressed by prior administration of clindamycin. But it increases the IL-6 production [84].

Peptides

Peptide antibiotics are a broad family comprising the bacillus antibiotics (tyrocidins, gramicidins, and bacitracin) predominantly active against gram-positive bacteria and polypeptides (polymyxins, streptogramins, daptomycin, teicoplanin and vancomycin [an antistaphylococcal glycopeptide]) active against gram-negative bacteria. The mechanisms underlying the antibacterial activity of these drugs differ but most are bactericidal. In general, peptide antibiotics do not significantly alter phagocyte functions at therapeutic concentrations.

Polymyxin B is the most extensively studied drug in this respect. Polymyxin B inhibits PKC [85]. It also decreased endotoxin levels in rats after cecal ligation and puncture [86]. It also attenuated NO and TNF-α production from Kupffer cells after LPS stimulation, probably to the ability of polymyxin B to bind the lipid A portion of LPS. This drug is widely used *in vitro* to neutralize possible LPS contamination [86, 87]. The amounts of intact drug or the reactive side chain necessary to achieve anti-LPS activity, however, are toxic and preclude systemic use of polymyxin in humans [38, 88]. Polymyxin B by itself is able to stimulate some cellular functions, for example, monocyte production of IL-1, IL-6, GM-CSF, and complement components [89, 90].

Vancomycin and teicoplanin have been reported to depress some PMN functions but only at very high, clinically irrelevant, concentrations. At a concentration of 50 mg/liter, teicoplanin also increased the production of TNF-α, IL-1, and IL-6 by concanavalin A-stimulated human monocytes [91]. Teicoplanin was able to bind and neutralize endotoxin. After incubation of teicoplanin with LPS for 3 h, it reduced *in vitro* reactivity and lethality of D-galactosamine-sensitized mice challenged intraperitoneally with *Salmonella enterica* LPS [92]. Vancomycin showed no inhibitory influence on the endotoxin-induced IL-6 and TNF-α production of human PMNs [46].

Quinolones

Quinolones are synthetic antibacterial compounds, whose first representative (nalidixic acid) was synthesized in 1962. Since then, thousands of compounds have been made, of which the 6-fluorinated molecules (fluoroquinolones) represent a breakthrough. The antibacterial activity of fluoroquinolones stems from their inhibitory effect on bacterial DNA gyrase and topoisomerase IV and thus on DNA replication. Fluoroquinolones might also affect mammalian DNA metabolism by inhibiting the topoisomerase II. The selectivity of fluoroquinolones for the bacterial topoisomerases is up to 1000-fold that of mammalian counterparts [93]. Effects of these drugs on the immune system under *in vivo* or clinically relevant *in vitro* conditions have not been well demonstrated. Therefore, this synopsis is limited to a review of published work describing effects of fluoroquinolones on cytokine synthesis under experimental conditions.

In vitro experiments

Ciprofloxacin decreased cytokine synthesis concentration dependently in LPS stimulated human monocytes [94, 95]. The inhibition of cytokine synthesis was statistically significant at high ciprofloxacin concentrations only. Lower ciprofloxacin concentrations ranging from 1–30 mg/L showed an inhibitory effect that did not reach statistical significance [96].

The effect of moxifloxacin on secretion of cytokines by human monocytes obtained from 10 healthy volunteers was studied following stimulation with either LPS or pansorbin (heat killed *Staphylococcus aureus*, Cowan strain). Exposure of LPS-stimulated monocytes to three clinically achievable moxifloxacin concentrations resulted in a significant inhibition of secretion of IL-1α, IL-1β, IL-8 and of TNF-α. Secretion of IL-4, IL-6 and IL-12 by LPS-stimulated monocytes was not significantly inhibited by either of the three moxifloxacin concentrations [94, 97]. Although exposure of LPS-stimulated monocytes to moxifloxacin inhibited cytokine-secretion, the same experimental procedure had no effect on cytokine secretion by pansorbin-stimulated monocytes [97]. These discrepant findings may be explained by the fact that LPS and heat killed *Staphylococcus aureus* preparations use different pathways to induce cytokine synthesis by human monocytes [98]. Alternatively, moxifloxacin may directly interact with LPS, and/or its receptors and/or its stimulatory pathway(s) to inhibit cytokine secretion. Ciprofloxacin had a concentration dependent inhibition of LPS activity [99]. Moxifloxacin has not yet been studied in this respect. However, neither ciprofloxacin or moxifloxacin are incorporated into LPS or displace cations from LPS [100].

Trovafloxacin significantly inhibits the secretion of IL-1α, IL-1β, IL-6, IL-10, granulocyte-macrophage colony stimulating factor (GM-CSF) and TNF-α by monocytes stimulated by either LPS or pansorbin [101]. Trovafloxacin exerted this effect on the monocytes obtained from all 10 volunteers; it had no demonstrable cytotoxicity under the experimental conditions studied. This finding is in contrast to the observations of the same group with moxifloxacin [97]. These differences may be due to the day-to-day variations observed in monocyte samples obtained from the same volunteer on different days [101].

Grepafloxacin at concentrations ranging from 1 to 30 mg/l inhibited the production of IL-1α and IL-1β but stimulated the synthesis of IL-2 by human LPS-stimulated monocytes [102]. Grepafloxacin inhibited the expression of IL-1α, IL-1β, TNF-α, IL-6 and IL-8 mRNA, indicating that the inhibitory effect of grepafloxacin is exerted, at least in part, at the gene transcription level [103]. Similarly, grepafloxacin inhibits TNF-α induced IL-8 expression in human airway epithelial cells [104]. Pretreatment of epithelial cells with grepafloxacin (1 to 25 mg/l) 1 h before TNF-α-stimulation resulted in a concentration dependent reduction of IL-8 synthesis; grepafloxacin concentrations of 1.0, 2.5, 5.0, 10 and 25 mg/l inhibited IL-8 production by 0%, 40%, 59%, 70% and 83%, respectively. This phenomenon was due to grepafloxacin mediated inhibition of TNF-α induced IL-8 mRNA expression [104].

Levofloxacin concentrations ranging from 5 to 100 mg/l stimulated IL-2 production by monocytes in a concentration dependent manner, with levofloxacin concentrations ≥ 10 mg/l causing a significant increase. By contrast, IL-1β production in LPS stimulated monocytes was concentration dependently decreased whereas TNF-α production was affected at a concentration of 100 mg levofloxacin/l only. IL-8 production was negligibly affected by levofloxacin [105].

These *in vitro* data indicate that most fluoroquinolone derivatives superinduce IL-2 synthesis. By contrast, they inhibit synthesis of IL-1 and TNF-α. However, diverse effects were reported, indicating a variation between different cells and/or stimuli studied and *in vitro* methods used.

Ex vivo studies

To investigate the *in vivo* effect of orally administered ciprofloxacin (25 mg/kg) on the capacity of peripheral blood monocytes from healthy human volunteers to produce IL-1α, IL-1β, IL-6 and TNF-α *ex vivo* in response to endotoxin stimulation was determined [106].

Eight patients received ciprofloxacin (25 mg/kg) orally twice daily for 7 days corresponding to a usual treatment. Peripheral blood was collected the day before the treatment (DO), 2 h after the last administration of ciprofloxacin (D7) and 7 days after the end of the treatment (D14). Extracellular and cell-associated TNF-α and IL-1α and extracellular IL-1 activity and cell-associated IL-6 production were significantly enhanced after 7 days administration of ciprofloxacin. Other cytokines measured were at the limit of significance. At D14 the capacity for cytokine production decreased compared to D7, showing that the elevation observed at D7 was probably related to ciprofloxacin treatment. Extracellular TNF-α and IL-6 production was even significantly lower at D14 than at D0 [106].

These data do not match with the observations of its *in vitro* studies [96, 107, 108] possibly due to *in vivo* cellular interactions that could modify monocytic reactivity to a secondary challenge of LPS. Another possibility is that ciprofloxacin was administered to healthy non-infected volunteers harbouring non-stimulated cells. The immunomodulatory effects of fluoroquinolones are noted in stimulated cells only.

In vivo studies

The evaluation of the effects of ciprofloxacin and rufloxacin in an intra-abdominal infection model represents one of the first attempts to determine whether fluoroquinolones *in vivo* alter the T cell response and cytokine production [109]. These two fluoroquinolones were studied as they are inactive against *B. fragilis in vitro* (MIC = 4 mg/l), so that their *in vivo* efficacy is most likely due to immunomodulatory effects [110].

Treatment of mice with rufloxacin or ciprofloxacin resulted in an elimination of *B. fragilis* from 66.6% and 63.5% of the animals, respectively. This therapeutic efficacy coincided with a modulation of TNF-α production *in vivo*. Other fluoroquinolones, like difloxacin [111] or temafloxacin [112], were found to be effective against *B. fragilis in vivo*, too, despite their lack of *in vitro* activity.

These results indicate that the *in vivo* efficacies of these four fluoroquinolones may be related to their ability to modulate TNF-α production. Morphological changes of *B. fragilis* triggered by sub-MIC levels making the bacteria more vulnerable to phagocytosis and intraleukocytic killing could be an alternative explanation. Both possibilities are not mutually exclusive.

The *in vivo* efficacy of trovafloxacin was studied in the same experimental model of intra-abdominal abscesses in rats. The decrease in mortality rate, elimination of infection and reduction of TNF-α concentrations was dose proportional [113, 114]. The protective effect of trovafloxacin was probably due to the modulation of TNF-α concentrations and not due to its antibacterial efficacy [115] since trovafloxacin was seen to be effective at subtherapeutic doses and in animals challenged with heat-killed bacteria [113].

Fluoroquinolones have been shown to protect mice from LPS induced death and from sublethal LPS-challenge [116, 117]. A set of additional *in vitro* and *in vivo* studies lends support to the above observations. Ciprofloxacin and trovafloxacin were found to modulate the inflammatory response of macrophages to LPS [47, 118] by reducing significantly the TNF-α response by stimulated with LPS or living *Pneumococcus aeruginosa*.

These data suggest that ciprofloxacin protects mice by decreasing TNF-α or IL-12 concentrations and by increasing IL-10 concentrations. Accentuation of IL-10 production and diminished IL-12 production were most pronounced in sublethally challenged animals. Both IL-10 and IL-12 are considered to play an important role in the functional differentiation of immunocompetent cells and trigger the initiation of the acquired immune response. The data suggest that the fluoroquinolones studied may affect cellular and humural immunity by attenuating cytokine responses in addition to their antibacterial activity.

Sulfones and sulfonamides

Dapsone (4,4'-diaminophenyl sulfone) was synthesized by Framm and Whitman in 1908. The antibacterial activity of sulfonamides was discovered in the early 1930s. Modification of the active derivative (sulfanilamide) has generated many compounds. The antibacterial action of all compounds is the same, i.e., inhibition of dihydropteroate synthase. The most frequently used antibacterial sulfonamide is sulfamethoxazole in combination with trimethoprim (cotrimoxazole). Sulfonamides exert an inhibitory effect on phagocyte functions, and many agents in this class have been switched from infections to inflammatory diseases, i.e., sulfasalazine and sulphapyridine. The mechanisms underlying the effects are unclear.

Dapsone inhibits neutrophil functions such as chemotaxis and oxidant production. It also irreversibly inhibits myeloperoxidase (MPO) and impairs the production of HOCl by converting MPO into its inactive compound II (ferryl) form [119]. It inhibits the adherence of neutrophils to antibodies, bound to the basement membrane in a dose-dependent manner. This may be related to an effect directly with the antibodies. This inhibition may contribute to the clinical efficacy of dapsone in antibody-mediated diseases [120].

Dapsone impairs the production of prostaglandin E_2 by neutrophils, a possible explanation for dapsone-induced potentiation of cell-mediated immunity [121].

The hematologic toxicity of dapsone is linked to its oxidative metabolism.

Clinical significance

Low-dose macrolide therapy has greatly increased survival in patients with diffuse panbronchiolotis, a chronic inflammatory airway disease that is relatively frequent in the Far East, with a high mortality during conventional treatment [122]. This has led to investigations concerning macrolide use in cystic fibrosis, bronchiectasis, and asthma. Preliminary results in patients with cystic fibrosis are encouraging [123].

For other indications, at present there is no place for attempts to modulate the immune response by antibacterials in clinical routine. Several effects, however, deserve attention for future research in humans.

References

1 Christensen GD, Beachey EH (1984) The molecular basis for the localization of bacterial infections. *Adv Intern Med* 30: 79–112

2 Nau R, Eiffert H (2002) Modulation of release of proinflammatory bacterial compounds by antibacterials: potential impact on course of inflammation and outcome in sepsis and meningitis. *Clin Microbiol Rev* 15(1): 95–110

3 Martinez JL, Baquero F (2002) Interactions among strategies associated with bacterial infection: pathogenicity, epidemicity, and antibiotic resistance. *Clin Microbiol Rev* 15(4): 647–79

4 Edlund C, Nord CE (1999) Effect of quinolones on intestinal ecology. *Drugs* 58 (Suppl) 2: 65–70

5 Dugas B, Mercenier A, Lenoir-Wijnkoop I, Arnaud C, Dugas N, Postaire E (1999) Immunity and probiotics. *Immunol Today* 20(9): 387–90

6 Pulverer G, Ko HL, Beuth J (1993) Immunomodulating effects of antibiotics influencing digestive flora. *Pathol Biol (Paris)* 41(8 Pt 2): 753–8

7 Milatovic D (1983) Antibiotics and phagocytosis. *Eur J Clin Microbiol* 2(5): 414–25

8 Bergeron Y, Deslauriers AM, Ouellet N, Gauthier MC, Bergeron MG (1999) Influence of cefodizime on pulmonary inflammatory response to heat-killed *Klebsiella pneumoniae* in mice. *Antimicrob Agents Chemother* 43(9): 2291–4

9 Carlier MB, Scorneaux B, Zenebergh A, Desnottes JF, Tulkens PM (1990) Cellular uptake, localization and activity of fluoroquinolones in uninfected and infected macrophages. *J Antimicrob Chemother* 26 (Suppl B): 27–39

10 Tulkens PM (1990) Intracellular pharmacokinetics and localization of antibiotics as predictors of their efficacy against intraphagocytic infections. *Scand J Infect Dis* (Suppl) 74: 209–17

11 Tulkens PM (1991) Intracellular distribution and activity of antibiotics. *Eur J Clin Microbiol Infect Dis* 10(2): 100–6

12 Neftel KA, Muller MR, Widmer U, Hugin AW (1986) Beta-lactam antibiotics inhibit human *in vitro* granulopoiesis and proliferation of some other cell types. *Cell Biol Toxicol* 2(4): 513–21

13 Lubran MM (1989) Hematologic side effects of drugs. Ann Clin Lab Sci 19(2): 114–21

14 Labro MT, Abdelghaffar H (2001) Immunomodulation by macrolide antibiotics. *J Chemother* 13(1): 3–8

15 Riesbeck K (2002) Immunomodulating activity of quinolones: review. *J Chemother* 14(1): 3–12
16 Labro MT (2000) Interference of antibacterial agents with phagocyte functions: immunomodulation or "immuno-fairy tales"? *Clin Microbiol Rev* 13(4): 615–50
17 Dalhoff A, Shalit I (2003) Immunomodulatory effects of quinolones. *Lancet Infect Dis* 3(6): 359–71
18 Brom C, Brom J, Konig W (1992) Neomycin induces stimulatory and inhibitory effects on leukotriene generation, guanine triphosphatase activity, and actin polymerization within human neutrophils. *Immunology* 75(1): 150–6
19 Mandell LA (1982) Effects of antimicrobial and antineoplastic drugs on the phagocytic and microbicidal function of the polymorphonuclear leukocyte. *Rev Infect Dis* 4(3): 683–97
20 Utili R, Adinolfi LE, Dilillo M, Tripodi MF, Marrone A, Ruggiero G (1991) Activity of aminoglycosides against phagocytosed bacteria. *J Antimicrob Chemother* 28(6): 897–904
21 Van Den BC, de Neeling AJ, Schot CS, Hustinx WN, Wemer J, de Wildt DJ (1992) Delayed antibiotic-induced lysis of *Escherichia coli in vitro* is correlated with enhancement of LPS release. *Scand J Infect Dis* 24(5): 619–27
22 Kadurugamuwa JL, Beveridge TJ (1997) Natural release of virulence factors in membrane vesicles by *Pseudomonas aeruginosa* and the effect of aminoglycoside antibiotics on their release. *J Antimicrob Chemother* 40(5): 615–21
23 Jaffuel D, Demoly P, Gougat C, Mautino G, Bousquet J, Mathieu M (1999) Rifampicin is not an activator of the glucocorticoid receptor in A549 human alveolar cells. *Mol Pharmacol* 55(5): 841–6
24 Herr AS, Wochnik GM, Rosenhagen MC, Holsboer F, Rein T (2000) Rifampicin is not an activator of glucocorticoid receptor. *Mol Pharmacol* 57(4): 732–7
25 Gupta S, Grieco MH, Siegel I (1975) Suppression of T-lymphocyte rosettes by rifampin. Studies in normals and patients with tuberculosis. *Ann Intern Med* 82(4): 484–8
26 Hauser WE Jr, Remington JS (1982) Effect of antibiotics on the immune response. *Am J Med* 72(5): 711–6
27 Tentori L, Graziani G, Porcelli SA, Sugita M, Brenner MB, Madaio R, Bonmassar E, Giuliani A, Aquino A (1998) Rifampin increases cytokine-induced expression of the CD1b molecule in human peripheral blood monocytes. *Antimicrob Agents Chemother* 42(3): 550–4
28 Malhotra V, Shanley TP, Pittet JF, Welch WJ, Wong HR (2001) Geldanamycin inhibits NF-kappaB activation and interleukin-8 gene expression in cultured human respiratory epithelium. *Am J Respir Cell Mol Biol* 25(1): 92–7
29 Wax S, Piecyk M, Maritim B, Anderson P (2003) Geldanamycin inhibits the production of inflammatory cytokines in activated macrophages by reducing the stability and translation of cytokine transcripts. *Arthritis Rheum* 48(2): 541–50
30 Bisht KS, Bradbury CM, Mattson D, Kaushal A, Sowers A, Markovina S, Ortiz KL, Sieck LK, Isaacs JS, Brechbiel MW et al (2003) Geldanamycin and 17-allylamino-17-

demethoxygeldanamycin potentiate the *in vitro* and *in vivo* radiation response of cervical tumor cells *via* the heat shock protein 90-mediated intracellular signaling and cytotoxicity. *Cancer Res* 63(24): 8984–95
31 Oleske JM, de la Cruz A, Ahdieh H, Sorvino D, La Braico J, Cooper R, Singh R, Lin R, Minnefor A (1983) Effects of antibiotics on polymorphonuclear leukocyte chemiluminescence and chemotaxis. *J Antimicrob Chemother* 12 (Suppl C): 35–8
32 Braga PC, Dal Sasso M, Maci S, Bondiolotti G, Fonti E, Reggio S (1996) Penetration of brodimoprim into human neutrophils and intracellular activity. *Antimicrob Agents Chemother* 40(10): 2392–8
33 Perry DK, Hand WL, Edmondson DE, Lambeth JD (1992) Role of phospholipase D-derived diradylglycerol in the activation of the human neutrophil respiratory burst oxidase. Inhibition by phosphatidic acid phosphohydrolase inhibitors. *J Immunol* 149(8): 2749–58
34 Labro MT (1990) Cefodizime as a biological response modifier: a review of its *in-vivo*, *ex-vivo* and *in-vitro* immunomodulatory properties. *J Antimicrob Chemother* 26 (Suppl C): 37–47
35 Barradell LB, Brogden RN (1992) Cefodizime. A review of its antibacterial activity, pharmacokinetic properties and therapeutic use. *Drugs* 44(5): 800–34
36 Labro MT (1992) Immunological evaluation of cefodizime: a unique molecule among cephalosporins. *Infection* 20 (Suppl 1): S45–S47
37 Limbert M, Bartlett RR, Dickneite G, Klesel N, Schorlemmer HU, Seibert G, Winkler I, Schrinner E (1984) Cefodizime, an aminothiazolyl cephalosporin. IV. Influence on the immune system. *J Antibiot (Tokyo)* 37(12): 1719–26
38 Ritts RE (1990) Antibiotics as biological response modifiers. *J Antimicrob Chemother* 26 (Suppl C): 31–6
39 Bergeron Y, Ouellet N, Deslauriers AM, Simard M, Olivier M, Bergeron MG (1998) Reduction by cefodizime of the pulmonary inflammatory response induced by heat-killed *Streptococcus pneumoniae* in mice. *Antimicrob Agents Chemother* 42(10): 2527–33
40 Alkharfy KM, Kellum JA, Frye RF, Matzke GR (2000) Effect of ceftazidime on systemic cytokine concentrations in rats. *Antimicrob Agents Chemother* 44(11): 3217–19
41 Schorlemmer HU, Dickneite G, Blumbach J, Durckheimer W, Sedlacek HH (1989) Immunomodulation by the new synthetic thiazole derivative tiprotimod. 2nd communication: immunopharmacological activity. *Arzneimittelforschung* 39(9): 1085–9
42 Shin WS, Min CK, Kim YR, Yoo JH, Kang MW (1996) *In-vitro* effects of cefodizime on leucocyte functions and colony formation from granulocyte-monocyte progenitors. *J Antimicrob Chemother* 37(1): 93–103
43 Wenisch C, Parschalk B, Hasenhundl M, Wiesinger E, Graninger W (1995) Effect of cefodizime and ceftriaxone on phagocytic function in patients with severe infections. *Antimicrob Agents Chemother* 39(3): 672–6
44 Gurer US, Cevikbas A, Johansson C, Derici K, Yardimci T (1999) Effect of fluconazole

on human polymorphonuclear leucocyte functions *ex vivo* against *Candida albicans. Chemotherapy* 45(4): 277–83

45 Gurer US, Palanduz S, Cevikbas A, Derici K, Johansson C, Ozturk S (1999) Effect of cefodizime, ofloxacin, ciprofloxacin and interferon alpha-2a, alone and in combination, on phagocytic and candidacidal functions of leucocytes from patients with chronic renal failure. *Medical Science Research* 27(5): 315–18

46 Krehmeier U, Bardenheuer M, Voggenreiter G, Obertacke U, Schade FU, Majetschak M (2002) Effects of antimicrobial agents on spontaneous and endotoxin-induced cytokine release of human peripheral blood mononuclear cells. *J Infect Chemother* 8(2): 194–7

47 Nwariaku FE, McIntyre KL, Sikes PJ, Mileski WJ (1997) The effect of antimicrobial agents on the induction of tumour necrosis factor by alveolar macrophages *in vitro* in response to endotoxin. *J Antimicrob Chemother* 39(2): 265–7

48 Chen M, Ona VO, Li M, Ferrante RJ, Fink KB, Zhu S, Bian J, Guo L, Farrell LA, Hersch SM et al (2000) Minocycline inhibits caspase-1 and caspase-3 expression and delays mortality in a transgenic mouse model of Huntington disease. *Nat Med* 6(7): 797–801

49 Yrjanheikki J, Tikka T, Keinanen R, Goldsteins G, Chan PH, Koistinaho J (1999) A tetracycline derivative, minocycline, reduces inflammation and protects against focal cerebral ischemia with a wide therapeutic window. *Proc Natl Acad Sci USA* 96(23): 13496–500

50 Golub LM, Lee HM, Ryan ME, Giannobile WV, Payne J, Sorsa T (1998) Tetracyclines inhibit connective tissue breakdown by multiple non-antimicrobial mechanisms. *Adv Dent Res* 12(2): 12–26

51 Petrinec D, Liao S, Holmes DR, Reilly JM, Parks WC, Thompson RW (1996) Doxycycline inhibition of aneurysmal degeneration in an elastase-induced rat model of abdominal aortic aneurysm: preservation of aortic elastin associated with suppressed production of 92 kD gelatinase. *J Vasc Surg* 23(2): 336–46

52 Prall AK, Longo GM, Mayhan WG, Waltke EA, Fleckten B, Thompson RW, Baxter BT (2002) Doxycycline in patients with abdominal aortic aneurysms and in mice: comparison of serum levels and effect on aneurysm growth in mice. *J Vasc Surg* 35(5): 923–9

53 Thompson RW, Baxter BT (1999) MMP inhibition in abdominal aortic aneurysms. Rationale for a prospective randomized clinical trial. *Ann NY Acad Sci* 878: 159–78

54 Baxter BT, Pearce WH, Waltke EA, Littooy FN, Hallett JW Jr, Kent KC, Upchurch GR Jr, Chaikof EL, Mills JL, Fleckten B et al (2002) Prolonged administration of doxycycline in patients with small asymptomatic abdominal aortic aneurysms: report of a prospective (Phase II) multicenter study. *J Vasc Surg* 36(1): 1–12

55 Meng Q, Xu J, Goldberg ID, Rosen EM, Greenwald RA, Fan S (2000) Influence of chemically modified tetracyclines on proliferation, invasion and migration properties of MDA-MB-468 human breast cancer cells. *Clin Exp Metastasis* 18(2): 139–46

56 Davies SR, Cole AA, Schmid TM (1996) Doxycycline inhibits type X collagen synthesis in avian hypertrophic chondrocyte cultures. *J Biol Chem* 271(42): 25966–70

57 Lokeshwar BL (1999) MMP inhibition in prostate cancer. *Ann NY Acad Sci* 878: 271–89

58 Ingham E (1990) Modulation of the proliferative response of murine thymocytes stimulated by IL-1, and enhancement of IL-1 beta secretion from mononuclear phagocytes by tetracyclines. *J Antimicrob Chemother* 26(1): 61–70
59 Gabler WL, Creamer HR (1991) Suppression of human neutrophil functions by tetracyclines. *J Periodontal Res* 26(1): 52–8
60 Gabler WL (1991) Fluxes and accumulation of tetracyclines by human blood cells. *Res Commun Chem Pathol Pharmacol* 72(1): 39–51
61 Jain A, Sangal L, Basal E, Kaushal GP, Agarwal SK (2002) Anti-inflammatory effects of erythromycin and tetracycline on *Propionibacterium acnes* induced production of chemotactic factors and reactive oxygen species by human neutrophils. *Dermatol Online* J 8(2): 2
62 Lauhio A, Leirisalo-Repo M, Lahdevirta J, Saikku P, Repo H (1991) Double-blind, placebo-controlled study of three-month treatment with lymecycline in reactive arthritis, with special reference to Chlamydia arthritis. *Arthritis Rheum* 34(1): 6–14
63 Palazzi C, Olivieri I, D'Amico E, Pennese E, Petricca A (2004) Management of reactive arthritis. *Expert Opin Pharmacother* 5(1): 61–70
64 Borderie D, Hernvann A, Hilliquin P, Lemarchal H, Kahan A, Ekindjian OG (2001) Tetracyclines inhibit nitrosothiol production by cytokine-stimulated osteoarthritic synovial cells. *Inflamm Res* 50(8): 409–14
65 Amin AR, Patel RN, Thakker GD, Lowenstein CJ, Attur MG, Abramson SB (1997) Post-transcriptional regulation of inducible nitric oxide synthase mRNA in murine macrophages by doxycycline and chemically modified tetracyclines. *FEBS Lett* 410 (2–3): 259–64
66 D'Agostino P, Arcoleo F, Barbera C, Di Bella G, La Rosa M, Misiano G, Milano S, Brai M, Cammarata G, Feo S et al (1998) Tetracycline inhibits the nitric oxide synthase activity induced by endotoxin in cultured murine macrophages. *Eur J Pharmacol* 346 (2–3): 283–90
67 Tilley BC, Alarcon GS, Heyse SP, Trentham DE, Neuner R, Kaplan DA, Clegg DO, Leisen JC, Buckley L, Cooper SM et al (1995) Minocycline in rheumatoid arthritis. A 48-week, double-blind, placebo-controlled trial. MIRA Trial Group. *Ann Intern Med* 122(2): 81–9
68 Campbell SM, Wernick R (1999) Update in rheumatology. *Ann Intern Med* 130(2): 135–42
69 Shapira L, Soskolne WA, Houri Y, Barak V, Halabi A, Stabholz A (1996) Protection against endotoxic shock and lipopolysaccharide-induced local inflammation by tetracycline: correlation with inhibition of cytokine secretion. *Infect Immun* 64(3): 825–8
70 Matsumoto T, Tateda K, Miyazaki S, Furuya N, Ohno A, Ishii Y, Hirakata Y, Yamaguchi K (1997) Immunomodulating effect of fosfomycin on gut-derived sepsis caused by *Pseudomonas aeruginosa* in mice. *Antimicrob Agents Chemother* 41(2): 308–13
71 Morikawa K, Nonaka M, Torii I, Morikawa S (2003) Modulatory effect of fosfomycin on acute inflammation in the rat air pouch model. *Int J Antimicrob Agents* 21(4): 334–9
72 Morikawa K, Oseko F, Morikawa S, Sawada M (1993) Immunosuppressive activity of

fosfomycin on human T-lymphocyte function *in vitro*. *Antimicrob Agents Chemother* 37(12): 2684–7
73 Morikawa K, Oseko F, Morikawa S (1993) Immunomodulatory effect of fosfomycin on human B-lymphocyte function. *Antimicrob Agents Chemother* 37(2): 270–5
74 Morikawa K, Watabe H, Araake M, Morikawa S (1996) Modulatory effect of antibiotics on cytokine production by human monocytes *in vitro*. *Antimicrob Agents Chemother* 40(6): 1366–70
75 Matsumoto T, Tateda K, Miyazaki S, Furuya N, Ohno A, Ishii Y, Hirakata Y, Yamaguchi K (1999) Fosfomycin alters lipopolysaccharide-induced inflammatory cytokine production in mice. *Antimicrob Agents Chemother* 43(3): 697–8
76 Marchese A, Bozzolasco M, Gualco L, Debbia EA, Schito GC, Schito AM (2003) Effect of fosfomycin alone and in combination with N-acetylcysteine on *E. coli* biofilms. *Int J Antimicrob Agents* 22 (Suppl) 2: 95–100
77 Van der AP, Husson M, Fruhling J (1987) Influence of various antibiotics on phagocytosis of *Staphylococcus aureus* by human polymorphonuclear leucocytes. *J Antimicrob Chemother* 20(3): 399–404
78 Luhrmann A, Tholke J, Behn I, Schumann J, Tiegs G, Hauschildt S (1998) Immunomodulating properties of the antibiotic novobiocin in human monocytes. *Antimicrob Agents Chemother* 42(8): 1911–16
79 Eder JP, Teicher BA, Holden SA, Cathcart KN, Schnipper LE, Frei E III (1989) Effect of novobiocin on the antitumor activity and tumor cell and bone marrow survivals of three alkylating agents. *Cancer Res* 49(3): 595–8
80 Shiozawa K, Oka M, Soda H, Yoshikawa M, Ikegami Y, Tsurutani J, Nakatomi K, Nakamura Y, Doi S, Kitazaki T et al (2004) Reversal of breast cancer resistance protein (BCRP/ABCG2)-mediated drug resistance by novobiocin, a coumermycin antibiotic. *Int J Cancer* 108(1): 146–51
81 Rappa G, Murren JR, Johnson LM, Lorico A, Sartorelli AC (2000) Novobiocin-induced VP-16 accumulation and MRP expression in human leukemia and ovarian carcinoma cells. *Anticancer Drug Des* 15(2): 127–34
82 Murren JR, DiStasio SA, Lorico A, McKeon A, Zuhowski EG, Egorin MJ, Sartorelli AC, Rappa G (2000) Phase I and pharmacokinetic study of novobiocin in combination with VP-16 in patients with refractory malignancies. *Cancer* J 6(4): 256–65
83 Boe NM, Dellinger EP, Minshew BH (1983) Effect of clindamycin on growth and haemolysin production by *Escherichia coli*. *J Antimicrob Chemother* 12 (Suppl C): 105–16
84 Nakano T, Hiramatsu K, Kishi K, Hirata N, Kadota J, Nasu M (2003) Clindamycin modulates inflammatory-cytokine induction in lipopolysaccharide-stimulated mouse peritoneal macrophages. *Antimicrob Agents Chemother* 47(1): 363–7
85 Aida Y, Pabst MJ, Rademacher JM, Hatakeyama T, Aono M (1990) Effects of polymyxin B on superoxide anion release and priming in human polymorphonuclear leukocytes. *J Leukoc Biol* 47(3): 283–91
86 Mayumi T, Takezawa J, Takahashi H, Kuwayama N, Fukuoka T, Shimizu K, Yamada

K, Kondo S, Aono K (1999) Low-dose intramuscular polymyxin B improves survival of septic rats. *Shock* 11(2): 82–6

87 Rifkind D (1967) Prevention by polymyxin B of endotoxin lethality in mice. *J Bacteriol* 93(4): 1463–4

88 Warren HS, Kania SA, Siber GR (1985) Binding and neutralization of bacterial lipopolysaccharide by colistin nonapeptide. *Antimicrob Agents Chemother* 28(1): 107–12

89 Damais C, Jupin C, Parant M, Chedid L (1987) Induction of human interleukin-1 production by polymyxin B. *J Immunol Methods* 101(1): 51–6

90 Hogasen AK, Abrahamsen TG (1995) Polymyxin B stimulates production of complement components and cytokines in human monocytes. *Antimicrob Agents Chemother* 39(2): 529–32

91 Tufano MA, Cipollaro dl, Ianniello R, Baroni A, Galdiero F (1992) Antimicrobial agents induce monocytes to release IL-1 alpha, IL-6, and TNF, and induce lymphocytes to release IL-4 and TNF tau. *Immunopharmacol Immunotoxicol* 14(4): 769–82

92 Foca A, Matera G, Berlinghieri MC, Liberto MC, De Sarro GB (1992) Teicoplanin reduces *in-vitro* reactivity and murine lethality of *Salmonella minnesota* R595 lipopolysaccharide. *J Antimicrob Chemother* 29(4): 443–6

93 Hussy P, Maass G, Tummler B, Grosse F, Schomburg U (1986) Effect of 4-quinolones and novobiocin on calf thymus DNA polymerase alpha primase complex, topoisomerases I and II, and growth of mammalian lymphoblasts. *Antimicrob Agents Chemother* 29(6): 1073–8

94 Williams AC, Galley HF, Webster NR (2001) The effect of moxifloxacin on release of interleukin-8 from human nutrophils. *Br J Anaesth* 87(4): 671–2

95 Roche Y, Gougerot-Pocidalo MA, Fay M, Etienne D, Forest N, Pocidalo JJ (1987) Comparative effects of quinolones on human mononuclear leucocyte functions. *J Antimicrob Chemother* 19(6): 781–90

96 Bailly S, Fay M, Roche Y, Gougerot-Pocidalo MA (1990) Effects of quinolones on tumor necrosis factor production by human monocytes. *Int J Immunopharmacol* 12(1): 31–6

97 Araujo FG, Slifer TL, Remington JS (2002) Effect of moxifloxacin on secretion of cytokines by human monocytes stimulated with lipopolysaccharide. *Clin Microbiol Infect* 8(1): 26–30

98 Rabehi L, Irinopoulou T, Cholley B, Haeffner-Cavaillon N, Carreno MP (2001) Gram-positive and gram-negative bacteria do not trigger monocytic cytokine production through similar intracellular pathways. *Infect Immun* 69(7): 4590–9

99 Nitsche D, Schulze C, Oesser S, Dalhoff A, Sack M (1996) Impact of different classes antimicrobial agents on plasma endotoxin activity. *Arch Surg* 131(2): 192–9

100 Lindner B, Wiese A, Brandenburg K, Seydel U, Dalhoff A (2002) Lack of interaction of fluoroquinolones with lipopolysaccharides. *Antimicrob Agents Chemother* 46(5): 1568–70

101 Khan AA, Slifer TR, Remington JS (1998) Effect of trovafloxacin on production of cytokines by human monocytes. *Antimicrob Agents Chemother* 42(7): 1713–17

102 Yamashita Y, Ashizawa T, Morimoto M, Hosomi J, Nakano H (1992) Antitumor quinolones with mammalian topoisomerase II mediated DNA cleavage activity. *Cancer Res* 52(10): 2818–22
103 Ono Y, Ohmoto Y, Ono K, Sakata Y, Murata K (2000) Effect of grepafloxacin on cytokine production *in vitro*. *J Antimicrob Chemother* 46(1): 91–4
104 Hashimoto S, Matsumoto K, Gon Y, Maruoka S, Hayashi S, Asai Y, Machino T, Horie T (2000) Grepafloxacin inhibits tumor necrosis factor-alpha-induced interleukin-8 expression in human airway epithelial cells. *Life Sci* 66(5): L–82
105 Yoshimura T, Kurita C, Usami E, Nakao T, Watanabe S, Kobayashi J, Yamazaki F, Nagai H (1996) Immunomodulatory action of levofloxacin on cytokine production by human peripheral blood mononuclear cells. *Chemotherapy* 42(6): 459–64
106 Bailly S, Pocidalo JJ, Fay M, Gougerot-Pocidalo MA (1993) *In vitro* and *in vivo* effects of quinolones on monocyte cytokine production. In: Ullmann U, Dalhoff A (eds): *Significance of cytokine in the treatment of infectious diseases*. Gustav Fischer Verlag, Stuttgart, New York 137–49
107 Roche Y, Fay M, Gougerot-Pocidalo MA (1987) Effects of quinolones on interleukin 1 production *in vitro* by human monocytes. *Immunopharmacology* 13(2): 99–109
108 Bailly S, Mahe Y, Ferrua B, Fay M, Tursz T, Wakasugi H, Gougerot-Pocidalo MA (1990) Quinolone-induced differential modification of IL-1 alpha and IL-1 beta production by LPS-stimulated human monocytes. *Cell Immunol* 128(1): 277–88
109 De Simone C, Baldinelli L, Ferrazzi M, De Santis S, Pugnaloni L, Sorice F (1986) Influence of ofloxacin, norfloxacin, nalidixic acid, pyromidic acid and pipemidic acid on human gamma-interferon production and blastogenesis. *J Antimicrob Chemother* 17(6): 811–14
110 Gollapudi SV, Chuah SK, Harvey T, Thadepalli HD, Thadepalli H (1993) *In vivo* effects of rufloxacin and ciprofloxacin on T-cell subsets and tumor necrosis factor production in mice infected with *Bacteroides fragilis*. *Antimicrob Agents Chemother* 37(8): 1711–12
111 Thadepalli H, Gollapudi SV, Chuah SK (1986) Therapeutic evaluation of difloxacin (A-56619) and A-56620 for experimentally induced *Bacteroides fragilis*-associated intra-abdominal abscess. *Antimicrob Agents Chemother* 30(4): 574–6
112 Thadepalli H, Hajji M, Perumal VK, Chuah SK, Gollapudi S (1992) Evaluation of temafloxacin in a rat model of intra-abdominal abscess. *J Antimicrob Chemother* 29(6): 687–92
113 Thadepalli H, Reddy U, Chuah SK, Thadepalli F, Malilay C, Polzer RJ, Hanna N, Esfandiari A, Brown P, Gollapudi S (1997) *in vivo* efficacy of trovafloxacin (CP-99,217), a new quinolone, in experimental intra-abdominal abscesses caused by *Bacteroides fragilis* and *Escherichia coli*. *Antimicrob Agents Chemother* 41(3): 583–6
114 Thadepalli H, Chuah SK, Reddy U, Hanna N, Clark R, Polzer RJ, Gollapudi S (1997) Efficacy of trovafloxacin for treatment of experimental *Bacteroides* infection in young and senescent mice. *Antimicrob Agents Chemother* 41(9): 1933–6

115 King A, May J, French G, Phillips I (2000) Comparative *in vitro* activity of gemifloxacin. *J Antimicrob Chemother* 45 (Suppl 1): 1–12

116 Khan AA, Slifer TR, Araujo FG, Suzuki Y, Remington JS (2000) Protection against lipopolysaccharide-induced death by fluoroquinolones. *Antimicrob Agents Chemother* 44(11): 3169–73

117 Purswani MU, Eckert SJ, Arora HK, Noel GJ (2002) Effect of ciprofloxacin on lethal and sublethal challenge with endotoxin and on early cytokine responses in a murine *in vivo* model. *J Antimicrob Chemother* 50(1): 51–8

118 Purswani M, Eckert S, Arora H, Johann-Liang R, Noel GJ (2000) The effect of three broad-spectrum antimicrobials on mononuclear cell responses to encapsulated bacteria: evidence for down-regulation of cytokine mRNA transcription by trovafloxacin. *J Antimicrob Chemother* 46(6): 921–9

119 van Zyl JM, Basson K, Kriegler A, van der Walt BJ (1991) Mechanisms by which clofazimine and dapsone inhibit the myeloperoxidase system. A possible correlation with their anti-inflammatory properties. *Biochem Pharmacol* 42(3): 599–608

120 Thuong-Nguyen V, Kadunce DP, Hendrix JD, Gammon WR, Zone JJ (1993) Inhibition of neutrophil adherence to antibody by dapsone: a possible therapeutic mechanism of dapsone in the treatment of IgA dermatoses. *J Invest Dermatol* 100(4): 349–55

121 Anderson JA, Adkinson NF Jr (1987) Allergic reactions to drugs and biologic agents. *JAMA* 258(20): 2891–9

122 Kudoh S, Azuma A, Yamamoto M, Izumi T, Ando M (1998) Improvement of survival in patients with diffuse panbronchiolitis treated with low-dose erythromycin. *Am J Respir Crit Care Med* 157(6 Pt 1): 1829–32

123 Jaffe A, Bush A (2001) Anti-inflammatory effects of macrolides in lung disease. *Pediatr Pulmonol* 31(6): 464–73

124 Wilschonski M, Yahav Y, Yaacov Y, Blau H, Bentur L, Rivlin J, Aviram M, Bdolah-Abram T, Bebok Z, Shushi L et al (2003) Gentamicin-induced correction of CFTR function in patients with cystic fibrosis and CFTR stop mutations. *N Engl J Med* 349(15): 1433–1441

Index

The PIR-Series
Progress in Inflammation Research

Homepage: http://www.birkhauser.ch

Up-to-date information on the latest developments in the pathology, mechanisms and therapy of inflammatory disease are provided in this monograph series. Areas covered include vascular responses, skin inflammation, pain, neuroinflammation, arthritis cartilage and bone, airways inflammation and asthma, allergy, cytokines and inflammatory mediators, cell signalling, and recent advances in drug therapy. Each volume is edited by acknowledged experts providing succinct overviews on specific topics intended to inform and explain. The series is of interest to academic and industrial biomedical researchers, drug development personnel and rheumatologists, allergists, pathologists, dermatologists and other clinicians requiring regular scientific updates.

Available volumes:

T Cells in Arthritis, P. Miossec, W. van den Berg, G. Firestein (Editors), 1998
Chemokines and Skin, E. Kownatzki, J. Norgauer (Editors), 1998
Medicinal Fatty Acids, J. Kremer (Editor), 1998
Inducible Enzymes in the Inflammatory Response,
D.A. Willoughby, A. Tomlinson (Editors), 1999
Cytokines in Severe Sepsis and Septic Shock, H. Redl, G. Schlag (Editors), 1999
Fatty Acids and Inflammatory Skin Diseases, J.-M. Schröder (Editor), 1999
Immunomodulatory Agents from Plants, H. Wagner (Editor), 1999
Cytokines and Pain, L. Watkins, S. Maier (Editors), 1999
In Vivo *Models of Inflammation*, D. Morgan, L. Marshall (Editors), 1999
Pain and Neurogenic Inflammation, S.D. Brain, P. Moore (Editors), 1999
Anti Inflammatory Drugs in Asthma, A.P. Sampson, M.K. Church (Editors), 1999
Novel Inhibitors of Leukotrienes, G. Folco, B. Samuelsson, R.C. Murphy (Editors), 1999
Vascular Adhesion Molecules and Inflammation, J.D. Pearson (Editor), 1999
Metalloproteinases as Targets for Anti-Inflammatory Drugs,
K.M.K. Bottomley, D. Bradshaw, J.S. Nixon (Editors), 1999
Free Radicals and Inflammation, P.G. Winyard, D.R. Blake, C.H. Evans (Editors), 1999
Gene Therapy in Inflammatory Diseases, C.H. Evans, P. Robbins (Editors), 2000
New Cytokines as Potential Drugs, S. K. Narula, R. Coffmann (Editors), 2000
High Throughput Screening for Novel Anti-inflammatories, M. Kahn (Editor), 2000
Immunology and Drug Therapy of Atopic Skin Diseases,
C.A.F. Bruijnzeel-Komen, E.F. Knol (Editors), 2000
Novel Cytokine Inhibitors, G.A. Higgs, B. Henderson (Editors), 2000
Inflammatory Processes. Molecular Mechanisms and Therapeutic Opportunities,
L.G. Letts, D.W. Morgan (Editors), 2000

Cellular Mechanisms in Airways Inflammation, C. Page, K. Banner, D. Spina (Editors), 2000
Inflammatory and Infectious Basis of Atherosclerosis, J.L. Mehta (Editor), 2001
Muscarinic Receptors in Airways Diseases, J. Zaagsma, H. Meurs, A.F. Roffel (Editors), 2001
TGF-β and Related Cytokines in Inflammation, S.N. Breit, S. Wahl (Editors), 2001
Nitric Oxide and Inflammation, D. Salvemini, T.R. Billiar, Y. Vodovotz (Editors), 2001
Neuroinflammatory Mechanisms in Alzheimer's Disease. Basic and Clinical Research, J. Rogers (Editor), 2001
Disease-modifying Therapy in Vasculitides, C.G.M. Kallenberg, J.W. Cohen Tervaert (Editors), 2001
Inflammation and Stroke, G.Z. Feuerstein (Editor), 2001
NMDA Antagonists as Potential Analgesic Drugs, D.J.S. Sirinathsinghji, R.G. Hill (Editors), 2002
Migraine: A Neuroinflammatory Disease? E.L.H. Spierings, M. Sanchez del Rio (Editors), 2002
Mechanisms and Mediators of Neuropathic pain, A.B. Malmberg, S.R. Chaplan (Editors), 2002
Bone Morphogenetic Proteins. From Laboratory to Clinical Practice, S. Vukicevic, K.T. Sampath (Editors), 2002
The Hereditary Basis of Allergic Diseases, J. Holloway, S. Holgate (Editors), 2002
Inflammation and Cardiac Diseases, G.Z. Feuerstein, P. Libby, D.L. Mann (Editors), 2003
Mind over Matter – Regulation of Peripheral Inflammation by the CNS, M. Schäfer, C. Stein (Editors), 2003
Heat Shock Proteins and Inflammation, W. van Eden (Editor), 2003
Pharmacotherapy of Gastrointestinal Inflammation, A. Guglietta (Editor), 2004
Arachidonate Remodeling and Inflammation, A.N. Fonteh, R.L. Wykle (Editors), 2004
Recent Advances in Pathophysiology of COPD, P.J. Barnes, T.T. Hansel (Editors), 2004
Cytokines and Joint Injury, W.B. van den Berg, P. Miossec (Editors), 2004
Cancer and Inflammation, D.W. Morgan, U. Forssmann, M.T. Nakada (Editors), 2004
Bone Morphogenetic Proteins: Bone Regeneration and Beyond, S. Vukicevic, K.T. Sampath (Editors), 2004